MATA REFERENCE MANUAL
VOLUME 2
[M-4]–[M-6]
RELEASE 10

A Stata Press Publication
StataCorp LP
College Station, Texas

Stata Press, 4905 Lakeway Drive, College Station, Texas 77845

Copyright © 1985–2007 by StataCorp LP
All rights reserved
Version 10
Typeset in TeX
Printed in the United States of America
10 9 8 7 6 5 4 3 2

ISBN-10: 1-59718-037-8 (volumes 1–2)
ISBN-10: 1-59718-035-1 (volume 1)
ISBN-10: 1-59718-036-X (volume 2)
ISBN-13: 978-1-59718-037-5 (volumes 1–2)
ISBN-13: 978-1-59718-035-1 (volume 1)
ISBN-13: 978-1-59718-036-8 (volume 2)

This manual is protected by copyright. All rights are reserved. No part of this manual may be reproduced, stored in a retrieval system, or transcribed, in any form or by any means—electronic, mechanical, photocopying, recording, or otherwise—without the prior written permission of StataCorp LP.

StataCorp provides this manual "as is" without warranty of any kind, either expressed or implied, including, but not limited to, the implied warranties of merchantability and fitness for a particular purpose. StataCorp may make improvements and/or changes in the product(s) and the program(s) described in this manual at any time and without notice.

The software described in this manual is furnished under a license agreement or nondisclosure agreement. The software may be copied only in accordance with the terms of the agreement. It is against the law to copy the software onto CD, disk, diskette, tape, or any other medium for any purpose other than backup or archival purposes.

The automobile dataset appearing on the accompanying media is Copyright © 1979 by Consumers Union of U.S., Inc., Yonkers, NY 10703-1057 and is reproduced by permission from CONSUMER REPORTS, April 1979.

Stata and Mata are registered trademarks and NetCourse is a trademark of StataCorp LP.

Other product names are registered trademarks or trademarks of their respective companies.

For copyright information about the software, type `help copyright` within Stata.

The suggested citation for this software is

StataCorp. 2007. *Stata Statistical Software: Release 10*. College Station, TX: StataCorp LP.

[M-4] Index and guide to functions

Title

[M-4] **intro** — Index and guide to functions

Contents

[M-4] Entry	Description
Mathematical	
matrix	Matrix functions
solvers	Matrix solvers and inverters
scalar	Scalar functions
statistical	Statistical functions
mathematical	Other important functions
Utility & manipulation	
standard	Standard matrices
utility	Matrix utility functions
manipulation	Matrix manipulation functions
Stata interface	
stata	Stata interface functions
String, I/O, & programming	
string	String manipulation functions
io	I/O functions
programming	Programming functions

Description

The entries in this section provides an index to the functions, grouped according to purpose.

Remarks

The next section, section [M-5], presents all the Mata functions, in alphabetical order.

Also See

[M-0] **intro** — Introduction to the Mata manual

Title

[M-4] **io** — I/O functions

Contents

[M-5] Manual entry	Function	Purpose
Console output		
printf( )	`printf()`	display
	`sprintf()`	display into string
errprintf( )	`errprintf()`	display error message
display( )	`display()`	display text interpreting SMCL
displayas( )	`displayas()`	set whether output is displayed
displayflush( )	`displayflush()`	flush terminal output buffer
liststruct( )	`liststruct()`	list structure's contents
more( )	`more()`	create --more-- condition
	`setmore()`	query or set more on or off
	`setmoreonexit()`	set more on or off on exit
File directories		
direxists( )	`direxists()`	whether directory exists
dir( )	`dir()`	file list
chdir( )	`pwd()`	obtain current working directory
	`chdir()`	change current working directory
	`mkdir()`	make new directory
	`rmdir()`	remove directory
File management		
findfile( )	`findfile()`	find file
fileexists( )	`fileexists()`	whether file exists
cat( )	`cat()`	read file into string matrix
unlink( )	`unlink()`	erase file
adosubdir( )	`adosubdir()`	obtain ado-subdirectory for file

File I/O

fopen( )
- fopen() — open file
- fclose() — close file
- fget() — read line of ASCII file
- fgetnl() — same, but include newline character
- fread() — read k bytes of binary file
- fput() — write line into ASCII file
- fwrite() — write k bytes into binary file
- fgetmatrix() — read matrix
- fputmatrix() — write matrix
- fstatus() — status of last I/O command
- ftell() — report location in file
- fseek() — seek to location in file
- ftruncate() — truncate file at current position

ferrortext( )
- ferrortext() — error text of file error code
- freturncode() — return code of file error code

bufio( )
- bufio() — initialize buffer
- bufbyteorder() — reset (specify) byte order
- bufmissingvalue() — reset (specify) missing-value encoding
- bufput() — copy into buffer
- bufget() — copy from buffer
- fbufput() — copy into and write buffer
- fbufget() — read and copy from buffer
- bufbfmtlen() — utility routine
- bufbfmtisnum() — utility routine

Filename & path manipulation

pathjoin( )
- pathjoin() — join paths
- pathsplit() — split paths
- pathbasename() — path basename
- pathsuffix() — file suffix
- pathrmsuffix() — remove file suffix
- pathisurl() — whether path is URL
- pathisabs() — whether path is absolute
- pathasciisuffix() — whether file is ASCII
- pathstatasuffix() — whether file is Stata
- pathlist() — process path list
- pathsubsysdir() — substitute for system directories
- pathsearchlist() — path list to search for file

Description

The above functions have to do with

1. Displaying output at the terminal.
2. Reading and writing data in a file.

Remarks

To display the contents of a scalar, vector, or matrix, it is sufficient merely to code the identity of the scalar, vector, or matrix:

```
: x
                1              2              3              4
      1    .1369840784    .643220668    .5578016951    .6047949435
```

You can follow this approach even in programs:

```
function example()
{
    ...
    "i am about to calculate the result"
    ...
    "the result is"
    b
}
```

On the other hand, display() and printf() (see [M-5] **display()** and [M-5] **printf()**) will allow you to exercise more control over how the output looks.

Changing the subject: you will find that many I/O functions come in two varieties: with and without an underscore in front of the name, such as _fopen() and fopen(). As always, functions that begin with an underscore are generally silent about their work and return flags indicating their success or failure. Functions that omit the underscore abort and issue the appropriate error message when things go wrong.

Also See

[M-4] **intro** — Index and guide to functions

Title

[M-4] manipulation — Matrix manipulation

Contents

[M-5] Manual entry	Function	Purpose
Transposition		
transposeonly( )	`transposeonly()`	transposition without conjugation
_transpose( )	`_transpose()`	transposition in place
Diagonals		
diag( )	`diag()`	create diagonal matrix from vector
_diag( )	`_diag()`	replace diagonal of matrix
diagonal( )	`diagonal()`	extract diagonal of matrix into vector
Triangular & symmetric		
lowertriangle( )	`lowertriangle()`	extract lower triangle
	`uppertriangle()`	extract upper triangle
makesymmetric( )	`makesymmetric()`	make matrix symmetric (Hermitian)
Sorting		
sort( )	`sort()`	sort rows of matrix
	`jumble()`	randomize order of rows of matrix
	`order()`	obtain permutation vector
	`_collate()`	order matrix on permutation vector
uniqrows( )	`uniqrows()`	sorted, unique rows

```
┌─ Editing ────────────────────────────────────────────────────────┐
_fillmissing( )      _fillmissing()      change matrix to contain missing values
editmissing( )       editmissing()       replace missing values in matrix
editvalue( )         editvalue()         replace values in matrix
edittozero( )        edittozero()        edit matrix for roundoff error (zeros)
                     edittozerotol()     same, absolute tolerance
edittoint( )         edittoint()         edit matrix for roundoff error (integers)
                     edittointtol()      same, absolute tolerance

┌─ Permutation vectors ────────────────────────────────────────────┐
invorder( )          invorder()          inverse of permutation vector
                     revorder()          reverse of permutation vector

┌─ Matrices into vectors & vice versa ─────────────────────────────┐
vec( )               vec()               convert matrix into column vector
                     vech()              convert symmetric matrix into column vector
                     invvech()           convert column vector into symmetric matrix
rowshape( )          rowshape()          reshape matrix to have r rows
                     colshape()          reshape matrix to have c columns
```

Description

The above functions manipulate matrices, such as extracting the diagonal and sorting.

Remarks

There is a thin line between manipulation and utility; also see

> [M-4] **utility**   Matrix utility functions

Also See

> [M-4] **intro** — Index and guide to functions

Title

> **[M-4] mathematical** — Important mathematical functions

Contents

[M-5] Manual entry	Function	Purpose
— Basics (also see [M-4] **scalar***) —*		
sum( )	rowsum( )	sum of each row
	colsum( )	sum of each column
	sum( )	overall sum
	quadrowsum( )	quad-precision sum of each row
	quadcolsum( )	quad-precision sum of each column
	quadsum( )	quad-precision overall sum
runningsum( )	runningsum( )	running sum of vector
	quadrunningsum( )	quad-precision runningsum( )
minmax( )	rowmin( )	minimum, by row
	colmin( )	minimum, by column
	min( )	minimum, overall
	rowmax( )	maximum, by row
	colmax( )	maximum, by column
	max( )	maximum, overall
	rowminmax( )	minimum and maximum, by row
	colminmax( )	minimum and maximum, by column
	minmax( )	minimum and maximum, overall
	rowmaxabs( )	rowmax(abs( ))
	colmaxabs( )	colmax(abs( ))
optimize( )	optimize( )	function maximization and minimization
	optimize_init( )	begin optimization
	optimize_init_*( )	set details
	optimize( )	perform optimization
	optimize_result_*( )	access results

[M-4] mathematical — Important mathematical functions

---- Fourier transform ----

fft( )
fft()	fast Fourier transform
invfft()	inverse fast Fourier transform
convolve()	convolution
deconvolve()	inverse of convolve()
Corr()	correlation
ftperiodogram()	power spectrum
ftpad()	pad to power-of-2 length
ftwrap()	convert to frequency-wraparound order
ftunwrap()	convert from frequency-wraparound order
ftretime()	change time scale of signal
ftfreqs()	frequencies of transform

---- Cubic splines ----

spline3( )
spline3()	fit cubic spline
spline3eval()	evaluate cubic spline

---- Polynomials ----

polyeval( )
polyeval()	evaluate polynomial
polysolve()	solve for polynomial
polytrim()	trim polynomial
polyderiv()	derivative of polynomial
polyinteg()	integral of polynomial
polyadd()	add polynomials
polymult()	multiply polynomials
polydiv()	divide polynomials
polyroots()	find roots of polynomial

---- Number-theoretic point sets ----

halton( )
halton()	generate a Halton or Hammersley set
ghalton()	generate a generalized Halton sequence

---- Base conversion ----

inbase( )
inbase()	convert to specified base
frombase()	convert from specified base

Description

The above functions are important mathematical functions that most people would not call either matrix functions or scalar functions, but that use matrices and scalars.

Remarks

For other mathematical functions, see

    [M-4] **matrix**   Matrix mathematical functions
    [M-4] **scalar**   Scalar mathematical functions
    [M-4] **statistical**   Statistical functions

Also See

[M-4] **intro** — Index and guide to functions

Title

[M-4] **matrix** — Matrix functions

Contents

[M-5] Manual entry	Function	Purpose
Characteristics		
trace( )	trace()	trace of matrix
det( )	det()	determinant
	dettriangular()	determinant of triangular matrix
norm( )	norm()	matrix and vector norms
cond( )	cond()	matrix condition number
rank( )	rank()	rank of matrix
Cholesky decomposition, solvers, & inverters		
cholesky( )	cholesky()	Cholesky square-root decomposition $A = GG'$
cholsolve( )	cholsolve()	solve $AX = B$ for X
cholinv( )	cholinv()	inverse of pos. def. symmetric matrix
invsym( )	invsym()	real symmetric matrix inversion
LU decomposition, solvers, & inverters		
lud( )	lud()	LU decomposition $A = PLU$
lusolve( )	lusolve()	solve $AX = B$ for X
luinv( )	luinv()	inverse of square matrix

(*Continued on next page*)

QR decomposition, solvers, & inverters

qrd( )
- `qrd()` — QR decomposition $A = QR$
- `qrdp()` — QR decomposition $A = QRP'$
- `hqrd()` — QR decomposition $A = f(H)R_1$
- `hqrdp()` — QR decomposition $A = f(H, tau)R_1 P'$
- `hqrdmultq()` — return QX or $Q'X$, $Q = f(H, tau)$
- `hqrdmultq1t()` — return $Q'_1 X$, $Q_1 = f(H, tau)$
- `hqrdq()` — return $Q = f(H, tau)$
- `hqrdq1()` — return $Q_1 = f(H, tau)$
- `hqrdr()` — return R
- `hqrdr1()` — return R_1

qrsolve( )
- `qrsolve()` — solve $AX = B$ for X

qrinv( )
- `qrinv()` — generalized inverse of matrix

Singular value decomposition, solvers, & inverters

svd( )
- `svd()` — singular value decomposition $A = UDV'$
- `svdsv()` — singular values s

fullsvd( )
- `fullsvd()` — singular value decomposition $A = USV'$
- `fullsdiag()` — convert s to S

svsolve( )
- `svsolve()` — solve $AX = B$ for X

pinv( )
- `pinv()` — Moore–Penrose pseudoinverse

Triangular solvers

solvelower( )
- `solvelower()` — solve $AX = B$ for X, A lower triangular
- `solveupper()` — solve $AX = B$ for X, A upper triangular

Eigensystems, powers, & transcendental

eigensystem( )	eigensystem()	eigenvalues and eigenvectors
	eigenvalues()	eigenvalues
	symeigensystem()	eigenvalues/eigenvectors of symmetric matrix
	symeigenvalues()	eigenvalues of symmetric matrix
	lefteigensystem()	eigenvalues and left eigenvectors
matpowersym( )	matpowersym()	powers of symmetric matrix
matexpsym( )	matexpsym()	exponentiation of symmetric matrix
	matlogsym()	logarithm of symmetric matrix

Equilibration

_equilrc( )	_equilrc()	row/column equilibration
	_equilr()	row equilibration
	_equilc()	column equilibration
	_perhapsequilrc()	row/column equilibration if necessary
	_perhapsequilr()	row equilibration if necessary
	_perhapsequilc()	column equilibration if necessary
	rowscalefactors()	row-scaling factors for equilibration
	colscalefactors()	column-scaling factors for equilibration

Description

The above functions are what most people would call mathematical matrix functions.

Remarks

For other mathematical functions, see

| [M-4] **scalar** | Scalar mathematical functions |
| [M-4] **mathematical** | Important mathematical functions |

Also See

[M-4] **intro** — Index and guide to functions

Title

[M-4] **programming** — Programming functions

Contents

[M-5] Manual entry	Function	Purpose
Argument and caller-preference processing		
args( )	args()	number of arguments
isfleeting( )	isfleeting()	whether argument is temporary
callersversion( )	callersversion()	obtain version number of caller
favorspeed( )	favorspeed()	whether speed or space is to be favored
Advanced parsing		
tokenget( )	tokeninit()	initialize parsing environment
	tokeninitstata()	initialize environment as Stata would
	tokenset()	set/reset string to be parsed
	tokengetall()	parse entire string
	tokenget()	parse next element of string
	tokenpeek()	peek at next tokenget() result
	tokenrest()	return yet-to-be-parsed portion
	tokenoffset()	query/reset offset in string
	tokenwchars()	query/reset white-space characters
	tokenpchars()	query/reset parsing characters
	tokenqchars()	query/reset quote characters
	tokenallownum()	query/reset number parsing
	tokenallowhex()	query/reset hex-number parsing
Accessing externals		
findexternal( )	findexternal()	find global
	crexternal()	create global
	rmexternal()	remove global
	nameexternal()	name of external
direxternal( )	direxternal()	obtain list of existing globals
valofexternal( )	valofexternal()	obtain value of global

Break key

setbreakintr( )
- setbreakintr() — turn off/on break-key interrupt
- querybreakintr() — whether break-key interrupt is off/on
- breakkey() — whether break key has been pressed
- breakkeyreset() — reset break key

Miscellaneous

assert( )
- assert() — abort execution if not true
- asserteq() — abort execution if not equal

c( )
- c() — access c() value

sizeof( )
- sizeof() — number of bytes consumed by object

swap( )
- swap() — interchange contents of variables

System info

byteorder( )
- byteorder() — byte order used by computer

stataversion( )
- stataversion() — version of Stata being used
- statasetversion() — version of Stata set

Exiting

exit( )
- exit() — terminate execution

error( )
- error() — issue standard Stata error message
- _error() — issue error message with traceback log

Also See

[M-4] **intro** — Index and guide to functions

Title

[M-4] **scalar** — Scalar mathematical functions

Contents

[M-5] Manual entry	Function	Purpose
Complex		
Re( )	Re( )	real part
	Im( )	imaginary part
C( )	C( )	make complex
Sign related		
abs( )	abs( )	absolute value (length if complex)
sign( )	sign( )	sign function
	quadrant( )	quadrant of value
dsign( )	sign( )	FORTRAN-like DSIGN function
conj( )	conj( )	complex conjugate
Transcendental & square root		
exp( )	exp( )	exponentiation
	ln( ), log( )	natural logarithm
	log10( )	base-10 logarithm
sqrt( )	sqrt( )	square root
sin( )	sin( )	sine
	cos( )	cosine
	tan( )	tangent
	asin( )	arcsine
	acos( )	arccosine
	atan( )	arctangent
	arg( )	arctangent of complex
	atan2( )	two-argument arctangent
	sinh( )	hyperbolic sine
	cosh( )	hyperbolic cosine
	tanh( )	hyperbolic tangent
	asinh( )	arc-hyperbolic sine
	acosh( )	arc-hyperbolic cosine
	atanh( )	arc-hyperbolic tangent
	pi( )	value of π

Factorial & Gamma

factorial( )
- `factorial()` — factorial
- `lnfactorial()` — natural logarithm of factorial
- `gamma()` — Gamma function
- `lngamma()` — natural logarithm of Gamma function
- `digamma()` — derivative of `lngamma()`
- `trigamma()` — second derivative of `lngamma()`

Modulus & integer rounding

mod( )
- `mod()` — modulus

trunc( )
- `trunc()` — truncate to integer
- `floor()` — round down to integer
- `ceil()` — round up to integer
- `round()` — round to closest integer or multiple

Dates

date( )
- `clock()` — %tc of string
- `mdyhms()` — %tc of month, day, year, hour, minute, and second
- `dhms()` — %tc of %td, hour, minute, and second
- `hms()` — %tc of hour, minute, and second
- `hh()` — hour of %tc
- `mm()` — minute of %tc
- `ss()` — second of %tc
- `dofc()` — %td of %tc

- `Cofc()` — %tC of %tc
- `Clock()` — %tC of string
- `Cmdyhms()` — %tC of month, day, year, hour, minute, and second
- `Cdhms()` — %tC of %td, hour, minute, and second
- `Chms()` — %tC of hour, minute, and second
- `hhC()` — hour of %tC
- `mmC()` — minute of %tC
- `ssC()` — second of %tC
- `dofC()` — %td of %tC

- `date()` — %td of string
- `mdy()` — %td of month, day, and year
- `yw()` — %tw of year and week
- `ym()` — %tm of year and month
- `yq()` — %tq of year and quarter
- `yh()` — %th of year and half
- `cofd()` — %tc of %td
- `Cofd()` — %tC of %td

date( ), *continued*	month()	month of %td
	day()	day-of-month of %td
	year()	year of %td
	dow()	day-of-week of %td
	week()	week of %td
	quarter()	quarter of %td
	halfyear()	half-of-year of %td
	doy()	day-of-year of %td
	yearly()	%ty of string
	yofd()	%ty of %td
	dofy()	%td of %ty
	halfyearly()	%th of string
	hofd()	%th of %td
	dofh()	%td of %th
	quarterly()	%tq of string
	qofd()	%tq of %td
	dofq()	%td of %tq
	monthly()	%tm of string
	mofd()	%tm of %td
	dofm()	%td of %tm
	weekly()	%tw of string
	wofd()	%tw of %td
	dofw()	%td of %tw
	hours()	hours of milliseconds
	minutes()	minutes of milliseconds
	seconds()	seconds of milliseconds
	msofhours()	milliseconds of hours
	msofminutes()	milliseconds of minutes
	msofseconds()	milliseconds of seconds

Description

With a few exceptions, the above functions are what most people would consider scalar functions, although in fact all will work with matrices, in an element-by-element fashion.

Remarks

For other mathematical functions, see

    [M-4] **matrix**   Matrix functions
    [M-4] **mathematical**   Important mathematical functions
    [M-4] **statistical**   Statistical functions

Also See

[M-4] **intro** — Index and guide to functions

Title

[M-4] **solvers** — Functions to solve AX=B and to obtain A inverse

Contents

[M-5] Manual entry	Function	Purpose
Solvers		
cholsolve( )	cholsolve()	A positive definite, symmetric or Hermitian
lusolve( )	lusolve()	A full rank, square, real or complex
qrsolve( )	qrsolve()	A general; $m \times n$, $m \geq n$, real or complex; least-squares generalized solution
svsolve( )	svsolve()	generalized; $m \times n$, real or complex; minimum norm, least-squares solution
Inverters		
invsym( )	invsym()	generalized; real symmetric
cholinv( )	cholinv()	positive definite; symmetric or Hermitian
luinv( )	luinv()	full rank; square; real or complex
qrinv( )	qrinv()	generalized; $m \times n$, $m \geq n$; real or complex
pinv( )	pinv()	generalized; $m \times n$, real or complex Moore–Penrose pseudoinverse

Description

The above functions solve $AX = B$ for X and solve for A^{-1}.

Remarks

Matrix solvers can be used to implement matrix inverters, and so the two nearly always come as a pair.

Solvers solve $AX = B$ for X. One way to obtain A^{-1} is to solve $AX = I$. If $f(A, B)$ solves $AX=B$, then $f(A, \text{I(rows}(A)))$ solves for the inverse. Some matrix inverters are in fact implemented this way, although usually custom code is written because memory savings are possible when it is known that $B = I$.

The pairings of inverter and solver are

inverter	solver
invsym()	(none)
cholinv()	cholsolve()
luinv()	lusolve()
qrinv()	qrsolve()
pinv()	svsolve()

Also See

[M-4] **intro** — Index and guide to functions

Title

[M-4] standard — Functions to create standard matrices

Contents

[M-5] Manual entry	Function	Purpose
Unit & constant matrices		
I( )	I()	identity matrix
e( )	e()	unit vectors
J( )	J()	matrix of constants
designmatrix( )	designmatrix()	design matrices
Block-diagonal matrices		
blockdiag( )	blockdiag()	block-diagonal matrix
Ranges		
range( )	range()	vector over specified range
unitcircle( )	unitcircle()	unit circle on complex plane
Random		
uniform( )	uniform()	uniformly distributed random numbers
	uniformseed()	obtain or set uniform() seed
Named matrices		
Hilbert( )	Hilbert()	Hilbert matrices
	invHilbert()	inverse Hilbert matrices
Toeplitz( )	Toeplitz()	Toeplitz matrices
Vandermonde( )	Vandermonde()	Vandermonde matrices

Description

The functions above create standard matrices such as the identity matrix, etc.

Remarks

For other mathematical functions, see

    [M-4] **matrix**   Matrix mathematical functions
    [M-4] **scalar**   Scalar mathematical functions
    [M-4] **mathematical**   Important mathematical functions

Also See

[M-4] **intro** — Index and guide to functions

Title

[M-4] **stata** — Stata interface functions

Contents

[M-5] Manual entry	Function	Purpose
Access to data		
st_nvar( )	st_nvar()	number of variables
	st_nobs()	number of observations
st_data( )	st_data()	load numeric data from Stata into matrix
	st_sdata()	load string data from Stata into matrix
st_store( )	st_store()	store numeric data in Stata dataset
	st_sstore()	store string data in Stata dataset
st_view( )	st_view()	make view onto num. data in Stata dataset
	st_sview()	make view onto string data in Stata dataset
st_subview( )	st_subview()	make view from view
st_viewvars( )	st_viewvars()	identify variables and observations
	st_viewobs()	corresponding to view
Variable names & indices		
st_varindex( )	st_varindex()	variable indices from variable names
st_varname( )	st_varname()	variable names from variable indices

Variable characteristics

st_varrename( )	st_varrename()	rename Stata variable
st_vartype( )	st_vartype()	storage type of Stata variable
	st_isnumvar()	whether variable is numeric
	st_isstrvar()	whether variable is string
st_varformat( )	st_varformat()	obtain/set format of Stata variable
	st_varlabel()	obtain/set variable label
	st_varvaluelabel()	obtain/set value label
st_vlexists( )	st_vlexists()	whether value label exists
	st_vldrop()	drop value
	st_vlmap()	map values
	st_vlsearch()	map text
	st_vlload()	load value label
	st_vlmodify()	create or modify value label

Temporary variables & time-series operators

st_tempname( )	st_tempname()	temporary variable name
	st_tempfilename()	temporary filename
st_tsrevar( )	st_tsrevar()	create time-series op.varname

Adding & removing variables & observations

st_addobs( )	st_addobs()	add observations to Stata dataset
st_addvar( )	st_addvar()	add variable to Stata dataset
st_dropvar( )	st_dropvar()	drop variables
	st_dropobsin()	drop specified observations
	st_dropobsif()	drop selected observations
	st_keepvar()	keep variables
	st_keepobsin()	keep specified observations
	st_keepobsif()	keep selected observations
st_updata( )	st_updata()	query/set data-have-changed flag

Executing Stata commands

stata( )	stata()	execute Stata command
st_macroexpand( )	st_macroexpand()	expand Stata macros

28   [M-4] stata — Stata interface functions

Accessing e(), r(), s(), macros, matrices, etc.

st_global( )	st_global()	obtain/set Stata global
st_local( )	st_local()	obtain/set local Stata macro
st_numscalar( )	st_numscalar()	obtain/set Stata numeric scalar
	st_strscalar()	obtain/set Stata string scalar
st_matrix( )	st_matrix()	obtain/set Stata matrix
	st_matrixrowstripe()	obtain/set row labels
	st_matrixcolstripe()	obtain/set column labels
	st_replacematrix()	replace existing Stata matrix
st_dir( )	st_dir()	obtain list of Stata objects
st_rclear( )	st_rclear()	clear r()
	st_eclear()	clear e()
	st_sclear()	clear s()

Parsing & verification

st_isname( )	st_isname()	whether valid Stata name
	st_islmname()	whether valid local macro name
st_isfmt( )	st_isfmt()	whether valid %*fmt*
	st_isnumfmt()	whether valid numeric %*fmt*
	st_isstrfmt()	whether valid string %*fmt*

Description

The above functions interface with Stata.

Remarks

The following manual entries have to do with getting data from or putting data into Stata:

[M-5] **st_data( )**	Load copy of current Stata dataset
[M-5] **st_view( )**	Make matrix that is a view onto current Stata dataset
[M-5] **st_store( )**	Modify values stored in current Stata dataset
[M-5] **st_nvar( )**	Numbers of variables and observations

In some cases, you may find yourself needing to translate variable names into variable indices and vice versa:

[M-5] **st_varname( )**	Obtain variable names from variable indices
[M-5] **st_varindex( )**	Obtain variable indices from variable names
[M-5] **st_tsrevar( )**	Create time-series op.varname variables

The other functions mostly have to do with getting and putting Stata's scalars, matrices, and returned results:

[M-5]	**st_local( )**	Obtain strings from and put strings into Stata
[M-5]	**st_global( )**	Obtain strings from and put strings into global macros
[M-5]	**st_numscalar( )**	Obtain values from and put values into Stata scalars
[M-5]	**st_matrix( )**	Obtain and put Stata matrices

The stata() function, documented in

[M-5]	**stata( )**	Execute Stata command

allows you to cause Stata to execute a command that you construct in a string.

Also See

[M-4] **intro** — Index and guide to functions

Title

[M-4] **statistical** — Statistical functions

Contents

[M-5] Manual entry	Function	Purpose
Random numbers		
uniform( )	uniform()	uniformly distributed random numbers
	uniformseed()	obtain or set uniform() seed
Means, variances, & correlations		
mean( )	mean()	mean
	variance()	variance
	quadvariance()	quad-precision variance
	meanvariance()	mean and variance
	quadmeanvariance()	quad-precision mean and variance
	correlation()	correlation
	quadcorrelation()	quad-precision correlation
cross( )	cross()	$X'X$, $X'Z$, $X'\text{diag}(w)Z$, etc.
corr( )	corr()	make correlation from variance matrix
crossdev( )	crossdev()	$(X{:}-x)'(X{:}-x)$, $(X{:}-x)'(Z{:}-z)$, etc.
quadcross( )	quadcross()	quad-precision cross()
	quadcrossdev()	quad-precision crossdev()
Factorial & combinations		
factorial( )	factorial()	factorial
	lnfactorial()	natural logarithm of factorial
	gamma()	Gamma function
	lngamma()	natural logarithm of Gamma function
	digamma()	derivative of lngamma()
	trigamma()	second derivative of lngamma()
comb( )	comb()	combinatorial function n choose k
cvpermute( )	cvpermutesetup()	permutation setup
	cvpermute()	return permutations, one at a time

Densities & distributions

normal( )

normalden()	normal density	
normal()	cumulative normal dist.	
invnormal()	inverse cumulative normal	
lnnormalden()	logarithm of the normal density	
lnnormal()	logarithm of the cumulative normal dist.	

binormal()	cumulative binormal dist.

chi2()	cumulative chi-squared dist.
chi2tail()	reverse cumulative chi-squared dist.
invchi2()	inverse cumulative chi-squared
invchi2tail()	inverse reverse cumulative chi-squared

nchi2()	noncentral cumulative chi-squared dist.
invnchi2()	inverse noncentral cumulative chi-squared
npnchi2()	noncentrality parameter of nchi2()

tden()	Student t density
ttail()	reverse cumulative t dist.
invttail()	inverse reverse cumulative t

Fden()	F density
F()	cumulative F dist.
Ftail()	reverse cumulative F dist.
invF()	inverse cumulative F
invFtail()	inverse reverse cumulative F

nFden()	noncentral F density
nFtail()	reverse cumulative noncentral F dist.
invnFtail()	inverse reverse cumulative noncentral F

binomial()	cumulative of binomial
binomialtail()	right cumulative of binomial
invbinomial()	inverse binomial (lower tail)
invbinomialtail()	inverse binomial (lower tail)

betaden()	beta density
ibeta()	cumulative beta dist.; a.k.a. incomplete beta function
ibetatail()	reverse cumulative beta dist.
invibeta()	inverse cumulative beta
invibetatail()	inverse reverse cumulative beta

normal( ), *continued*

nbetaden()	noncentral beta density
nibeta()	cumulative noncentral beta dist.
invnibeta()	inverse cumulative noncentral beta
gammaden()	Gamma density
gammap()	cumulative Gamma dist.; a.k.a. incomplete gamma function
gammaptail()	reverse cumulative Gamma dist.;
invgammap()	inverse cumulative Gamma
invgammaptail()	inverse reverse cumulative Gamma
dgammapda()	dg/da
dgammapdx()	dg/dx
dgammapdada()	d^2g/da^2
dgammapdadx()	$d^2g/dadx$
dgammapdxdx()	d^2g/dx^2

Maximization & minimization

optimize( )

optimize()	function maximization and minimization
optimize_init()	begin optimization
optimize_init_*()	set details
optimize()	perform optimization
optimize_result_*()	access results

Logits, odds, & related

logit( )

logit()	log of the odds ratio
invlogit()	inverse log of the odds ratio
cloglog()	complementary log-log
invcloglog()	inverse complementary log-log

Multivariate normal

ghk( )
ghkfast( )

ghk()	GHK multivariate normal (MVN) simulator
ghkfastsetup()	GHK MVN initialization
ghkfast()	GHK MVN simulated probabilities

Description

The above functions are statistical, probabilistic, or designed to work with data matrices.

Remarks

Concerning data matrices, see

    [M-4] **stata**   Stata interface functions

and especially

    [M-5] **st_data( )**   Load copy of current Stata dataset
    [M-5] **st_view( )**   Make matrix that is a view onto current Stata dataset

For other mathematical functions, see

    [M-4] **matrix**   Matrix mathematical functions
    [M-4] **scalar**   Scalar mathematical functions
    [M-4] **mathematical**   Important mathematical functions

Also See

[M-4] **intro** — Index and guide to functions

Title

[M-4] **string** — String manipulation functions

Contents

[M-5] Manual entry	Function	Purpose
Parsing		
tokens( )	tokens()	obtain tokens (words) from string
invtokens( )	invtokens()	concatenate string vector into string scalar
strmatch( )	strmatch()	pattern matching
tokenget( )	...	advanced parsing
Length & position		
strlen( )	strlen()	length of string
fmtwidth( )	fmtwidth()	width of %*fmt*
strpos( )	strpos()	find substring within string
indexnot( )	indexnot()	find character not in list
Editing		
substr( )	substr()	extract substring
strupper( )	strupper()	convert to uppercase
	strlower()	convert to lowercase
	strproper()	convert to proper case
strtrim( )	stritrim()	replace multiple, consecutive internal blanks with one blank
	strltrim()	remove leading blanks
	strrtrim()	remove trailing blanks
	strtrim()	remove leading and trailing blanks
subinstr( )	subinstr()	substitute text
	subinword()	substitute word
_substr( )	_substr()	substitute into string
strdup( )	*	duplicate string
strreverse( )	strreverse()	reverse string

[M-4] string — String manipulation functions

```
┌─ ASCII translation ─────────────────────────────────────────┐
strofreal( )      strofreal()      convert real to string
strtoreal( )      strtoreal()      convert string to real
ascii( )          ascii()          obtain ASCII codes of string
                  char()           make string from ASCII codes
```

Description

The above functions are for manipulating strings. Strings in Mata are strings of ASCII characters, usually the printable characters, but Mata enforces no such restriction. In particular, strings may contain binary 0.

Remarks

In addition to the above functions, two operators are especially useful for dealing with strings.

The first is +. Addition is how you concatenate strings:

```
: "abc" + "def"
abcdef
: command = "list"
: args = "mpg weight"
: result = command + " " + args
: result
list mpg weight
```

The second is *. Multiplication is how you duplicate strings:

```
: 5*"a"
aaaaa
: "b"*3
bbb
: indent = 20
: title = indent*" " + "My Title"
: title
                    My Title
```

Also See

[M-4] **intro** — Index and guide to functions

Title

[M-4] utility — Matrix utility functions

Contents

[M-5] Manual entry	Function	Purpose
Complex		
Re( )	Re()	real part
	Im()	imaginary part
C( )	C()	make complex
Shape & type		
rows( )	rows()	number of rows
	cols()	number of columns
	length()	number of elements of vector
eltype( )	eltype()	element type of object
	orgtype()	organizational type of object
isreal( )	isreal()	object is real matrix
	iscomplex()	object is complex matrix
	isstring()	object is string matrix
	ispointer()	object is pointer matrix
isrealvalues( )	isrealvalues()	whether matrix contains only real values
isview( )	isview()	whether matrix is view
Properties		
issymmetric( )	issymmetric()	whether matrix is symmetric (Hermitian)
	issymmetriconly()	whether matrix is mechanically symmetric
isdiagonal( )	isdiagonal()	whether matrix is diagonal
diag0cnt( )	diag0cnt()	count 0s on diagonal
Selection		
select( )	select()	select rows or columns
	st_select()	select rows or columns of view

[M-4] utility — Matrix utility functions

Missing values
---

missing()	missing()	count of missing values
	rowmissing()	count of missing values, by row
	colmissing()	count of missing values, by column
	nonmissing()	count of nonmissing values
	rownonmissing()	count of nonmissing values, by row
	colnonmissing()	count of nonmissing values, by column
missingof()	missingof()	appropriate missing value

Range, sums, & cross products
---

minmax()	rowmin()	minimum, by row
	colmin()	minimum, by column
	min()	minimum, overall
	rowmax()	maximum, by row
	colmax()	maximum, by column
	max()	maximum, overall
	rowminmax()	minimum and maximum, by row
	colminmax()	minimum and maximum, by column
	minmax()	minimum and maximum, overall
	rowmaxabs()	rowmax(abs())
	colmaxabs()	colmax(abs())
minindex()	minindex()	indices of minimums
	maxindex()	indices of maximums
sum()	rowsum()	sum of each row
	colsum()	sum of each column
	sum()	overall sum
	quadrowsum()	quad-precision sum of each row
	quadcolsum()	quad-precision sum of each column
	quadsum()	quad-precision overall sum
runningsum()	runningsum()	running sum of vector
	quadrunningsum()	quad-precision runningsum()
cross()	cross()	$X'X$, $X'Z$, etc.
crossdev()	crossdev()	$(X:-x)'(X:-x)$, $(X:-x)'(Z:-z)$, etc.
quadcross()	quadcross()	quad-precision cross()
	quadcrossdev()	quad-precision crossdev()

[M-4] utility — Matrix utility functions

――― Programming ―――

reldif( )	reldif()	relative difference
	mreldif()	max. relative difference between matrices
	mreldifsym()	max. relative difference from symmetry
	mreldifre()	max. relative difference from real
all( )	all()	sum(!L)==0
	any()	sum(L)!=0
	allof()	all(P:==s)
	anyof()	any(P:==s)
panelsetup( )	panelsetup()	initialize panel-data processing
	panelstats()	summary statistics on panels
	panelsubmatrix()	obtain matrix for panel i
	panelsubview()	obtain view matrix for panel i

――― Constants & tolerances ―――

mindouble( )	mindouble()	minimum nonmissing value
	maxdouble()	maximum nonmissing value
	smallestdouble()	smallest $e > 0$
epsilon( )	epsilon()	unit roundoff error
floatround( )	floatround()	round to float precision
solve_tol( )	solve_tol()	tolerance used by solvers and inverters

Description

Matrix utility functions tell you something about the matrix, such as the number of rows or whether it is diagonal.

Remarks

There is a thin line between utility and manipulation; also see

    [M-4] **manipulation**   Matrix manipulation functions

Also See

[M-4] **intro** — Index and guide to functions

[M-5] Mata functions

Title

[M-5] **intro** — Mata functions

Contents

See [M-4] **intro**

Description

The following entries, alphabetically arranged, document all the Mata functions.

Remarks

See [M-4] **intro** for an index, grouped logically, of the functions presented in this section.

Also See

[M-4] **intro** — Index and guide to functions

[M-0] **intro** — Introduction to the Mata manual

Title

[M-5] **abs( )** — Absolute value (length)

Syntax

>*real matrix* abs(*numeric matrix Z*)

Description

For Z real, abs(Z) returns the elementwise absolute values of Z.

For Z complex, abs(Z) returns the elementwise length of each element. If $Z = a + bi$, returned is $\operatorname{sqrt}(a^2 + b^2)$, although the calculation is not made in that way. The method actually used prevents overflow.

Conformability

abs(Z):
>>Z: $r \times c$
>>*result*: $r \times c$

Diagnostics

abs(.) returns . (missing).

Also See

[M-4] **scalar** — Scalar mathematical functions

Title

[M-5] **adosubdir( )** — Determine ado-subdirectory for file

Syntax

>   *string scalar* adosubdir(*string scalar filename*)

Description

adosubdir(*filename*) returns the subdirectory in which Stata would search for *filename*. Typically, the subdirectory will be simply the first letter of *filename*. However, certain files may result in a different subdirectory, depending on their extension.

Remarks

adosubdir("xyz.ado") returns "x" because Stata ado-files are located in subdirectories with name given by their first letter.

adosubdir("xyz.style") returns "style" because Stata style files are located in subdirectories named style.

Conformability

>   adosubdir(*filename*):
>   *filename*: 1×1
>   *result*: 1×1

Diagnostics

adosubdir() returns the first letter of the filename if the filetype is unknown to Stata, thus treating unknown filetypes as if they were ado-files.

adosubdir() aborts with error if the filename is too long for the operating system; nothing else causes abort with error.

Also See

[M-4] **io** — I/O functions

Title

[M-5] **all( )** — Element comparisons

Syntax

>*real scalar* `all(`*real matrix L*`)`

>*real scalar* `any(`*real matrix L*`)`

>*real scalar* `allof(`*transmorphic matrix P, transmorphic scalar s*`)`

>*real scalar* `anyof(`*transmorphic matrix P, transmorphic scalar s*`)`

Description

`all(`*L*`)` is equivalent to `sum(!`*L*`)==0` but is significantly faster.

`any(`*L*`)` is equivalent to `sum(`*L*`)!=0` but is slightly faster.

`allof(`*P*, *s*`)` returns 1 if every element of *P* equals *s* and returns 0 otherwise. `allof(`*P*, *s*`)` is faster and consumes less memory than the equivalent construction `all(`*P*`:==`*s*`)`.

`anyof(`*P*, *s*`)` returns 1 if any element of *P* equals *s* and returns 0 otherwise. `anyof(`*P*, *s*`)` is faster and consumes less memory than the equivalent `any(`*P*`:==`*s*`)`.

Remarks

These functions are fast, so their use is encouraged over alternative constructions.

`all()` and `any()` are typically used with logical expressions to detect special cases, such as

>   if (any(x :< 0)) {
>   ...
>   }

or

>   if (all(x :>= 0)) {
>   ...
>   }

`allof()` and `anyof()` are used to look for special values:

>   if (allof(x, 0)) {
>   ...
>   }

or

```
        if (anyof(x, 0)) {
                . . .
        }
```

Do not use allof() and anyof() to check for missing values—e.g., anyof(x, .)—because to really check, you would have to check not only . but also .a, .b, ..., .z. Instead use missing(); see [M-5] **missing( )**.

Conformability

all(*L*), any(*L*):
    L: $r \times c$
    result: 1×1

allof(*P*, *s*), anyof(*P*, *s*):
    P: $r \times c$
    s: 1×1
    result: 1×1

Diagnostics

all(*L*) and any(*L*) treat missing values in *L* as true.

all(*L*) and any(*L*) return 0 (false) if *L* is $r \times 0$, $0 \times c$, or 0×0.

allof(*P*, *s*) and anyof(*P*, *s*) return 0 (false) if *P* is $r \times 0$, $0 \times c$, or 0×0.

Also See

[M-4] **utility** — Matrix utility functions

Title

[M-5] **args( )** — Number of arguments

Syntax

> *real scalar* `args()`

Description

`args()` returns the number of arguments actually passed to the function; see [M-2] **optargs**.

Conformability

`args()`:
>*result:* 1×1

Diagnostics

None.

Also See

[M-4] **programming** — Programming functions

Title

[M-5] **ascii( )** — Manipulate ASCII codes

Syntax

> *real rowvector* `ascii(`*string scalar s*`)`

> *string scalar* `char(`*real rowvector c*`)`

Description

`ascii(`*s*`)` returns a row vector containing the ASCII codes corresponding to *s*. For instance, `ascii("abc")` returns (97, 98, 99).

`char(`*c*`)` returns a string consisting of the specified ASCII codes. For instance, `char((97, 98, 99))` returns `"abc"`.

Conformability

`ascii(`*s*`)`:
    s: 1 × 1
    result: 1 × `strlen(`*s*`)`

`char(`*c*`)`:
    c: 1 × *n*, *n* ≥ 0
    result: 1 × 1

Diagnostics

`ascii(`*s*`)` returns `J(1,0,.)` if `strlen(`*s*`)==0`.

In `char(`*c*`)`, if any element of *c* is outside the range 0 to 255, the returned string is terminated at that point. For instance, `char((97,98,99,1000,97,98,99))="abc"`.

`char(J(1,0,.))` returns `""`.

Also See

[M-4] **string** — String manipulation functions

Title

[M-5] **assert( )** — Abort execution if false

Syntax

> *void* `assert(`*real scalar r*`)`
>
> *void* `asserteq(`*transmorphic matrix A*, *transmorphic matrix B*`)`

Description

`assert(`*r*`)` produces the error message "assertion is false" and aborts with error if $r == 0$.

`asserteq(`*A, B*`)` is logically equivalent to `assert(`*A==B*`)`. If the assertion is false, however, information is presented on the number of mismatches.

Remarks

In the midst of complicated code, you know that a certain calculation must produce a result greater than 0, but you worry that perhaps you have an error in your code:

```
...
assert(n>0)
...
```

In another spot, you have produced matrix A and know every element of A should be positive or zero:

```
...
assert(A:>=0)
...
```

Once you are convinced that your function works, these verifications should be removed. In a third part of your code, however, the problem is different if the number of rows *r* exceed the number of columns *c*. In all the cases you need to use it, however, *r* will be less than *c*, so you are not much interested in programming the alternative solution:

```
...
assert(rows(PROBLEM) < cols(PROBLEM))
...
```

Leave that one in.

Conformability

```
assert(r):
            r:      1 × 1
       result:      void
```

asserteq(A, B):
        A: $r_1 \times c_1$
        B: $r_2 \times c_2$
   result: void

Diagnostics

assert(r) aborts with error if $r == 0$.

asserteq(A, B) aborts with error if $A \neq B$.

Also See

[M-4] **programming** — Programming functions

Title

[M-5] blockdiag( ) — Block-diagonal matrix

Syntax

>   *numeric matrix* `blockdiag(`*numeric matrix* Z_1, *numeric matrix* Z_2`)`

Description

`blockdiag`(Z_1, Z_2) returns a block-diagonal matrix with Z_1 in the upper-left corner and Z_2 in the lower right, i.e.,

$$\begin{bmatrix} Z_1 & \mathbf{0} \\ \mathbf{0} & Z_2 \end{bmatrix}$$

Z_1 and Z_2 may be either real or complex and need not be of the same type.

Remarks

To create a block diagonal matrix of Z_1, Z_2, Z_3, code

>   : `blockdiag(`*Z1*, `blockdiag(`*Z2*,*Z3*`))`

Conformability

`blockdiag`(Z_1, Z_2):
$\quad\quad Z_1: \quad r_1 \times c_1$
$\quad\quad Z_2: \quad r_2 \times c_2$
$\quad\ result: \quad r_1 + r_2 \times c_1 + c_2$

Diagnostics

None. Either or both Z_1 and Z_2 may be void.

Also See

[M-4] **standard** — Functions to create standard matrices

Title

> **[M-5] bufio( )** — Buffered (binary) I/O

Syntax

colvector C =	bufio()
real scalar	bufbyteorder(*C*)
void	bufbyteorder(*C*, *real scalar byteorder*)
real scalar	bufmissingvalue(*C*)
void	bufmissingvalue(*C*, *real scalar version*)
void	bufput(*C*, *B*, *offset*, *bfmt*, *X*)
scalar	bufget(*C*, *B*, *offset*, *bfmt*)
rowvector	bufget(*C*, *B*, *offset*, *bfmt*, *c*)
matrix	bufget(*C*, *B*, *offset*, *bfmt*, *r*, *c*)
void	fbufput(*C*, *fh*, *bfmt*, *X*)
scalar	fbufget(*C*, *fh*, *bfmt*)
rowvector	fbufget(*C*, *fh*, *bfmt*, *c*)
matrix	fbufget(*C*, *fh*, *bfmt*, *r*, *c*)
real scalar	bufbfmtlen(*string scalar bfmt*)
real scalar	bufbfmtisnum(*string scalar bfmt*)

where

C:	*colvector* returned by bufio()
B:	*string scalar* (buffer)
offset:	*real scalar* (buffer position, starts at 0)
fh:	file handle returned by fopen()
bfmt:	*string scalar* (binary format; see below)
r:	*string scalar*
c:	*string scalar*
X:	value to be written; see *Remarks*

bfmt may contain

bfmt	meaning
%{ 8|4 }z	8-byte double or 4-byte double
%{ 4|2|1 }b[s|u]	4-, 2-, or 1-byte integer; Stata, signed, or unsigned
%#s	text string
%#S	binary string

Description

These functions manipulate buffers (string scalars) containing binary data and, optionally, perform I/O.

bufio() returns a control vector, C, that you pass to the other buffer functions. C specifies the byte order of the buffer and specifies how missing values are to be encoded. Despite its name, bufio() opens no files and performs no I/O. bufio() merely returns a vector of default values for use with the remaining buffer functions.

bufbyteorder() and bufmissingvalue() allow changing the defaults in C.

bufput() and bufget() copy elements into and out of buffers. No I/O is performed. Buffers can then be written by using fwrite() and read by using fread(); see [M-5] **fopen( )**.

fbufput() and fbufget() do the same, and they perform the corresponding I/O by using fwrite() or fread().

bufbfmtlen(*bfmt*) and bufbfmtisnum(*bfmt*) are utility routines for processing *bfmts*; they are rarely used. bufbfmtlen(*bfmt*) returns the implied length, in bytes, of the specified *bfmt*, and bufbfmtisnum(*bfmt*) returns 1 if the *bfmt* is numeric, 0 if string.

Remarks

If you wish simply to read and write matrices, etc., see fgetmatrix() and fputmatrix() and the other functions in [M-5] **fopen( )**.

The functions documented here are of interest if

1. you wish to create your own binary-data format because you are writing routines in low-level languages such as FORTRAN or C and need to transfer data between your new routines and Stata, or

2. you wish to write a program to read and write the binary format of another software package.

These are advanced and tedious programming projects.

Remarks are presented under the following headings:

> *Basics*
> *Argument C*
> *Arguments B and offset*
> *Argument fh*
> *Argument bfmt*
> *bfmts for numeric data*
> *bfmts for string data*
> *Argument X*
> *Arguments r and c*
> *Advanced issues*

Basics

Let's assume that you wish to write a matrix to disk so you can move it back and forth from FORTRAN. You settle on a file format in which the number of rows and number of columns are first written as 4-byte integers, and then the values of the matrix are written as 8-byte doubles, by row:

One solution to writing matrices in such a format is

```
fh = fopen("filename", "w")
C  = bufio()
fbufput(C, fh, "%4b", rows(X))
fbufput(C, fh, "%4b", cols(X))
fbufput(C, fh, "%8z", X)
fclose(fh)
```

The code to read the matrix back is

```
fh   = fopen("filename", "r")
C    = bufio()
rows = fbufget(C, fh, "%4b")
cols = fbufget(C, fh, "%4b")
X    = fbufget(C, fh, "%8z", rows, cols)
fclose(fh)
```

Another solution, which would be slightly more efficient, is

```
fh = fopen("filename", "w")
C  = bufio()
buf = 8*char(0)
bufput(C, buf, 0, "%4b", rows(X))
bufput(C, buf, 4, "%4b", cols(X))
fwrite(C, buf)
fbufput(C, fh, "%8z", X)
fclose(fh)
```

and

```
fh   = fopen("filename", "r")
C    = bufio()
buf  = fread(fh, 8)
rows = bufget(C, buf, 0, "%4b")
cols = bufget(C, buf, 4, "%4b")
X    = fbufget(C, fh, "%8z", rows, cols)
fclose(fh)
```

What makes the above approach more efficient is that, rather than writing 4 bytes (the number of rows), and 4 bytes again (the number of columns), we created one 8-byte buffer and put the two 4-byte fields in it, and then we wrote all 8 bytes at once. We did the reverse when we read back the data: we read all 8 bytes and then broke out the fields. The benefit is minuscule here but, in general, writing longer buffers results in faster I/O.

In all the above examples, we wrote and read the entire matrix with one function call,

```
fbufput(C, fh, "%8z", X)
```

and

```
X    = fbufget(C, fh, "%8z", rows, cols)
```

Perhaps you would have preferred our having coded

```
for (i=1; i<=rows(X); i++) {
        for (j=1; j<=cols(X); j++) {
                fbufput(C, fh, "%8z", X[i,j])
        }
}
```

and perhaps you would have preferred our having coded something similar to read back the matrix. Had we done so, the results would have been the same.

If you are familiar with FORTRAN, you know that it records matrices in column-dominant order, rather than the row-dominant order used by Mata. It would be a little easier to code the FORTRAN side of things if we changed our file-format design to write columns first:

# rows	# cols	X[1,1]	X[2,1]	...
4 bytes	4 bytes	8 bytes	8 bytes	

One way we could do that would be to write the loops ourselves:

```
fh = fopen("filename", "w")
C  = bufio()
fbufput(C, fh, "%4b", rows(X))
fbufput(C, fh, "%4b", cols(X))
for (j=1; j<=cols(X); i++) {
        for (i=1; i<=rows(X); j++) {
                fbufput(C, fh, "%8z", X[i,j])
        }
}
```

and
```
fh   = fopen("filename", "r")
C    = bufio()
rows = fbufget(C, fh, "%4b")
cols = fbufget(C, fh, "%4b")
X    = J(rows, cols, .)
for (j=1; j<=cols(X); i++) {
        for (i=1; i<=rows(X); j++) {
                X[i,j] = fbufget(C, fh, "%8z")
        }
}
```

We could do that, but there are more efficient and easier ways to proceed. For instance, we could simply transpose the matrix before writing and after reading, and if we do that transposition in place, our code will use no extra memory:

```
fh = fopen("filename", "w")
C  = bufio()
fbufput(C, fh, "%4b", rows(X))
fbufput(C, fh, "%4b", cols(X))
_transpose(X)
fbufput(C, fh, "%8z", X)
_transpose(X)
fclose(fh)
```

The code to read the matrices back is

```
fh   = fopen("filename", "r")
C    = bufio()
rows = fbufget(C, fh, "%4b")
cols = fbufget(C, fh, "%4b")
X    = fbufget(C, fh, "%8z", cols, rows)
_transpose(X)
fclose(fh)
```

Argument C

Argument C in

> bufput(C, B, *offset*, *bfmt*, X),
>
> bufget(C, B, *offset*, *bfmt*, ...),
>
> fbufput(C, *fh*, *bfmt*, X), and
>
> fbufget(C, *fh*, *bfmt*, ...)

specifies the control vector. You obtain C by coding

```
C = bufio()
```

bufio() returns C, which is nothing more than a vector filled in with default values. The other buffer routines know how to interpret the vector. The vector contains two pieces of information:

1. The byte order to be used
2. The missing-value coding scheme to be used

Some computer hardware writes numbers left to right (e.g., Sun), and other computer hardware writes numbers right to left (e.g., Intel); see [M-5] **byteorder()**. If you are going to write binary files, and if you want your files to be readable on all computers, you must write code to deal with this issue.

Many programmers ignore the issue because the programs they write are intended for one computer or on computers like the one they use. If that is the case, you can ignore the issue, too. The default byte order filled in by `bufio()` is the byte order of the computer you are using.

If you intend to read and write files across different computers, however, you will need to concern yourself with byte order, and how you do that is described in *Advanced issues* below.

The other issue you may need to consider is missing values. If you are writing a binary format that is intended to be used outside Stata, it is best if the values you write simply do not include missing values. Not all packages have them, and the packages that do don't agree on how they are encoded. In such cases, if the data you are writing include missing values, change the values to another value such as −1, 99, 999, or −9999.

If, however, you are writing binary files in Stata to be read back in Stata, you can allow Stata's missing values ., .a, .b, ..., .z. No special action is required. The missing-value scheme in C specifies how those missing values are encoded, and there is only one way right now, so there is in fact no issue at all. C includes the extra information in case Stata ever changes the way it encodes missing values so that you will have a way to read and write old-format files. How this process works is described in *Advanced issues*.

Arguments B and offset

Functions

>   bufput(*C*, *B*, *offset*, *bfmt*, *X*) and
>
>   bufget(*C*, *B*, *offset*, *bfmt*, ...)

do not perform I/O; they copy values into and out of the buffer. *B* specifies the buffer, and *offset* specifies the position within it.

B is a string scalar.

offset is an integer scalar specifying the position within *B*. Offset 0 specifies the first byte of *B*.

For `bufput()`, *B* must already exist and be long enough to receive the result, and it is usual to code something like

```
B = (4 + 4 + rows(X)*cols(X)*8) * char(0)
bufput(C, 0, "%4b", rows(X))
bufput(C, 4, "%4b", cols(X))
bufput(C, 8, "%8z", X)
```

Argument fh

Argument *fh* in

>fbufput(*C*, *fh*, *bfmt*, *X*) and

>fbufget(*C*, *fh*, *bfmt*, ...)

plays the role of arguments *B* and *offset* in bufput() and bufget(). Rather than copy into or out of a buffer, data are written to, or read from, file *fh*. *fh* is obtained from fopen(); see [M-5] **fopen( )**.

Argument bfmt

Argument *bfmt* in

>bufput(*C*, *B*, *offset*, *bfmt*, *X*),

>bufget(*C*, *B*, *offset*, *bfmt*, ...),

>fbufput(*C*, *fh*, *bfmt*, *X*), and

>fbufget(*C*, *fh*, *bfmt*, ...)

specifies how the elements are to be written or read.

bfmts for numeric data

The numeric *bfmts* are

bfmt	interpretation
%8z	8-byte floating point
%4z	4-byte floating point
%4bu	4-byte unsigned integer
%4bs	4-byte signed integer
%4b	4-byte Stata integer
%2bu	2-byte unsigned integer
%2bs	2-byte signed integer
%2b	2-byte Stata integer
%1bu	1-byte unsigned integer
%1bs	1-byte signed integer
%1b	1-byte Stata integer

A Stata integer is the same as a signed integer, except that the largest 27 values are given the interpretation ., .a, .b, ..., .z.

bfmts for string data

The string *bfmts* are

bfmt	interpretation
%#s	text string
%#S	binary string

where # represents the length of the string field. Examples include %8s and %639876S.

When writing, it does not matter whether you use %#s or %#S, the same actions are taken:

1. If the string being written is shorter than #, the field is padded with char(0).

2. If the string being written is longer than #, only the first # bytes of the string are written.

When reading, the distinction between %#s and %#S is important:

1. When reading with %#s, if char(0) appears within the first # bytes, the returned result is truncated at that point.

2. When reading with %#S, a full # bytes are returned in all cases.

Argument X

Argument X in

$$\texttt{bufput}(C, B, \textit{offset}, \textit{bfmt}, X) \text{ and}$$

$$\texttt{fbufput}(C, \textit{fh}, \textit{bfmt}, X)$$

specifies the value to be written. X may be real or string and may be a scalar, vector, or matrix. If X is a vector, the elements are written one after the other. If X is a matrix, the elements of the first row are written one after the other, followed by the elements of the second row, and so on.

In

$$X = \texttt{bufget}(C, B, \textit{offset}, \textit{bfmt}, \ldots) \text{ and}$$

$$X = \texttt{fbufget}(C, \textit{fh}, \textit{bfmt}, \ldots)$$

X is returned.

Arguments r and c

Arguments r and c are optional in the following:

$$X = \texttt{bufget}(C, B, \textit{offset}, \textit{bfmt}),$$

$$X = \texttt{bufget}(C, B, \textit{offset}, \textit{bfmt}, c),$$

$$X = \texttt{bufget}(C, B, \textit{offset}, \textit{bfmt}, r, c),$$

$$X = \texttt{fbufget}(C, \textit{fh}, \textit{bfmt}),$$

$X = \text{fbufget}(C, \mathit{fh}, \mathit{bfmt}, c)$, and

$X = \text{fbufget}(C, \mathit{fh}, \mathit{bfmt}, r, c)$.

If r is not specified, results are as if $r = 1$.

If c is not specified, results are as if $c = 1$.

Thus

$X = \text{bufget}(C, B, \mathit{offset}, \mathit{bfmt})$ and

$X = \text{fbufget}(C, \mathit{fh}, \mathit{bfmt})$

read one element and return it, whereas

$X = \text{bufget}(C, B, \mathit{offset}, \mathit{bfmt}, c)$ and

$X = \text{fbufget}(C, \mathit{fh}, \mathit{bfmt}, c)$

read c elements and return them in a column vector, and

$X = \text{bufget}(C, B, \mathit{offset}, \mathit{bfmt}, r, c)$ and

$X = \text{fbufget}(C, \mathit{fh}, \mathit{bfmt}, r, c)$

read $r * c$ elements and return them in an $r \times c$ matrix.

Advanced issues

A properly designed binary-file format includes a signature line first thing in the file:

```
fh = fopen(filename, "w")
fwrite(fh, "MyFormat For Mats v. 1.0")
        /* ----+----1----+----2---- */
```

and

```
fh = fopen(filename, "r")
if (fread(fh, 24) != "MyFormat For Mats v. 1.0") {
        errprintf("%s not My-Format file\n", filename)
        exit(610)
}
```

If you are concerned with byte order and mapping of missing values, you should write the byte order and missing-value mapping in the file, write in natural byte order, and be prepared to read back in either byte order.

The code for writing is

```
fh = fopen(filename, "w")
fwrite(fh, "MyFormat For Mats v. 1.0")

C = bufio()
fbufput(C, fh, "%1bu", bufbyteorder(C))
fbufput(C, fh, "%2bu", bufmissingvalue(C))
```

and the corresponding code for reading the file is

```
fh = fopen(filename, "r")
if (fread(fh, 24) != "MyFormat For Mats v. 1.0") {
        errprintf("%s not My-Format file\n", filename)
        exit(610)
}

C = bufio()
bufbyteorder(C, fbufget(C, "%1bu"))
bufmissingvalue(C, fbufget(C, "%2bu"))
```

All we write in the file before recording the byte order are strings and bytes. This way, when we read the file later, we can set the byte order before reading any 2-, 4-, or 8-byte fields.

bufbyteorder(*C*) — bufbyteorder() with one argument — returns the byte-order encoding recorded in *C*. It returns 1 (meaning HILO) or 2 (meaning LOHI).

bufbyteorder(*C*, *value*) — bufbyteorder() with two arguments — resets the byte order recorded in *C*. Once reset, all buffer functions will automatically reverse bytes if necessary.

bufmissingvalue() works the same way. With one argument, it returns a code for the encoding scheme recorded in *C* (said code being the Stata release number multiplied by 100). With two arguments, it resets the code. Once the code is reset, all buffer routines used will automatically take the appropriate action.

Conformability

bufio():
      result: *colvector*

bufbyteorder(*C*):
      C: *colvector*   made by bufio()
      result: 1×1 containing 1 (HILO) or 2 (LOHI)

bufbyteorder(*C*, *byteorder*):
      C: *colvector*   made by bufio()
      byteorder: 1×1 containing 1 (HILO) or 2 (LOHI)
      result: *void*

bufmissingvalue(*C*):
      C: *colvector*   made by bufio()
      result: 1×1

bufmissingvalue(*C*, *version*):
      C: *colvector*   made by bufio()
      version: 1×1
      result: *void*

bufput(*C*, *B*, *offset*, *bfmt*, *X*):
        C: *colvector*   made by bufio()
        B: 1×1
    offset: 1×1
     bfmt: 1×1
        X: $r \times c$
    result: *void*

bufget(*C*, *B*, *offset*, *bfmt*):
        C: *colvector*   made by bufio()
        B: 1×1
    offset: 1×1
     bfmt: 1×1
    result: 1×1

bufget(*C*, *B*, *offset*, *bfmt*, *r*):
        C: *colvector*   made by bufio()
        B: 1×1
    offset: 1×1
     bfmt: 1×1
        r: 1×1
    result: $1 \times c$

bufget(*C*, *B*, *offset*, *bfmt*, *r*, *c*):
        C: *colvector*   made by bufio()
        B: 1×1
    offset: 1×1
     bfmt: 1×1
        r: 1×1
        c: 1×1
    result: $r \times c$

fbufput(*C*, *fh*, *bfmt*, *X*):
        C: *colvector*   made by bufio()
       fh: 1×1
     bfmt: 1×1
        X: $r \times c$
    result: *void*

fbufget(*C*, *fh*, *bfmt*):
        C: *colvector*   made by bufio()
       fh: 1×1
     bfmt: 1×1
    result: 1×1

fbufget(*C*, *fh*, *bfmt*, *r*):
        C: *colvector*   made by bufio()
       fh: 1×1
     bfmt: 1×1
        r: 1×1
    result: $1 \times c$

fbufget(*C*, *fh*, *bfmt*, *r*, *c*):
    C: colvector   made by bufio()
    fh: 1×1
    bfmt: 1×1
    r: 1×1
    c: 1×1
    result: $r \times c$

bufbfmtlen(*bfmt*):
    bfmt: 1×1
    result: 1×1

bufbfmtisnum(*bfmt*):
    bfmt: 1×1
    result: 1×1

Diagnostics

bufio() cannot fail.

bufbyteorder(*C*) cannot fail. bufbyteorder(*C*, *byteorder*) aborts with error if *byteorder* is not 1 or 2.

bufmissingvalue(*C*) cannot fail. bufmissingvalue(*C*, *version*) aborts with error if *version* < 100 or *version* > stataversion().

bufput(*C*, *B*, *offset*, *bfmt*, *X*) aborts with error if *B* is too short to receive the result, *offset* is out of range, *bfmt* is invalid, or *bfmt* is a string format and *X* is numeric or vice versa. Putting a void matrix results in 0 bytes being inserted into the buffer and is not an error.

bufget(*C*, *B*, *offset*, *bfmt*, ...) aborts with error if *B* is too short, *offset* is out of range, or *bfmt* is invalid. Reading zero rows or columns results in a void returned result and is not an error.

fbufput(*C*, *fh*, *bfmt*, *X*) aborts with error if *fh* is invalid, *bfmt* is invalid, or *bfmt* is a string format and *X* is numeric or vice versa. Putting a void matrix results in 0 bytes being written and is not an error. I/O errors are possible; use fstatus() to detect them.

fbufget(*C*, *fh*, *bfmt*, ...) aborts with error if *fh* is invalid or *bfmt* is invalid. Reading zero rows or columns results in a void returned result and is not an error. End-of-file and I/O errors are possible; use fstatus() to detect them.

bufbfmtlen(*bfmt*) and bufbfmtisnum(*bfmt*) abort with error if *bfmt* is invalid.

Also See

[M-5] **fopen()** — File I/O

[M-4] **io** — I/O functions

Title

[M-5] byteorder( ) — Byte order used by computer

Syntax

>   *real scalar* byteorder()

Description

byteorder() returns 1 if the computer is HILO (records most significant byte first) and returns 2 if LOHI (records least significant byte first).

Remarks

Pretend that the values 00 and 01 are recorded at memory positions 58 and 59 and that you know what is recorded there is a 2-byte integer. How should the 2-byte number be interpreted: as 0001 (meaning 1) or 0100 (meaning 256 in decimal)? For different computers, the answer varies. For HILO computers, the number is to be interpreted as 0001. For LOHI computers, the number is interpreted as 0100.

Regardless, it does not matter because the computer is consistent with itself. An issue arises, however, when we write binary files that may be read on computers using a different byte order or when we read files from computers that used a different byte order.

Stata and Mata automatically handle these problems for you, so you may wish to stop reading. byteorder(), however, is the tool on which the solution is based. If you intend to write code based on your own binary-file format or to write code to process the binary files of others, then you may need to use it.

There are two popular solutions to the byte-order problem: (1) write the file in a known byte order or (2) write the file by using whichever byte order is convenient and record the byte order used in the file. StataCorp tends to use the second, but others who have worried about this problem have used both solutions.

In solution (1), it is usually agreed that the file will be written in HILO order. If you are using a HILO computer, you write and read files the ordinary way, ignoring the problem altogether. If you are on a LOHI computer, however, you must reverse the bytes before placing them in the file. If you are writing code designed to execute on both kinds of computers, you must write code for both circumstances, and you must consider the problem when both reading and writing the files.

In solution (2), files are written LOHI or HILO, depending solely on the computer being used. Early in the file, however, the byte order is recorded. When reading the file, you compare the order in which the file is recorded with the order of the computer and, if they are different, you reverse bytes.

Mata-buffered I/O utilities will automatically reverse bytes for you. See [M-5] **bufio( )**.

Conformability

byteorder():
    result: 1×1

Diagnostics

None.

Also See

[M-5] **bufio( )** — Buffered (binary) I/O

[M-5] **stataversion( )** — Version of Stata being used

[M-4] **programming** — Programming functions

Title

[M-5] **C( )** — Make complex

Syntax

>   *complex matrix*   C(*numeric matrix A*)

>   *complex matrix*   C(*real matrix R, real matrix I*)

Description

C(*A*) returns *A* converted to complex. C(*A*) returns *A* if *A* is already complex. If *A* is real, C(*A*) returns *A*+0*i*—*A* cast up to complex. Coding C(*A*) is thus how you ensure that the matrix is treated as complex.

C(*R*, *I*) returns the complex matrix *R*+*Ii* and is faster than the alternative *R* + *I*:*1i.

Remarks

Many of Mata's functions are overloaded, meaning they return a real when given real arguments and a complex when given complex arguments. Given real arguments, if the result cannot be expressed as a real, missing value is returned. Thus sqrt(-1) evaluates to missing, whereas sqrt(-1+0i) is 1*i*.

C( ) is the fast way to make arguments that might be real into complex. You can code

>   result = sqrt(C(*x*))

If *x* already is complex, C( ) does nothing; if *x* is real, C(*x*) returns the complex equivalent.

The two-argument version of C( ) is less frequently used. C(*R*, *I*) is literally equivalent to *R* :+ *I**1i, meaning that *R* and *I* need only be c-conformable.

For instance, C(1, (1,2,3)) evaluates to (1+1i, 1+2i, 1+3i).

Conformability

C(*A*):

> A: $r \times c$
> *result*: $r \times c$

C(*R*, *I*):

> R: $r_1 \times c_1$
> I: $r_2 \times c_2$,   *R* and *I* c-conformable
> *result*: $\max(r_1,r_2) \times \max(c_1,c_2)$

Diagnostics

C(Z), if Z is complex, literally returns Z and not a copy of Z. This makes execution of C() applied to complex arguments instant.

In C(R, I), the i,j element of the result will be missing anywhere $R[i,j]$ or $I[i,j]$ is missing. For instance, C((1,3,.), (.,2,4)) results in (., 3+2i, .). If $R[i,j]$ and $I[i,j]$ are both missing, then the $R[i,j]$ value will be used; e.g., C(.a, .b) results in .a.

Also See

[M-5] **Re( )** — Extract real or imaginary part

[M-4] **scalar** — Scalar mathematical functions

[M-4] **utility** — Matrix utility functions

Title

> [M-5] **c( )** — Access c( ) value

Syntax

>   *scalar* c(*string scalar name*)

returned is either a real or string scalar, depending on the value of *name*.

Description

c(*name*) returns Stata's c-class value; see [P] **creturn**.

Do not confuse c() with C(), which makes complex out of real arguments; see [M-5] **C()**.

Remarks

See [P] **creturn** or, in Stata, type

        . creturn list

to see what is stored in c(). Among the especially useful c() values are

>   *string* c("current_date")
>   *string* c("current_time")
>   *string* c("os")
>   *string* c("dirsep")

Conformability

c(*name*):
>   *name*: 1×1
>   *result*: 1×1

Diagnostics

c(*name*) returns a string or real depending on the particular c() value requested. If *name* is an invalid name or contains a name for which no c() value is defined, returned is "".

Also See

[M-4] **programming** — Programming functions

Title

[M-5] **callersversion( )** — Obtain version number of caller

Syntax

>   *real scalar*   callersversion()

Description

callersversion() returns the version set by the caller (see [M-2] **version**), or if the caller did not set the version, it returns the version of Stata under which the caller was compiled.

Remarks

callersversion() is how [M-2] **version** is made to work. Say that you have written function

>   *real matrix* useful(*real matrix A*, *real scalar k*)

and assume that useful() aborts with error if *A* is void. You wrote useful() in the days of Stata 9. For Stata 10, you want to change useful() so that it returns J(0,0,.) if *A* is void, but you want to maintain the current behavior for old Stata 9 callers and programs. You do that as follows:

```
real matrix useful(real matrix A, real scalar k)
{
        ...
        if (callersversion()>=10) {
                if (rows(A)==0 | cols(A)==0) return(J(0,0,.))
        }
        ...
}
```

Conformability

callersversion():
>   *result*: 1×1

Diagnostics

None.

Also See

[M-2] **version** — Version control

[M-4] **programming** — Programming functions

Title

[M-5] **cat( )** — Load file into string matrix

Syntax

> *string colvector* cat(*string scalar filename* [, *real scalar line1* [, *real scalar line2*]])

Description

cat(*filename*) returns a column vector containing the lines from ASCII file *filename*.

cat(*filename*, *line1*) returns a column vector containing the lines from ASCII file *filename* starting with line number *line1*.

cat(*filename*, *line1*, *line2*) returns a column vector containing the lines from ASCII file *filename* starting with line number *line1* and ending with line number *line2*.

Remarks

cat(*filename*) removes new-line characters at the end of lines.

Conformability

cat(*filename*, *line1*, *line2*):
    filename: 1×1
    line1: 1×1 (optional)
    line2: 1×1 (optional)
    result: $r \times 1$, $r \geq 0$

Diagnostics

cat(*filename*) aborts with error if *filename* does not exist.

cat() returns a 0×1 result if *filename* contains 0 bytes.

Also See

[M-4] **io** — I/O functions

Title

[M-5] chdir( ) — Manipulate directories

Syntax

string scalar   pwd()

void   chdir(*string scalar dirpath*)

real scalar   _chdir(*string scalar dirpath*)

void   mkdir(*string scalar dirpath*)

void   mkdir(*string scalar dirpath, real scalar public*)

real scalar   _mkdir(*string scalar dirpath*)

real scalar   _mkdir(*string scalar dirpath, real scalar public*)

void   rmdir(*string scalar dirpath*)

real scalar   _rmdir(*string scalar dirpath*)

Description

pwd() returns the full name (path) of the current working directory.

chdir(*dirpath*) changes the current working directory to *dirpath*. chdir() aborts with error if the directory does not exist or the operating system cannot change to it.

_chdir(*dirpath*) does the same thing but returns 170 (a return code) when chdir() would abort. _chdir() returns 0 if it is successful.

mkdir(*dirpath*) and mkdir(*dirpath, public*) create directory *dirpath*. mkdir() aborts with error if the directory already exists or cannot be created. If *public* $\neq$ 0 is specified, the directory is given permissions so that everyone can read it; otherwise, it is given the usual permissions.

_mkdir(*dirpath*) and _mkdir(*dirpath, public*) do the same thing but return 693 (a return code) when mkdir() would abort. _mkdir() returns 0 if it is successful.

rmdir(*dirpath*) removes directory *dirpath*. rmdir() aborts with error if the directory does not exist, is not empty, or the operating system refuses to remove it.

_rmdir(*dirpath*) does the same thing but returns 693 (a return code) when rmdir() would abort. _rmdir() returns 0 if it is successful.

Conformability

pwd():
    result: 1×1

chdir(*dirpath*):
    dirpath: 1×1
    result: *void*

_chdir(*dirpath*):
    dirpath: 1×1
    result: 1×1

mkdir(*dirpath*, *public*):
    dirpath: 1×1
    public: 1×1 (optional)
    result: *void*

_mkdir(*dirpath*, *public*):
    dirpath: 1×1
    public: 1×1 (optional)
    result: 1×1

rmdir(*dirpath*):
    dirpath: 1×1
    result: *void*

_rmdir(*dirpath*):
    dirpath: 1×1
    result: 1×1

Diagnostics

pwd() never aborts with error, but it can return "" if the operating system does not know or does not have a name for the current directory (which happens when another process removes the directory in which you are working).

chdir(*dirpath*) aborts with error if the directory does not exist or the operating system cannot change to it.

_chdir(*dirpath*) never aborts with error; it returns 0 on success and 170 on failure.

mkdir(*dirpath*) and mkdir(*dirpath*, *public*) abort with error if the directory already exists or the operating system cannot change to it.

_mkdir(*dirpath*) and _mkdir(*dirpath*, *public*) never abort with error; they return 0 on success and 693 on failure.

rmdir(*dirpath*) aborts with error if the directory does not exist, is not empty, or the operating system cannot remove it.

_rmdir(*dirpath*) never aborts with error; it returns 0 on success and 639 on failure.

Also See

[M-4] **io** — I/O functions

Title

[M-5] cholesky( ) — Cholesky square-root decomposition

Syntax

> *numeric matrix*   cholesky(*numeric matrix A*)
>
> *void*   _cholesky(*numeric matrix A*)

Description

cholesky(*A*) returns the Cholesky decomposition *G* of symmetric (Hermitian), positive-definite matrix *A*. cholesky() returns a lower-triangular matrix of missing values if *A* is not positive definite.

_cholesky(*A*) does the same thing, except that it overwrites *A* with the Cholesky result.

Remarks

The Cholesky decomposition *G* of a symmetric, positive-definite matrix *A* is

$$A = GG'$$

where *G* is lower triangular. When *A* is complex, *A* must be Hermitian, and *G'*, of course, is the conjugate transpose of *G*.

Decomposition is performed via [M-1] **LAPACK**.

Conformability

cholesky(*A*):
> *A*: $n \times n$
> *result*: $n \times n$

_cholesky(*A*):
> *input*:
> > *A*: $n \times n$
>
> *output*:
> > *A*: $n \times n$

Diagnostics

cholesky() returns a lower-triangular matrix of missing values if *A* contains missing values or if *A* is not positive definite.

_cholesky(*A*) overwrites *A* with a lower-triangular matrix of missing values if *A* contains missing values or if *A* is not positive definite.

Both functions use the elements from the lower triangle of *A* without checking whether *A* is symmetric or, in the complex case, Hermitian.

> André-Louis Cholesky (1875–1918) was born near Bordeaux in France. He studied at the Ecole Polytechnique and then joined the French army. Cholesky served in Tunisia and Algeria and then worked in the Geodesic Section of the Army Geographic Service, where he invented his now-famous method. In the war of 1914–1918, he served in the Vosges and in Romania but after return to the Eastern front was fatally wounded. Cholesky's method was written up posthumously by one of his fellow officers but attracted little attention until the 1940s.

Reference

Chabert, J.-L., ed. 1999. *A History of Algorithms: From the Pebble to the Microchip*. Berlin: Springer.

Also See

[M-5] **lud( )** — LU decomposition

[M-4] **matrix** — Matrix functions

Title

[M-5] cholinv( ) — Symmetric, positive-definite matrix inversion

Syntax

numeric matrix	cholinv(*numeric matrix A*)
numeric matrix	cholinv(*numeric matrix A, real scalar tol*)
void	_cholinv(*numeric matrix A*)
void	_cholinv(*numeric matrix A, real scalar tol*)

Description

cholinv(*A*) and cholinv(*A, tol*) return the inverse of real or complex, symmetric (Hermitian), positive-definite, square matrix *A*.

_cholinv(*A*) and _cholinv(*A, tol*) do the same thing except that, rather than returning the inverse matrix, they overwrite the original matrix *A* with the inverse.

In all cases, optional argument *tol* specifies the tolerance for determining singularity; see *Remarks* below.

Remarks

These routines calculate the inverse of a symmetric, positive-definite square matrix *A*. See [M-5] **luinv( )** for the inverse of a general square matrix.

A is required to be square and positive definite. See [M-5] **qrinv( )** and [M-5] **pinv( )** for generalized inverses of nonsquare or rank-deficient matrices. See [M-5] **invsym( )** for generalized inverses of real, symmetric matrices.

cholinv(*A*) is logically equivalent to cholsolve(*A*, I(rows(*A*))); see [M-5] **cholsolve( )** for details and for use of the optional *tol* argument.

Conformability

cholinv(*A, tol*):
        A: $n \times n$
     tol: 1×1 (optional)
   result: $n \times n$

_cholinv(*A, tol*):
  input:
        A: $n \times n$
     tol: 1×1 (optional)
  output:
        A: $n \times n$

Diagnostics

The inverse returned by these functions is real if A is real and is complex if A is complex. If you use these functions with a non–positive definite matrix, or a matrix that is too close to singularity, returned will be a matrix of missing values. The determination of singularity is made relative to *tol*. See *Tolerance* under *Remarks* in [M-5] **cholsolve( )** for details.

cholinv(A) and _cholinv(A) return a result containing all missing values if A is not positive definite or if A contains missing values.

_cholinv(A) aborts with error if A is a view.

See [M-5] **cholsolve( )** and [M-1] **tolerance** for information on the optional *tol* argument.

Both functions use the elements from the lower triangle of A without checking whether A is symmetric or, in the complex case, Hermitian.

Also See

[M-5] **invsym( )** — Symmetric real matrix inversion

[M-5] **luinv( )** — Square matrix inversion

[M-5] **qrinv( )** — Generalized inverse of matrix via QR decomposition

[M-5] **pinv( )** — Moore–Penrose pseudoinverse

[M-5] **cholsolve( )** — Solve AX=B for X using Cholesky decomposition

[M-4] **matrix** — Matrix functions

[M-4] **solvers** — Functions to solve AX=B and to obtain A inverse

Title

[M-5] cholsolve( ) — Solve AX=B for X using Cholesky decomposition

Syntax

numeric matrix   cholsolve(*numeric matrix A*, *numeric matrix B*)

numeric matrix   cholsolve(*numeric matrix A*, *numeric matrix B*, *real scalar tol*)

void   _cholsolve(*numeric matrix A*, *numeric matrix B*)

void   _cholsolve(*numeric matrix A*, *numeric matrix B*, *real scalar tol*)

Description

cholsolve(*A*, *B*) solves $AX = B$ and returns X for symmetric (Hermitian), positive-definite A. cholsolve() returns a matrix of missing values if A is not positive definite or if A is singular.

cholsolve(*A*, *B*, *tol*) does the same thing; it allows you to specify the tolerance for declaring that A is singular; see *Tolerance* under *Remarks* below.

_cholsolve(*A*, *B*) and _cholsolve(*A*, *B*, *tol*) do the same thing except that, rather than returning the solution X, they overwrite B with the solution, and in the process of making the calculation, they destroy the contents of A.

Remarks

The above functions solve $AX = B$ via Cholesky decomposition and are accurate. When A is not symmetric and positive definite, [M-5] **lusolve( )**, [M-5] **qrsolve( )**, and [M-5] **svsolve( )** are alternatives based on the LU decomposition, the QR decomposition, and the singular value decomposition (SVD). The alternatives differ in how they handle singular A. Then the LU-based routines return missing values, whereas the QR-based and SVD-based routines return generalized (least-squares) solutions.

Remarks are presented under the following headings:

> *Derivation*
> *Relationship to inversion*
> *Tolerance*

Derivation

We wish to solve for X

$$AX = B \qquad (1)$$

when A is symmetric and positive definite. Perform the Cholesky decomposition of A so that we have $A = GG'$. Then (1) can be written

$$GG'X = B \qquad (2)$$

Define
$$Z = G'X \qquad (3)$$
Then (2) can be rewritten
$$GZ = B \qquad (4)$$
It is easy to solve (4) for Z because G is a lower-triangular matrix. Once Z is known, it is easy to solve (3) for X because G' is upper triangular.

Relationship to inversion

See *Relationship to inversion* in [M-5] **lusolve( )** for a discussion of the relationship between solving the linear system and matrix inversion.

Tolerance

The default tolerance used is
$$\eta = \frac{(1\text{e-}13)*\texttt{trace(abs}(G))}{n}$$

where G is the lower-triangular Cholesky factor of A: $n \times n$. A is declared to be singular if `cholesky()` (see [M-5] **cholesky( )**) finds that A is not positive definite, or if A is found to be positive definite, if any diagonal element of G is less than or equal to η. Mathematically, positive definiteness implies that the matrix is not singular. In the numerical method used, two checks are made: `cholesky()` makes one and then the η rule is applied to ensure numerical stability in the use of the result `cholesky()` returns.

If you specify *tol* > 0, the value you specify is used to multiply η. You may instead specify *tol* ≤ 0 and then the negative of the value you specify is used in place of η; see [M-1] **tolerance**.

See [M-5] **lusolve( )** for a detailed discussion of the issues surrounding solving nearly singular systems. The main point to keep in mind is that if A is ill conditioned, then small changes in A or B can lead to radically large differences in the solution for X.

Conformability

cholsolve(*A*, *B*, *tol*):
   input:
        A: $n \times n$
        B: $n \times k$
        tol: 1×1 (optional)
   output:
        result: $n \times k$

_cholsolve(*A*, *B*, *tol*):
   input:
        A: $n \times n$
        B: $n \times k$
        tol: 1×1 (optional)
   output:
        A: 0×0
        B: $n \times k$

Diagnostics

cholsolve(A, B, ...), and _cholsolve(A, B, ...) return a result of all missing values if A is not positive definite or if A contains missing values.

_cholsolve(A, B, ...) also aborts with error if A or B is a view.

All functions use the elements from the lower triangle of A without checking whether A is symmetric or, in the complex case, Hermitian.

Also See

[M-5] **cholesky()** — Cholesky square-root decomposition

[M-5] **cholinv()** — Symmetric, positive-definite matrix inversion

[M-5] **solvelower()** — Solve AX=B for X, A triangular

[M-5] **lusolve()** — Solve AX=B for X using LU decomposition

[M-5] **qrsolve()** — Solve AX=B for X using QR decomposition

[M-5] **svsolve()** — Solve AX=B for X using singular value decomposition

[M-5] **solve_tol()** — Tolerance used by solvers and inverters

[M-4] **matrix** — Matrix functions

[M-4] **solvers** — Functions to solve AX=B and to obtain A inverse

Title

[M-5] comb( ) — Combinatorial function

Syntax

> *real matrix* comb(*real matrix n*, *real matrix k*)

Description

comb(*n*, *k*) returns the elementwise combinatorial function *n*-choose-*k*, the number of ways to choose *k* items from *n* items, regardless of order.

Conformability

comb(*n*, *k*):
        n: $r_1 \times c_1$
        k: $r_2 \times c_2$,   *n* and *k* r-conformable
   result: $\max(r_1,r_2) \times \max(c_1,c_2)$

Diagnostics

comb(*n*, *k*) returns missing when either argument is missing or when the result would be larger than 10^{300}.

Also See

[M-4] **statistical** — Statistical functions

Title

[M-5] **cond( )** — Condition number

Syntax

>   *real scalar*   cond(*numeric matrix A*)

>   *real scalar*   cond(*numeric matrix A, real scalar p*)

Description

cond(*A*) returns cond(*A*, 2).

cond(*A, p*) returns the value of the condition number of *A* for the specified norm *p*, where *p* may be 0, 1, 2, or . (missing).

Remarks

The condition number of a matrix *A* is

$$cond = \text{norm}(A, p) \times \text{norm}(A^{-1}, p)$$

These functions return missing when *A* is singular.

Values near 1 indicate that the matrix is well conditioned, and large values indicate ill conditioning.

Conformability

cond(*A*):
>   *A*: $r \times c$
>   *result*: 1×1

cond(*A, p*):
>   *A*: $r \times c$
>   *p*: 1×1
>   *result*: 1×1

Diagnostics

cond(*A, p*) aborts with error if *p* is not 0, 1, 2, or . (missing).

cond(*A*) and cond(*A, p*) return missing when *A* is singular or if *A* contains missing values.

cond(*A*) and cond(*A, p*) return 1 when *A* is void.

cond(*A*) and cond(*A*, 2) return missing if the SVD algorithm fails to converge, which is highly unlikely; see [M-5] **svd( )**.

Also See

[M-5] **norm( )** — Matrix and vector norms

[M-4] **matrix** — Matrix functions

Title

[M-5] **conj( )** — Complex conjugate

Syntax

>   *numeric matrix* conj(*numeric matrix Z*)

Description

conj(*Z*) returns the elementwise complex conjugate of *Z*, i.e., conj(*a+bi*) = *a* − *bi*. conj() may be used with real or complex matrices. If *Z* is real, *Z* is returned unmodified.

Remarks

Given $m \times n$ matrix *Z*, conj(*Z*) returns an $m \times n$ matrix; it does not return the transpose. To obtain the conjugate transpose matrix, also known as the adjoint matrix, adjugate matrix, Hermitian adjoin, or Hermitian transpose, code

>   *Z'*

See [M-2] **op_transpose**.

A matrix equal to its conjugate transpose is called Hermitian or self-adjoint, although in this manual, we often use the term symmetric.

Conformability

    conj(Z):
            Z: r × c
         result: r × c

Diagnostics

conj(*Z*) returns a real matrix if *Z* is real and a complex matrix if *Z* is complex.

conj(*Z*), if *Z* is real, returns *Z* itself and not a copy. This makes conj() execute instantly when applied to real matrices.

Also See

[M-5] **_transpose( )** — Transposition in place

[M-4] **scalar** — Scalar mathematical functions

Title

> [M-5] **corr( )** — Make correlation matrix from variance matrix

Syntax

>   *real matrix*   corr(*real matrix V*)
>
>   *void*   _corr(*real matrix V*)

Description

corr(*V*) returns the correlation matrix corresponding to variance matrix *V*.

_corr(*V*) changes the contents of *V* from being a variance matrix to being a correlation matrix.

Remarks

See function variance() in [M-5] **mean( )** for obtaining a variance matrix from data.

Conformability

corr(*V*):
>   *input*:
>   >   *V*: $k \times k$
>
>   *output*:
>   >   *result*: $k \times k$

_corr(*V*):
>   *input*:
>   >   *V*: $k \times k$
>
>   *output*:
>   >   *V*: $k \times k$

Diagnostics

corr() and _corr() abort with error if *V* is not square. *V* should also be symmetric, but this is not checked.

Also See

[M-5] **mean( )** — Means, variances, and correlations

[M-4] **stata** — Stata interface functions

[M-4] **statistical** — Statistical functions

Title

[M-5] **cross( )** — Cross products

Syntax

>   *real matrix* cross(*X*, *Z*)
>
>   *real matrix* cross(*X*, *w*, *Z*)
>
>   *real matrix* cross(*X*, *xc*, *Z*, *zc*)
>
>   *real matrix* cross(*X*, *xc*, *w*, *Z*, *zc*)

where

X:	*real matrix X*
xc:	*real scalar xc*
w:	*real vector w*
Z:	*real matrix Z*
zc:	*real scalar zc*

Description

cross() makes calculations of the form

$$X'X$$
$$X'Z$$
$$X' \operatorname{diag}(w) X$$
$$X' \operatorname{diag}(w) Z$$

cross() is designed for making calculations that often arise in statistical formulas. In one sense, cross() does nothing that you cannot easily write out in standard matrix notation. For instance, cross(*X*, *Z*) calculates $X'Z$. cross(), however, has the following differences and advantages over the standard matrix-notation approach:

1. cross() omits the rows in *X* and *Z* that contain missing values, which amounts to dropping observations with missing values.

2. cross() uses less memory and is especially efficient when used with views.

3. cross() watches for special cases and makes calculations in those special cases more efficiently. For instance, if you code cross(*X*, *X*), cross() observes that the two matrices are the same and makes the calculation for a symmetric matrix result.

cross(*X*, *Z*) returns $X'Z$. Usually rows(*X*)==rows(*Z*), but *X* is also allowed to be a scalar, which is then treated as if J(rows(*Z*), 1, 1) were specified. Thus cross(1, *Z*) is equivalent to colsum(*Z*).

cross(*X*, *w*, *Z*) returns $X' \operatorname{diag}(w) Z$. Usually, rows(*w*)==rows(*Z*) or cols(*w*)==rows(*Z*), but *w* is also allowed to be a scalar, which is treated as if J(rows(*Z*), 1, *w*) were specified. Thus cross(*X*,1,*Z*) is the same as cross(*X*,*Z*). *Z* may also be a scalar, just as in the two-argument case.

cross(X, xc, Z, zc) is similar to cross(X, Z) in that $X'Z$ is returned. In the four-argument case, however, X is augmented on the right with a column of 1s if xc!=0 and Z is similarly augmented if zc!=0. cross(X, 0, Z, 0) is equivalent to cross(X, Z). Z may be specified as a scalar.

cross(X, xc, w, Z, zc) is similar to cross(X, w, Z) in that $X'\text{diag}(w)Z$ is returned. As with the four-argument cross(), X is augmented on the right with a column of 1s if xc!=0 and Z is similarly augmented if zc!=0. Both Z and w may be specified as scalars. cross(X, 0, 1, Z, 0) is equivalent to cross(X, Z).

Remarks

In the following examples, we are going to calculate linear regression coefficients using $b = (X'X)^{-1}X'y$, which means using $\sum x/n$, and variances using $n/(n-1) \times (\sum x^2/n - mean^2)$. See [M-5] **crossdev()** for examples of the same calculations made in a more numerically stable way.

The examples use the automobile data. Since we are using the absolute form of the calculation equations, it would be better if all variables had values near 1 (in which case the absolute form of the calculation equations are perfectly adequate). Thus we suggest

```
. sysuse auto
. replace weight = weight/1000
```

Some of the examples use a weight w. For that, you might try

```
. gen w = int(4*uniform())+1
```

▷ Example 1: Linear regression, the traditional way

```
: y = X = .
: st_view(y, ., "mpg")
: st_view(X, ., ("weight", "foreign"))
:
: X = X, J(rows(X),1,1)
: b = invsym(X'X)*X'y
```

Comments: Does not handle missing values and uses much memory if X is large.

◁

▷ Example 2: Linear regression using cross()

```
: y = X = .
: st_view(y, ., "mpg")
: st_view(X, ., ("weight", "foreign"))
:
: XX = cross(X,1 , X,1)
: Xy = cross(X,1 , y,0)
: b  = invsym(XX)*Xy
```

Comments: There is still an issue with missing values; mpg might not be missing everywhere weight and foreign are missing.

◁

▷ Example 3: Linear regression using cross( ) and one view

```
: // We will form
: //
: //    (y X)'(y X)  =  (y'y, y'X \ X'y, X'X)
:
: M = .
: st_view(M, ., ("mpg", "weight", "foreign"), 0)
:
: CP = cross(M,1 , M,1)
: XX = CP[|2,2 \ .,.|]
: Xy = CP[|2,1 \ .,1|]
: b  = invsym(XX)*Xy
```

Comments: Using one view handles all missing-value issues (we specified fourth argument 0 to st_view(); see [M-5] **st_view( )**).

◁

▷ Example 4: Linear regression using cross( ) and subviews

```
: M = X = y = .
: st_view(M, ., ("mpg", "weight", "foreign"), 0)
: st_subview(y, M, ., 1)
: st_subview(X, M, ., (2\.))
:
: XX = cross(X,1 , X,1)
: Xy = cross(X,1 , y,0)
: b  = invsym(XX)*Xy
```

Comments: Using subviews also handles all missing-value issues; see [M-5] **st_subview( )**. The subview approach is a little less efficient than the previous solution but is perhaps easier to understand. The efficiency issue concerns only the extra memory required by the subviews y and X, which is not much.

Also, this subview solution could be used to handle the missing-value problems of calculating linear regression coefficients the traditional way, shown in example 1:

```
: M = X = y = .
: st_view(M, ., ("mpg", "weight", "foreign"), 0)
: st_subview(y, M, ., 1)
: st_subview(X, M, ., (2\.))
:
: X = X, J(rows(X), 1, 1)
: b = invsym(X'X)*X'y
```

◁

▷ Example 5: Weighted linear regression, the traditional way

```
: M = w = y = X = .
: st_view(M, ., ("w", "mpg", "weight", "foreign"), 0)
: st_subview(w, M, ., 1)
: st_subview(y, M, ., 2)
: st_subview(X, M, ., (3\.))
:
: X = X, J(rows(X), 1, 1)
: b = invsym(X'diag(w)*X)*X'diag(w)'y
```

Comments: The memory requirements are now truly impressive because diag(w) is an $N \times N$ matrix! That is, the memory requirements are truly impressive when N is large. Part of the power of Mata is that you can write things like invsym(X'diag(w)*X)*X'diag(w)'y and obtain solutions. We do not mean to be dismissive of the traditional approach; we merely wish to emphasize its memory requirements and note that there are alternatives.

◁

▷ Example 6: Weighted linear regression using cross()

```
: M = w = y = X = .
: st_view(M, ., ("w", "mpg", "weight", "foreign"), 0)
: st_subview(w, M, ., 1)
: st_subview(y, M, ., 2)
: st_subview(X, M, ., (3\.))
:
: XX = cross(X,1 ,w, X,1)
: Xy = cross(X,1 ,w, y,0)
: b  = invsym(XX)*Xy
```

Comments: The memory requirements here are no greater than they were in example 4, which this example closely mirrors. We could also have mirrored the logic of example 3:

```
: M = w = M2 = .
: st_view(M, ., ("w", "mpg", "weight", "foreign"), 0)
: st_subview(w, M, ., 1)
: st_subview(M2, M, ., (2\.))
:
: CP = cross(M,1 ,w, M,1)
: XX = CP[|3,3 \ .,.|]
: Xy = CP[|3,1 \ .,2|]
: b  = invsym(XX)*Xy
```

Note how similar these solutions are to their unweighted counterparts. The only important difference is the appearance of w as the middle argument of cross(). Since specifying the middle argument as a scalar 1 is also allowed and produces unweighted estimates, the above code could be modified to produce unweighted or weighted estimates, depending on how w is defined.

◁

▷ Example 7: Mean of one variable

```
: x = .
: st_view(x, ., "mpg", 0)
:
: CP = cross(1,0 , x,1)
: mean = CP[1]/CP[2]
```

Comments: An easier and every bit as good a solution would be

```
: x = .
: st_view(x, ., "mpg", 0)
:
: mean = mean(x,1)
```

mean() is implemented in terms of cross(). Actually, mean() is implemented using the quad-precision version of cross(); see [M-5] **quadcross()**. We could implement our solution in terms of quadcross():

```
: x = .
: st_view(x, ., "mpg", 0)
:
: CP = quadcross(1,0 , x,1)
: mean = CP[1]/CP[2]
```

quadcross() returns a double-precision result just as does cross(). The difference is that quadcross() uses quad precision internally in calculating sums.

◁

▷ Example 8: Means of multiple variables

```
: X = .
: st_view(X, ., ("mpg", "weight", "displ"), 0)
:
: CP = cross(1,0 , X,1)
: n   = cols(CP)
: means = CP[|1\n-1|] :/ CP[n]
```

Comments: The above logic will work for one variable, too. With mean(), the solution would be

```
: X = .
: st_view(X, ., ("mpg", "weight", "displ"), 0)
:
: means = mean(X, 1)
```

◁

▷ Example 9: Weighted means of multiple variables

```
: M = w = X = .
: st_view(M, ., ("w", "mpg", "weight, "displ"), 0)
: st_subview(w, M, ., 1)
: st_subview(X, M, ., (2\.))
:
: CP = cross(1,0, w,  X,1)
: n  = cols(CP)
: means = CP[|1\n-1|] :/ CP[n]
```

Comments: Note how similar this solution is to the unweighted solution: w now appears as the middle argument of cross(). The line CP = cross(1,0, w, X,1) could also be coded CP = cross(w,0, X,1); it would make no difference.

The mean() solution to the problem is

```
: M = w = X = .
: st_view(M, ., ("w", "mpg", "weight", "displ"), 0)
: st_subview(w, M, ., 1)
: st_subview(X, M, ., (2\.))
:
: means = mean(X, w)
```

◁

▷ Example 10: Variance matrix, traditional approach 1

```
: X = .
: st_view(X, ., ("mpg", "weight", "displ"), 0)
:
: n     = rows(X)
: means = mean(X, 1)
: cov   = (X'X/n - means'means)*(n/(n-1))
```

Comments: This above is not 100% traditional since we used mean() to obtain the means, but that does make the solution more understandable. The solution requires calculating X', requiring that the data matrix be duplicated. Also, we have used a numerically poor calculation formula.

◁

▷ Example 11: Variance matrix, traditional approach 2

```
: X = .
: st_view(X, ., ("mpg", "weight", "displ"), 0)
:
: n     = rows(X)
: means = mean(X, 1)
: cov   = (X:-means)'(X:-means) :/ (n-1)
```

Comments: We use a better calculation formula and, in the process, increase our memory usage substantially.

◁

▷ Example 12: Variance matrix using cross()

```
: X = .
: st_view(X, ., ("mpg", "weight, "displ"), 0)
:
: n     = rows(X)
: means = mean(X, 1)
: XX    = cross(X, X)
: cov   = ((XX:/n)-means'means)*(n/(n-1))
```

Comments: The above solution conserves memory but uses the numerically poor calculation formula. A related function, crossdev(), will calculate deviation crossproducts:

```
: X = .
: st_view(X, ., ("mpg", "weight, "displ"), 0)
:
: n     = rows(X)
: means = mean(X, 1)
: xx    = crossdev(X,means, X,means)
: cov   = xx:/(n-1)
```

See [M-5] **crossdev( )**. The easiest solution, however, is

```
: X = .
: st_view(X, ., ("mpg", "weight, "displ"), 0)
:
: cov = variance(X, 1)
```

See [M-5] **mean()** for a description of the variance() function. variance() is implemented in terms of crossdev().

◁

▷ Example 13: Weighted variance matrix, traditional approaches

```
: M = w = X = .
: st_view(M, ., ("w", "mpg", "weight", "displ"), 0)
: st_subview(w, M, ., 1)
: st_subview(X, M, ., (2\.))
:
: n     = colsum(w)
: means = mean(X, w)
: cov   = (X'diag(w)*X:/n - means'means)*(n/(n-1))
```

Comments: Above we use the numerically poor formula. Using the better deviation formula, we would have

```
: M = w = X = .
: st_view(M, ., ("w", "mpg", "weight", "displ"), 0)
: st_subview(w, M, ., 1)
: st_subview(X, M, ., (2\.))
:
: n     = colsum(w)
: means = mean(X, w)
: cov   = (X:-means)'diag(w)*(X:-means) :/ (n-1)
```

The memory requirements include making a copy of the data with the means removed and making an $N \times N$ diagonal matrix.

◁

▷ Example 14: Weighted variance matrix using cross()

```
: M = w = X = .
: st_view(M, ., ("w", "mpg", "weight", "displ"), 0)
: st_subview(w, M, ., 1)
: st_subview(X, M, ., (2\.))
:
: n     = colsum(w)
: means = mean(X, w)
: cov   = (cross(X,w,X):/n - means'means)*(n/(n-1))
```

Comments: As in example 12, the above solution conserves memory but uses a numerically poor calculation formula. Better is to use crossdev():

```
: M = w = X = .
: st_view(M, ., ("w", "mpg", "weight", "displ"), 0)
: st_subview(w, M, ., 1)
: st_subview(X, M, ., (2\.))
:
: n     = colsum(w)
: means = mean(X, w)
: cov   = crossdev(X,means, w, X,means) :/ (n-1)
```

and easiest is to use `variance()`:

```
: M = w = X = .
: st_view(M, ., ("w", "mpg", "weight", "displ"), 0)
: st_subview(w, M, ., 1)
: st_subview(X, M, ., (2\.))
:
: cov = variance(X, w)
```

See [M-5] **crossdev( )** and [M-5] **mean( )**.

◁

Comment concerning cross( ) and missing values

`cross()` automatically omits rows containing missing values in making its calculation. Depending on this feature, however, is considered bad style because so many other Mata functions do not provide that feature and it is easy to make a mistake.

The right way to handle missing values is to exclude them when constructing views and subviews, as we have done above. When we constructed a view, we invariably specified fourth argument 0 to `st_view()`. In formal programming situations, you will probably specify the name of the `touse` variable you have previously constructed in your ado-file that calls your Mata function.

Conformability

$\text{cross}(X, xc, w, Z, zc)$:

X:	$n \times v_1$ or 1×1,	1×1 treated as if $n \times 1$	
xc:	1×1		(optional)
w:	$n \times 1$ or $1 \times n$ or 1×1		(optional)
Z:	$n \times v_2$		
zc:	1×1		(optional)
$result$:	$(v_1+(xc!=0)) \times (v_2+(zc!=0))$		

Diagnostics

$\text{cross}(X, xc, w, Z, zc)$ omits rows in X and Z that contain missing values.

Also See

[M-5] **crossdev( )** — Deviation cross products

[M-5] **quadcross( )** — Quad-precision cross products

[M-5] **mean( )** — Means, variances, and correlations

[M-4] **statistical** — Statistical functions

[M-4] **utility** — Matrix utility functions

Title

[M-5] crossdev( ) — Deviation cross products

Syntax

>*real matrix* crossdev(*X, x, Z, z*)

>*real matrix* crossdev(*X, x, w, Z, z*)

>*real matrix* crossdev(*X, xc, x, Z, zc, z*)

>*real matrix* crossdev(*X, xc, x, w, Z, zc, z*)

where

X:	*real matrix X*
xc:	*real scalar xc*
x:	*real rowvector x*
w:	*real vector w*
Z:	*real matrix Z*
zc:	*real scalar zc*
z:	*real rowvector z*

Description

crossdev( ) makes calculations of the form

$$(X:-x)'(X:-x)$$
$$(X:-x)'(Z:-z)$$
$$(X:-x)'\operatorname{diag}(w)(X:-x)$$
$$(X:-x)'\operatorname{diag}(w)(Z:-z)$$

crossdev( ) is a variation on [M-5] **cross( )**. crossdev( ) mirrors cross( ) in every respect except that it has two additional arguments: *x* and *z*. *x* and *z* record the amount by which *X* and *Z* are to be deviated. *x* and *z* usually contain the (appropriately weighted) column means of *X* and *Z*.

Remarks

x usually contains the same number of rows as *X* but, if $xc \neq 0$, *x* may contain an extra element on the right recording the amount from which the constant 1 should be deviated.

The same applies to *z*: it usually contains the same number of rows as *Z* but, if $zc \neq 0$, *z* may contain an extra element on the right.

▷ Example 1: Linear regression using one view

```
: M = .
: st_view(M, ., ("mpg", "weight", "foreign"), 0)
:
: means = mean(M, 1)
: CP = crossdev(M,means , M,means)
: xx = CP[|2,2 \ .,.|]
: xy = CP[|2,1 \ .,1|]
: b  = invsym(xx)*xy
: b  = b \ means[1]-means[|2\.|]*b
```

Compare this solution with example 3 in [M-5] **cross( )**.

◁

▷ Example 2: Linear regression using subviews

```
: M = X = y = .
: st_view(M, ., ("mpg", "weight", "foreign"), 0)
: st_subview(y, M, ., 1)
: st_subview(X, M, ., (2\.))
:
: xmean = mean(X, 1)
: ymean = mean(y, 1)
: xx    = crossdev(X,xmean , X,xmean)
: xy    = crossdev(X,xmean , y,ymean)
: b     = invsym(xx)*xy
: b     = b \ ymean-xmean*b
```

Compare this solution with example 4 in [M-5] **cross( )**.

◁

▷ Example 3: Weighted linear regression

```
: M = X = y = w = .
: st_view(M, ., ("w", "mpg", "weight", "foreign"), 0)
: st_subview(w, M, ., 1)
: st_subview(y, M, ., 2)
: st_subview(X, M, ., (3\.))
:
: xmean = mean(X, w)
: ymean = mean(y, w)
: xx    = crossdev(X,xmean, w, X,xmean)
: xy    = crossdev(X,xmean, w, y,ymean)
: b     = invsym(xx)*xy
: b     = b \ ymean-xmean*b
```

Compare this solution with example 6 in [M-5] **cross( )**.

◁

▷ Example 4: Variance matrix

```
: X = .
: st_view(X, ., ("mpg", "weight", "displ"), 0)
:
: n     = rows(X)
: means = mean(X, 1)
: xx    = crossdev(X,means , X,means)
: cov   = xx:/(n-1)
```

This is exactly what variance() does; see [M-5] **mean( )**. Compare this solution with example 12 in [M-5] **cross( )**.

◁

▷ Example 5: Weighted variance matrix

```
: M = w = X = .
: st_view(M, ., ("w", "mpg", "weight", "displ"), 0)
: st_subview(w, M, ., 1)
: st_subview(X, M, ., (2\.))
:
: n     = colsum(w)
: means = mean(X, w)
: cov   = crossdev(X,means, w, X,means) :/ (n-1)
```

This is exactly what variance() does with weighted data; see [M-5] **mean( )**. Compare this solution with example 14 in [M-5] **cross( )**.

◁

Conformability

crossdev(X, xc, x, w, Z, zc, z):

X:	$n \times v_1$ or 1×1, 1×1 treated as if $n \times 1$	
xc:	1×1	(optional)
x:	$1 \times v_1$ or $1 \times v_1 + (xc \neq 0)$	
w:	$n \times 1$ or $1 \times n$ or 1×1	(optional)
Z:	$n \times v_2$	
zc:	1×1	(optional)
z:	$1 \times v_2$ or $1 \times v_2 + (zc \neq 0)$	
result:	$(v_1 + (xc \neq 0)) \times (v_2 + (zc \neq 0))$	

Diagnostics

crossdev(X, xc, x, w, Z, zc, z) omits rows in X and Z that contain missing values.

Also See

[M-5] **cross()** — Cross products

[M-5] **quadcross()** — Quad-precision cross products

[M-4] **stata** — Stata interface functions

[M-4] **utility** — Matrix utility functions

[M-4] **statistical** — Statistical functions

Title

[M-5] **cvpermute( )** — Obtain all permutations

Syntax

> *info* = cvpermutesetup(*real colvector V* [, *real scalar unique*])

> *real colvector* cvpermute(*info*)

where *info* should be declared *transmorphic*.

Description

cvpermute() returns all permutations of the values of column vector *V*, one at a time. If $V = (1\backslash 2\backslash 3)$, there are six permutations and they are $(1\backslash 2\backslash 3)$, $(1\backslash 3\backslash 2)$, $(2\backslash 1\backslash 3)$, $(2\backslash 3\backslash 1)$, $(3\backslash 1\backslash 2)$, and $(3\backslash 2\backslash 1)$. If $V = (1\backslash 2\backslash 1)$, there are three permutations and they are $(1\backslash 1\backslash 2)$, $(1\backslash 2\backslash 1)$, and $(2\backslash 1\backslash 1)$.

Vector *V* is specified by calling cvpermutesetup(),

> *info* = cvpermutesetup(*V*)

info holds information that is needed by cvpermute() and it is *info*, not *V*, that is passed to cvpermute(). To obtain the permutations, repeated calls are made to cvpermute() until it returns J(0,1,.):

> *info* = cvpermutesetup(*V*)
> while ((*p*=cvpermute(*info*)) != J(0,1,.)) {
>   ... *p* ...
> }

Column vector *p* will contain a permutation of *V*.

cvpermutesetup() may be specified with one or two arguments:

> *info* = cvpermutesetup(*V*)

> *info* = cvpermutesetup(*V*, *unique*)

unique is usually not specified. If *unique* is specified, it should be 0 or 1. Not specifying *unique* is equivalent to specifying *unique* = 0. Specifying *unique* = 1 states that the elements of *V* are unique or, at least, are to be treated that way.

When the arguments of *V* are unique—for instance, $V = (1\backslash 2\backslash 3)$—specifying *unique* = 1 will make cvpermute() run faster. The same permutations will be returned, although usually in a different order.

When the arguments of *V* are not unique—for instance, $V = (1\backslash 2\backslash 1)$—specifying *unique* = 1 will make cvpermute() treat them as if they were unique. With *unique* = 0, there are three permutations of $(1\backslash 2\backslash 1)$. With *unique* = 1, there are six permutations, just as there are with $(1\backslash 2\backslash 3)$.

Remarks

▷ Example 1

You have the following data:

```
    v1      v2
    22      29
    17      33
    21      26
    20      32
    16      35
```

You wish to do an exact permutation test for the correlation between v1 and v2.

That is, you wish to (1) calculate the correlation between v1 and v2—call that value r—and then (2) calculate the correlation between v1 and v2 for all permutations of v1, and count how many times the result is more extreme than r.

For the first step,

```
: X = (22, 29 \
>      17, 33 \
>      21, 26 \
>      20, 32 \
>      16, 35)
:
: correlation(X)
[symmetric]
                      1              2
    1             1
    2      -.8468554653            1
```

The correlation is $-.846855$. For the second step,

```
: V1 = X[,1]
: V2 = X[,2]
: num = den = 0
: info = cvpermutesetup(V1)
: while ((V1=cvpermute(info)) != J(0,1,.)) {
>         rho = correlation((V1,V2))[2,1]
>         if (rho<=-.846 | rho>=.846) num++
>         den++
> }
: (num, den, num/den)
              1         2          3
    1        13       120   .1083333333
```

Of the 120 permutations, 13 (10.8%) were outside $.846855$ or $-.846855$.

[M-5] cvpermute( ) — Obtain all permutations

▷ Example 2

You now wish to do the same thing but using the Spearman rank-correlation coefficient. Mata has no function that will calculate that, but Stata has a command that does—see [R] **spearman**—so we will use the Stata command as our subroutine.

This time, we will assume that the data have been loaded into a Stata dataset:

```
. list

        var1   var2
     ┌──────────────┐
  1. │  22     29   │
  2. │  17     33   │
  3. │  21     26   │
  4. │  20     32   │
  5. │  16     35   │
     └──────────────┘
```

For the first step,

```
. spearman var1 var2

 Number of obs =      5
 Spearman's rho =    -0.9000

Test of Ho: var1 and var2 are independent
     Prob > |t| =     0.0374
```

For the second step,

```
. mata
                                                     ──── mata (type end to exit) ────
: V1 = st_data(., "var1")
: info = cvpermutesetup(V1)
: num = den = 0
: while ((V1=cvpermute(info)) != J(0,1,.)) {
>         st_store(., "var1", V1)
>         stata("quietly spearman var1 var2")
>         rho = st_numscalar("r(rho)")
>         if (rho<=-.9 | rho>=.9) num++
>         den++
> }
: (num, den, num/den)
           1       2           3
      ┌──────────────────────────────┐
   1  │   2      120    .0166666667  │
      └──────────────────────────────┘
```

Only two of the permutations resulted in a rank correlation of at least .9 in magnitude.

In the code above, we obtained the rank correlation from r(rho) which, we learned from [R] **spearman**, is where spearman stores it.

Also note how we replaced the contents of var1 by using st_store(). Our code leaves the dataset changed and so could be improved.

◁

Conformability

cvpermutesetup(*V*, *unique*):
        V: $n \times 1$
   unique: 1×1 (optional)
    result: $1 \times L$

cvpermute(*info*):
     info: $1 \times L$
    result: $n \times 1$ or 0×1

where

$$L = \begin{cases} 3 & \text{if } n = 0 \\ 4 & \text{if } n = 1 \\ (n+3)(n+2)/2 - 6 & \text{otherwise} \end{cases}$$

The value of L is not important except that the *info* vector returned by cvpermutesetup() and then passed to cvpermute() consumes memory. For instance,

n	L	Total memory ($8 * L$)
5	22	176 bytes
10	72	576
50	1,372	10,560
100	5,247	41,976
1,000	502,497	4,019,976

Diagnostics

cvpermute() returns J(0,1,.) when there are no more permutations.

Also See

[M-4] **statistical** — Statistical functions

Title

[M-5] **date( )** — Date and time manipulation

Syntax

tc = clock(*strdatetime*, *pattern* [, *year*])

tc = mdyhms(*month*, *day*, *year*, *hour*, *minute*, *second*)

tc = dhms(*td*, *hour*, *minute*, *second*)

tc = hms(*hour*, *minute*, *second*)

hour = hh(*tc*)

minute = mm(*tc*)

second = ss(*tc*)

td = dofc(*tc*)

tC = Cofc(*tc*)

tC = Clock(*strdatetime*, *pattern* [, *year*])

tC = Cmdyhms(*month*, *day*, *year*, *hour*, *minute*, *second*)

tC = Cdhms(*td*, *hour*, *minute*, *second*)

tC = Chms(*hour*, *minute*, *second*)

hour = hhC(*tC*)

minute = mmC(*tC*)

second = ssC(*tC*)

td = dofC(*tC*)

tc = cofC(*tC*)

td = date(*strdate*, *dpattern* [, *year*])

td = mdy(*month*, *day*, *year*)

tw = yw(*year*, *week*)

tm = ym(*year*, *month*)

tq = yq(*year*, *quarter*)

th = yh(*year*, *half*)

tc = cofd(*td*)

tC = Cofd(*td*)

month = month(*td*)

day = day(*td*)

year = year(*td*)

dow = dow(*td*)

week = week(*td*)

quarter = quarter(*td*)

half = halfyear(*td*)

doy = doy(*td*)

ty = yearly(*strydate*, *ypattern* [, *year*])

ty = yofd(*td*)

td = dofy(*ty*)

th = halfyearly(*strhdate*, *hpattern* [, *year*])

th = hofd(*td*)

td = dofh(*th*)

tq = quarterly(*strqdate*, *qpattern* [, *year*])

tq = qofd(*td*)

td = dofq(*tq*)

tm = monthly(*strmdate*, *mpattern* [, *year*])

tm = mofd(*td*)

td = dofm(*tm*)

tw = weekly(*strwdate*, *wpattern* [, *year*])

tw = wofd(*td*)

td = dofw(*tw*)

hours = hours(*ms*)

minutes = minutes(*ms*)

seconds = seconds(*ms*)

ms = msofhours(*hours*)

ms = msofminutes(*minutes*)

ms = msofseconds(*seconds*)

where

tc: number of milliseconds from 01jan1960 00:00:00.000, unadjusted for leap seconds

tC: number of milliseconds from 01jan1960 00:00:00.000, adjusted for leap seconds

strdatetime: string-format date, time, or date/time, e.g., "15jan1992", "1/15/1992", "15-1-1992", "January 15, 1992", "9:15", "13:42", "1:42 p.m.", "1:42:15.002 pm", "15jan1992 9:15", "1/15/1992 13:42", "15-1-1992 1:42 p.m.", "January 15, 1992 1:42:15.002 pm"

pattern: order of month, day, year, hour, minute, and seconds in *strdatetime*, plus optional default century, e.g., "DMYhms" (meaning day, month, year, hour, minute, second), "DMYhm", "MDYhm", "hmMDY", "hms", "hm", "MDY", "MD19Y", "MDY20Yhm"

td: number of days from 01jan1960

strdate: string-format date, e.g., "15jan1992", "1/15/1992", "15-1-1992", "January 15, 1992"

dpattern: order of month, day, and year in *strdate*, plus optional default century, e.g., "DMY" (meaning day, month, year), "MDY", "MD19Y"

hour: hour, 0–23

minutes: minutes, 0–59

seconds: seconds, 0.000–59.999 (maximum 60.999 in case of leap second)

month: month number, 1–12

day: day-of-month number, 1–31

year: year, e.g., 1942, 1995, 2008

week: week within year, 1–52

quarter: quarter within year, 1–4

half: half within year, 1–2

dow: day of week, 0–6, 0 = Sunday

doy: day within year, 1–366

ty: calendar year

strydate: string-format calendar year, e.g., "1980", "80"

ypattern: pattern of *strydate*, e.g., "Y", "19Y"

th: number of halves from 1960h1

strhdate: string-format *hdate*, e.g., "1982-1", "1982h2", "2 1982"

hpattern: pattern of *strhdate*, e.g., "YH", "19YH", "HY"

tq:	number of quarters from 1960q1
strqdate:	string-format *qdate*, e.g., "1982-3", "1982q2", "3 1982"
qpattern:	pattern of *strqdate*, e.g., "YQ", "19YQ", "QY"
tm:	number of months from 1960m1
strmdate:	string-format *mdate*, e.g., "1982-3", "1982m2", "3/1982"
mpattern:	pattern of *strmdate*, e.g., "YM", "19YM", "MR"
tw:	number of weeks from 1960w1
strwdate:	string-format *wdate*, e.g., "1982-3", "1982w2", "1982-15"
wpattern:	pattern of *strwdate*, e.g., "YW", "19YW", "WY"
hours:	interval of time in hours (positive or negative, real)
minutes:	interval of time in minutes (positive or negative, real)
seconds:	interval of time in seconds (positive or negative, real)
ms:	interval of time in milliseconds (positive or negative, integer)

Functions return an element-by-element result. Functions are usually used with scalars.

All variables are *real matrix* except the *str** and **pattern* variables, which are *string matrix*.

Description

These functions mirror Stata's date functions; see [D] **dates and times**.

Conformability

clock(*strdatetime*, *pattern*, *year*), Clock(*strdatetime*, *pattern*, *year*):

strdatetime:	$r_1 \times c_1$
pattern:	$r_2 \times c_2$ (c-conformable with *strdatetime*)
year:	$r_3 \times c_3$ (optional, c-conformable)
result:	$\max(r_1, r_2, r_3) \times \max(c_1, c_2, c_3)$

mdyhms(*month*, *day*, *year*, *hour*, *minute*, *second*),
Cmdyhms(*month*, *day*, *year*, *hour*, *minute*, *second*):

month:	$r_1 \times c_1$
day:	$r_2 \times c_2$
year:	$r_3 \times c_3$
hour:	$r_4 \times c_4$
minute:	$r_5 \times c_5$
second:	$r_6 \times c_6$ (all variables c-conformable)
result:	$\max(r_1, r_2, r_3, r_4, r_5, r_6) \times \max(c_1, c_2, c_3 c_4, c_5, c_6)$

hms(*hour*, *minute*, *second*), Chms(*hour*, *minute*, *second*):

hour:	$r_1 \times c_1$
minute:	$r_2 \times c_2$
second:	$r_3 \times c_3$
result:	$\max(r_1, r_2, r_3) \times \max(c_1, c_2, c_3)$

dhms(*td*, *hour*, *minute*, *second*), Cdhms(*td*, *hour*, *minute*, *second*):
  td: $r_1 \times c_1$
  hour: $r_2 \times c_2$
  minute: $r_3 \times c_3$
  second: $r_4 \times c_4$ (all variables c-conformable)
  result: $\max(r_1, r_2, r_3, r_4) \times \max(c_1, c_2, c_3 c_4)$

hh(*x*), mm(*x*), ss(*x*), hhC(*x*), mmC(*x*), ssC(*x*),
  x: $r \times c$
  result: $r \times c$

date(*strdate*, *dpattern*, *year*):
  strdate: $r_1 \times c_1$
  dpattern: $r_2 \times c_2$ (c-conformable with *strdate*)
  year: $r_3 \times c_3$ (optional, c-conformable)
  result: $\max(r_1, r_2, r_3) \times \max(c_1, c_2, c_3)$

mdy(*month*, *day*, *year*):
  month: $r_1 \times c_1$
  day: $r_2 \times c_2$
  year: $r_3 \times c_3$ (all variables c-conformable)
  result: $\max(r_1, r_2, r_3) \times \max(c_1, c_2, c_3)$

yw(*year*, *detail*), ym(*year*, *detail*), yq(*year*, *detail*), yh(*year*, *detail*):
  year: $r_1 \times c_1$
  detail: $r_2 \times c_2$ (c-conformable with *year*)
  result: $\max(r_1, r_2) \times \max(c_1, c_2)$

month(*td*), day(*td*), year(*td*), dow(*td*), week(*td*), quarter(*td*), halfyear(*td*), doy(*td*):
  td: $r \times c$
  result: $r \times c$

yearly(*str*, *pat*, *year*), halfyearly(*str*, *pat*, *year*), quarterly(*str*, *pat*, *year*), monthly(*str*, *pat*, *year*), weekly(*str*, *pat*, *year*):
  str: $r_1 \times c_1$
  pat: $r_2 \times c_2$ (c-conformable with *str*)
  year: $r_3 \times c_3$ (optional, c-conformable)
  result: $\max(r_1, r_2, r_3) \times \max(c_1, c_2, c_3)$

Cofc(*x*), cofC(*x*), dofc(*x*), dofC(*x*), cofd(*x*), Cofd(*x*), yofd(*x*), dofy(*x*), hofd(*x*), dofh(*x*), qofd(*x*), dofq(*x*), mofd(*x*), dofm(*x*), wofd(*x*), dofw(*x*):
  x: $r \times c$
  result: $r \times c$

hours(*x*), minutes(*x*), seconds(*x*), msofhours(*x*), msofminutes(*x*): msofseconds(*x*):
  x: $r \times c$
  result: $r \times c$

Diagnostics

None.

Also See

[M-4] **scalar** — Scalar mathematical functions

Title

[M-5] **designmatrix( )** — Design matrices

Syntax

>   *real matrix*   designmatrix(*real colvector v*)

Description

designmatrix(*v*) returns a rows(*v*) × colmax(*v*) matrix with ones in the indicated columns and zero everywhere else.

Remarks

designmatrix((1\2\3)) is equal to I(3), the 3 × 3 identity matrix.

Conformability

designmatrix(*v*):
>   *v*: $r \times 1$
>   *result*: $r \times$ colmax(*v*)   (0 × 0 if $r = 0$)

Diagnostics

designmatrix(*v*) aborts with error if any element of *v* is <1.

Also See

[M-4] **standard** — Functions to create standard matrices

Title

> **[M-5] det( )** — Determinant of matrix

Syntax

>> *numeric scalar* det(*numeric matrix A*)

>> *numeric scalar* dettriangular(*numeric matrix A*)

Description

det(*A*) returns the determinant of *A*.

dettriangular(*A*) returns the determinant of *A*, treating *A* as if it were triangular (even if it is not).

Remarks

Calculation of the determinant is made by obtaining the LU decomposition of *A* and then calculating the determinant of *U*:

$$\begin{aligned}\det(A) &= \det(PLU) \\ &= \det(P) \times \det(L) \times \det(U) \\ &= \pm 1 \times 1 \times \det(U) \\ &= \pm \det(U)\end{aligned}$$

Since *U* is (upper) triangular, det(*U*) is simply the product of its diagonal elements. See [M-5] **lud( )**.

Conformability

    det(A), dettriangular(A):
            A: n × n
       result: 1 × 1

Diagnostics

det(*A*) and dettriangular(*A*) return 1 if *A* is 0×0.

det(*A*) aborts with error if *A* is not square and returns missing if *A* contains missing values.

dettriangular(*A*) aborts with error if *A* is not square and returns missing if any element on the diagonal of *A* is missing.

Both det(*A*) and dettriangular(*A*) will return missing value if the determinant exceeds 8.99e+307.

Also See

[M-5] **lud( )** — LU decomposition

[M-4] **matrix** — Matrix functions

Title

[M-5] _diag( ) — Replace diagonal of a matrix

Syntax

> *void* _diag(*numeric matrix Z*, *numeric vector v*)

Description

_diag(*Z*, *v*) replaces the diagonal of the matrix *Z* with *v*. *Z* need not be square.

1. If *v* is a vector, the vector replaces the principal diagonal.
2. If *v* is 1×1, each element of the principal diagonal is replaced with *v*.
3. If *v* is a void vector (1×0 or 0×1), *Z* is left unchanged.

Conformability

_diag(*Z*, *v*):

input:
    Z: $n \times m, n \leq m$
    v: $1 \times 1, 1 \times n,$ or $n \times 1$

or

    Z: $n \times m, n > m$
    v: $1 \times 1, 1 \times m,$ or $m \times 1$

output:
    Z: $n \times m$

Diagnostics

_diag(*Z*, *v*) aborts with error if *Z* or *v* is a view.

Also See

[M-5] **diag( )** — Create diagonal matrix

[M-4] **manipulation** — Matrix manipulation

Title

[M-5] **diag( )** — Create diagonal matrix

Syntax

>　　*numeric matrix*　　diag(*numeric matrix Z*)
>
>　　*numeric matrix*　　diag(*numeric vector z*)

Description

diag() creates diagonal matrices.

diag(*Z*), *Z* a matrix, extracts the principal diagonal of *Z* to create a new matrix. *Z* must be square.

diag(*z*), *z* a vector, creates a new matrix with the elements of *z* on its diagonal.

Remarks

Do not confuse diag() with its functional inverse, diagonal(); see [M-5] **diagonal( )**. diag() creates a matrix from a vector (or matrix); diagonal() extracts the diagonal of a matrix into a vector.

Use of diag() should be avoided because it wastes memory. The colon operators will allow you to use vectors directly:

Desired calculation	Equivalent
diag(*v*)*X,	
v is a column	*v*:*X
v is a row	*v*':*X
v is a matrix	diagonal(*v*):*X
X*diag(*v*)	
v is a column	*X*:*v'
v is a row	*X*:*v
v is a matrix	*X*:*diagonal(*v*)'

In the above table, it is assumed that *v* is real. If *v* might be complex, the transpose operators that appear must be changed to transposeonly() calls, because we do not want the conjugate. For instance, *v*':*X would become transposeonly(*v*):*X.

Conformability

diag(*Z*):
    Z: $m \times n$
    result: $\min(m,n) \times \min(m,n)$

diag(*z*):
    z: $1 \times n$ or $n \times 1$
    result: $n \times n$

Diagnostics

None.

Also See

[M-5] **_diag( )** — Replace diagonal of a matrix

[M-5] **diagonal( )** — Extract diagonal into column vector

[M-5] **isdiagonal( )** — Whether matrix is diagonal

[M-4] **manipulation** — Matrix manipulation

Title

[M-5] diag0cnt() — Count zeros on diagonal

Syntax

>*real scalar* diag0cnt(*real matrix X*)

Description

diag0cnt(*X*) returns the number of principal diagonal elements of *X* that are 0.

Remarks

diag0cnt() is often used after [M-5] **invsym()** to count the number of columns dropped because of collinearity.

Conformability

diag0cnt(*X*):
>>*X*: $r \times c$
>>*result*: 1×1

Diagnostics

diag0cnt(*X*) returns 0 if *X* is void.

Also See

[M-5] **invsym()** — Symmetric real matrix inversion

[M-4] **utility** — Matrix utility functions

Title

[M-5] **diagonal( )** — Extract diagonal into column vector

Syntax

> *numeric colvector*   diagonal(*numeric matrix A*)

Description

diagonal(*A*) extracts the diagonal of *A* and returns it in a column vector.

Remarks

diagonal() may be used with nonsquare matrices.

Do not confuse diagonal() with its functional inverse, diag(); see [M-5] **diag( )**. diagonal() extracts the diagonal of a matrix into a vector; diag() creates a diagonal matrix from a vector.

Conformability

diagonal(*A*):
      A: $r \times c$
  result: $\min(r,c) \times 1$

Diagnostics

None.

Also See

[M-5] **diag( )** — Create diagonal matrix

[M-5] **isdiagonal( )** — Whether matrix is diagonal

[M-5] **blockdiag( )** — Block-diagonal matrix

[M-4] **manipulation** — Matrix manipulation

Title

[M-5] **dir( )** — File list

Syntax

>   *string colvector*   dir(*dirname*, *filetype*, *pattern*)
>
>   *string colvector*   dir(*dirname*, *filetype*, *pattern*, *prefix*)

where

dirname:	*string scalar*	containing directory name
filetype:	*string scalar*	containing "files", "dirs", or "other"
pattern:	*string scalar*	containing match pattern
prefix:	*real scalar*	containing 0 or 1

Description

dir(*dirname*, *filetype*, *pattern*) returns a column vector containing the names of the files in *dir* that match *pattern*.

dir(*dirname*, *filetype*, *pattern*, *prefix*) does the same thing but allows you to specify whether you want a simple list of files (*prefix* = 0) or a list of filenames prefixed with *dirname* (*prefix* ≠ 0). dir(*dirname*, *filetype*, *pattern*) is equivalent to dir(*dirname*, *filetype*, *pattern*, 0).

pattern is interpreted by [M-5] **strmatch( )**.

Remarks

Examples:

    dir(".", "dirs", "*")
        returns a list of all directories in the current directory.

    dir(".", "files", "*")
        returns a list of all regular files in the current directory.

    dir(".", "files", "*.sthlp")
        returns a list of all *.sthlp files found in the current directory.

Conformability

dir(*dirname*, *filetype*, *pattern*, *prefix*):

dirname:	1×1	
filetype:	1×1	
pattern:	1×1	
prefix:	1×1	(optional)
result:	$k \times 1$,	k number of files matching pattern

Diagnostics

dir(*dirname*, *filetype*, *pattern*, *prefix*) returns J(0,1,"") if

1. no files matching *pattern* are found,

2. directory *dirname* does not exist, or

3. *filetype* is misspecified (is not equal to "files", "dirs", or "others").

dirname may be specified with or without the directory separator on the end.

dirname = "" is interpreted the same as *dirname* = "."; the current directory is searched.

Also See

[M-4] **io** — I/O functions

Title

[M-5] direxists( ) — Whether directory exists

Syntax

>   *real scalar*   `direxists(`*string scalar dirname*`)`

Description

`direxists(`*dirname*`)` returns 1 if *dirname* contains a valid path to a directory and returns 0 otherwise.

Conformability

`direxists(`*dirname*`)`:
>   *dirname*: 1×1
>   *result*: 1×1

Diagnostics

None.

Also See

[M-4] **io** — I/O functions

Title

> **[M-5] direxternal( )** — Obtain list of existing external globals

Syntax

> *string colvector* direxternal(*string scalar pattern*)

Description

direxternal(*pattern*) returns a column vector containing the names matching *pattern* of the existing external globals. direxternal() returns J(0,1,"") if there are no such globals.

pattern is interpreted by [M-5] **strmatch( )**.

Remarks

See [M-5] **findexternal( )** for the definition of a global.

A list of all globals can be obtained by direxternal("*").

Conformability

direxternal(*pattern*):
    pattern: 1×1
    result: $n \times 1$

Diagnostics

direxternal(*pattern*) returns J(0,1,"") when there are no globals matching *pattern*.

Also See

[M-5] **findexternal( )** — Find, create, and remove external globals

[M-4] **programming** — Programming functions

Title

[M-5] **display( )** — Display text interpreting SMCL

Syntax

>   *void*   display(*string colvector s*)

>   *void*   display(*string colvector s, real scalar asis*)

Description

display(*s*) displays the string or strings contained in *s*.

display(*s, asis*) does the same thing but allows you to control how SMCL codes are treated. display(*s*, 0) is equivalent to display(*s*); any SMCL codes are honored.

display(*s, asis*), *asis* ≠ 0, displays the contents of *s* exactly as they are. For instance, when *asis* ≠ 0, "{it}" is just the string of characters {, i, t, and } and those characters are displayed; {it} is not given the SMCL interpretation of enter italic mode.

Remarks

When *s* is a scalar, the differences between coding

>   : display(*s*)

and coding

>   : *s*

are

1. display(*s*) will not indent *s*; *s* by itself causes *s* to be indented by two spaces.

2. display(*s*) will honor any SMCL codes contained in *s*; *s* by itself is equivalent to display(*s*, 1). For example,

    : s = "this is an {it:example}"
    : display(s)
    this is an *example*
    : s
      this is an {it:example}

3. When *s* is a vector, display(*s*) simply displays the lines, whereas *s* by itself adorns the lines with row and column numbers:

```
        : s = ("this is line 1" \ "this is line 2")
        : display(s)
        this is line 1
        this is line 2
        : s

            1 │ this is line 1
            2 │ this is line 2
```

Another alternative to display() is printf(); see [M-5] **printf**( ). When *s* is a scalar, display() and printf() do the same thing:

```
        : display("this is an {it:example}")
        this is an example
        : printf("%s\n", "this is an {it:example}")
        this is an example
```

printf(), however, will not allow *s* to be nonscalar; it has other capabilities.

Conformability

display(*s*, *asis*)
        s: $k \times 1$
    asis: 1×1 (optional)
  result: *void*

Diagnostics

None.

Also See

[M-5] **displayas**( ) — Set display level

[M-5] **displayflush**( ) — Flush terminal-output buffer

[M-5] **printf**( ) — Format output

[M-4] **io** — I/O functions

Title

[M-5] **displayas( )** — Set display level

Syntax

> *void* displayas(*string scalar level*)

where *level* may be

level	minimum abbreviation
"result"	"res"
"text"	"txt"
"error"	"err"
"input"	"inp"

Description

displayas(*level*) sets whether and how subsequent output is to be displayed.

Remarks

If this function is never invoked, then the output level is result. Say that Mata was invoked in such a way that all output except error messages is being suppressed (e.g., quietly was coded in front of the mata command or in front of the ado-file that called your Mata function). If output is being suppressed, then Mata output is being suppressed, including any output created by your program. Say that you reach a point in your program where you wish to output an error message. You coded

> printf("{err:you made a mistake}\n")

Even though you coded the SMCL directive {err:}, the error message will still be suppressed. SMCL determines how something is rendered, not whether it is rendered. What you need to code is

> displayas("err")
> printf("{err:you made a mistake}\n")

Actually, you could code

> displayas("err")
> printf("you made a mistake\n")

because, in addition to setting the output level (telling Stata that all subsequent output is of the specified level), it also sets the current SMCL rendering to what is appropriate for that kind of output. Hence, if you coded

> displayas("err")
> printf("{res:you made a mistake}\n")

the text you made a mistake would appear in the style of results despite any quietlys attempting to suppress output. Coding the above is considered bad style.

Conformability

    displayas(level):
            level: 1 × 1
            result: void

Diagnostics

displayas(level) aborts with error if *level* contains an inappropriate string.

Also See

[M-5] **printf( )** — Format output

[M-5] **display( )** — Display text interpreting SMCL

[M-4] **io** — I/O functions

Title

[M-5] **displayflush( )** — Flush terminal-output buffer

Syntax

>*void* displayflush()

Description

To achieve better performance, Stata buffers terminal output, so, within a program, output may not appear when a display() or printf() command is executed. The output might be held in a buffer and displayed later.

displayflush() forces Stata to display all pending output at the terminal. displayflush() is rarely used.

Remarks

See [M-5] **printf( )** for an example of the use of displayflush().

Use of displayflush() slows execution. Use displayflush() only when it is important that output be displayed at the terminal now, such as when providing messages indicating what your program is doing.

Diagnostics

None.

Also See

[M-5] **printf( )** — Format output

[M-5] **display( )** — Display text interpreting SMCL

[M-4] **io** — I/O functions

Title

[M-5] **dsign( )** — FORTRAN-like DSIGN( ) function

Syntax

> *real scalar* dsign(*real scalar a, real scalar b*)

Description

dsign(*a*, *b*) returns *a* with the sign of *b*, defined as $|a|$ if $b \geq 0$ and $-|a|$ otherwise.

This function is useful when translating FORTRAN programs.

The in-line construction

> (b >=0 ? abs(*a*) : -abs(*a*))

is clearer. Also, differentiate carefully between what dsign() returns (equivalent to the above construction) and signum(*b*)*abs(*a*), which is almost equivalent but returns 0 when *b* is 0 rather than abs(*a*). (Message: dsign() is not one of our favorite functions.)

Conformability

dsign(*a*, *b*):
>   *a*: 1×1
>   *b*: 1×1
> *result*: 1×1

Diagnostics

dsign(., *b*) returns . for all *b*.

dsign(*a*, .) returns abs(*a*) for all *a*.

Also See

[M-5] **sign( )** — Sign and complex quadrant functions

[M-4] **scalar** — Scalar mathematical functions

Title

[M-5] e( ) — Unit vectors

Syntax

> *real rowvector* e(*real scalar i, real scalar n*)

Description

e(i, n) returns a $1 \times n$ unit vector, a vector with all elements equal to zero except for the ith, which is set to one.

Conformability

e(i, n):
- i: 1×1
- n: 1×1
- *result*: $1 \times n$

Diagnostics

e(i, n) aborts with error if $n < 1$ or if $i < 1$ or if $i > n$. Arguments i and n are interpreted as trunc(i) and trunc(n).

Also See

[M-4] **standard** — Functions to create standard matrices

Title

[M-5] **editmissing( )** — Edit matrix for missing values

Syntax

> *numeric matrix* editmissing(*numeric matrix A*, *numeric scalar v*)
>
> *void* _editmissing(*numeric matrix a*, *numeric scalar v*)

Description

editmissing(*A*, *v*) returns *A* with any missing values changed to *v*.

_editmissing(*A*, *v*) replaces all missing values in *A* with *v*.

Remarks

editmissing( ) and _editmissing( ) are very fast.

If you want to change nonmissing values to other values, including missing, see [M-5] **editvalue( )**.

Conformability

editmissing(*A*, *v*):
        A: $r \times c$
        v: 1×1
    result: $r \times c$

_editmissing(*A*, *v*):
  input:
        A: $r \times c$
        v: 1×1
  output:
        A: $r \times c$

Diagnostics

editmissing(*A*, *v*) and _editmissing(*A*, *v*) change all missing elements to *v*, including not only . but also .a, .b, ..., .z.

Also See

[M-5] **editvalue( )** — Edit (change) values in matrix

[M-4] **manipulation** — Matrix manipulation

Title

[M-5] edittoint( ) — Edit matrix for roundoff error (integers)

Syntax

> *numeric matrix*   edittoint(*numeric matrix Z, real scalar amt*)
> *void*   _edittoint(*numeric matrix Z, real scalar amt*)
>
> *numeric matrix*   edittointtol(*numeric matrix Z, real scalar tol*)
> *void*   _edittointtol(*numeric matrix Z, real scalar tol*)

Description

These edit functions set elements of matrices to integers that are close to integers.

edittoint(*Z, amt*) and _edittoint(*Z, amt*) set

$$Z_{ij} = \text{round}(Z_{ij}) \quad \text{if } |Z_{ij} - \text{round}(Z_{ij})| \leq |tol|$$

for Z real and set

$$\text{Re}(Z_{ij}) = \text{round}(\text{Re}(Z_{ij})) \quad \text{if } |\text{Re}(Z_{ij}) - \text{round}(\text{Re}(Z_{ij}))| \leq |tol|$$

$$\text{Im}(Z_{ij}) = \text{round}(\text{Im}(Z_{ij})) \quad \text{if } |\text{Im}(Z_{ij}) - \text{round}(\text{Im}(Z_{ij}))| \leq |tol|$$

for Z complex, where in both cases

$$tol = \text{abs}(amt)*\text{epsilon}(\text{sum}(\text{abs}(Z))/(\text{rows}(Z)*\text{cols}(Z)))$$

edittoint() leaves Z unchanged and returns the edited matrix. _edittoint() edits Z in place.

edittointtol(*Z, tol*) and _edittointtol(*Z, tol*) do the same thing, except that *tol* is specified directly.

Remarks

These functions mirror the edittozero() functions documented in [M-5] **edittozero( )**, except that, rather than solely resetting to zero values close to zero, they reset to integer values close to integers.

See [M-5] **edittozero( )**. Whereas use of the functions documented there is recommended, use of the functions documented here generally is not. Although zeros commonly arise in real problems so that there is reason to suspect small numbers would be zero but for roundoff error, integers arise more rarely.

If you have reason to believe that integer values are likely, then by all means use these functions.

Conformability

edittoint(*Z*, *amt*):

              Z: $r \times c$
          amt: 1×1
        result: $r \times c$

_edittoint(*Z*, *amt*):
   input:

              Z: $r \times c$
          amt: 1×1
   output:
              Z: $r \times c$

edittointtol(*Z*, *tol*):
              Z: $r \times c$
           tol: 1×1
        result: $r \times c$

_edittointtol(*Z*, *tol*):
   input:

              Z: $r \times c$
           tol: 1×1
   output:
              Z: $r \times c$

Diagnostics

None.

Also See

[M-5] **edittozero( )** — Edit matrix for roundoff error (zeros)

[M-4] **manipulation** — Matrix manipulation

Title

[M-5] **edittozero( )** — Edit matrix for roundoff error (zeros)

Syntax

> *numeric matrix*   edittozero(*numeric matrix Z, real scalar amt*)
>
> *void*   _edittozero(*numeric matrix Z, real scalar amt*)
>
> *numeric matrix*   edittozerotol(*numeric matrix Z, real scalar tol*)
>
> *void*   _edittozerotol(*numeric matrix Z, real scalar tol*)

Description

These edit functions set elements of matrices to zero that are close to zero. edittozero(*Z, amt*) and _edittozero(*Z, amt*) set

$$Z_{ij} = 0 \quad \text{if } |Z_{ij}| \leq |tol|$$

for Z real and set

$$\text{Re}(Z_{ij}) = 0 \quad \text{if } |\text{Re}(Z_{ij})| \leq |tol|$$
$$\text{Im}(Z_{ij}) = 0 \quad \text{if } |\text{Im}(Z_{ij})| \leq |tol|$$

for Z complex, where in both cases

> tol = abs(*amt*)*epsilon(sum(abs(Z))/(rows(Z)*cols(Z)))

edittozero() leaves Z unchanged and returns the edited matrix. _edittozero() edits Z in place.

edittozerotol(*Z, tol*) and _edittozerotol(*Z, tol*) do the same thing, except that *tol* is specified directly.

Remarks

Remarks are presented under the following headings:

> Background
> Treatment of complex values
> Recommendations

Background

Numerical roundoff error leads to, among other things, numbers that should be zero being small but not zero, and so it is sometimes desirable to reset those small numbers to zero.

The problem is in identifying those small numbers. Is 1e−14 small? Is 10,000? The answer is that, given some matrix, 1e−14 might not be small because most of the values in the matrix are around 1e−14, and the small values are 1e−28, and given some other matrix, 10,000 might indeed be small because most of the elements are around 1e+18.

edittozero() makes an attempt to determine what is small. edittozerotol() leaves that determination to you. In edittozerotol(Z, tol), you specify *tol* and elements for which $|Z_{ij}| \leq tol$ are set to zero.

Using edittozero(Z, amt), however, you specify *amt* and then *tol* is calculated for you based on the size of the elements in Z and *amt*, using the formula

$$tol = amt * \texttt{epsilon}(average\ value\ of\ |Z_{ij}|)$$

epsilon() refers to machine precision, and epsilon(*x*) is the function that returns machine precision in units of *x*:

$$\texttt{epsilon}(x) = |x|*\texttt{epsilon}(1)$$

where epsilon(1) is approximately 2.22e−16 on most computers; see [M-5] **epsilon( )**.

Treatment of complex values

The formula

$$tol = amt * \texttt{epsilon}(average\ value\ of\ |Z_{ij}|)$$

is used for both real and complex values. For complex, $|Z_{ij}|$ refers to the modulus (length) of the complex element.

However, rather than applying the reset rule

$$Z_{ij} = 0 \quad \text{if } |Z_{ij}| \leq |tol|$$

as is done when Z is real, the reset rules are

$$\text{Re}(Z_{ij}) = 0 \quad \text{if } |\text{Re}(Z_{ij})| \leq |tol|$$
$$\text{Im}(Z_{ij}) = 0 \quad \text{if } |\text{Im}(Z_{ij})| \leq |tol|$$

The first rule, applied even for complex, may seem more appealing, but the use of the second has the advantage of mapping numbers close to being purely real or purely imaginary to purely real or purely imaginary results.

Recommendations

1. Minimal editing is performed by edittozero(Z,1). Values less than 2.22e−16 times the average would be set to zero.

2. It is often reasonable to code edittozero(Z, 1000), which sets to zero values less than 2.22e−13 times the average.

3. For a given calculation, the amount of roundoff error that arises with complex matrices (matrices with nonzero imaginary part) is greater than the amount that arises with real matrices (matrices with zero imaginary part even if stored as complex). That is because, in addition to all the usual sources of roundoff error, multiplication of complex values involves the addition operator, which introduces additional roundoff error. Hence, whatever is the appropriate value of *amt* or *tol* with real matrices, it is larger for complex matrices.

Conformability

edittozero(*Z, amt*):

        Z: $r \times c$
      amt: 1×1
      result: $r \times c$

_edittozero(*Z, amt*):
  input:
        Z: $r \times c$
      amt: 1×1
  output:
        Z: $r \times c$

edittozerotol(*Z, tol*):

        Z: $r \times c$
      tol: 1×1
      result: $r \times c$

_edittozerotol(*Z, tol*):
  input:
        Z: $r \times c$
      tol: 1×1
  output:
        Z: $r \times c$

Diagnostics

edittozero(*Z, amt*) and _edittozero(*Z, amt*) leave scalars unchanged because they base their calculation of the likely roundoff error on the average value of $|Z_{ij}|$.

Also See

[M-5] **edittoint( )** — Edit matrix for roundoff error (integers)

[M-4] **manipulation** — Matrix manipulation

Title

[M-5] editvalue( ) — Edit (change) values in matrix

Syntax

>   *matrix*   editvalue(*matrix A, scalar from, scalar to*)

>   *void*   _editvalue(*matrix A, scalar from, scalar to*)

where *A*, *from*, and *to* may be real, complex, or string.

Description

editvalue(*A, from, to*) returns *A* with all elements equal to *from* changed to *to*.

_editvalue(*A, from, to*) does the same thing but modifies *A* itself.

Remarks

editvalue() and _editvalue() are fast.

If you wish to change missing values to nonmissing values, it is better to use [M-5] **editmissing( )**. editvalue(*A*, ., 1) would change all . missing values to 1 but leave .a, .b, ..., .z unchanged. editmissing(*A*, 1) would change all missing values to 1.

Conformability

editvalue(*A, from, to*):

A:	$r \times c$
from:	1×1
to:	1×1
result:	$r \times c$

_editvalue(*A, from, to*):

input:

A:	$r \times c$
from:	1×1
to:	1×1

output:

A:	$r \times c$

Diagnostics

editvalue(*A, from, to*) returns a matrix of the same type as *A*.

editvalue(*A, from, to*) and _editvalue(*A, from, to*) abort with error if *from* and *to* are incompatible with *A*. That is, if *A* is real, *to* and *from* must be real. If *A* is complex, *to* and *from* must each be either real or complex. If *A* is string, *to* and *from* must be string.

_editvalue(*A, from, to*) aborts with error if *A* is a view.

Also See

[M-5] **editmissing( )** — Edit matrix for missing values

[M-4] **manipulation** — Matrix manipulation

Title

[M-5] eigensystem( ) — Eigenvalues and eigenvectors

Syntax

void	eigensystem(*A*, *X*, *L* [, *rcond* [, *nobalance*]])
void	lefteigensystem(*A*, *X*, *L* [, *rcond* [, *nobalance*]])
complex rowvector	eigenvalues(*A* [, *rcond* [, *nobalance*]])
void	symeigensystem(*A*, *X*, *L*)
real rowvector	symeigenvalues(*A*)
void	_eigensystem(*A*, *X*, *L* [, *rcond* [, *nobalance*]])
void	_lefteigensystem(*A*, *X*, *L* [, *rcond* [, *nobalance*]])
complex rowvector	_eigenvalues(*A* [, *rcond* [, *nobalance*]])
void	_symeigensystem(*A*, *X*, *L*)
real rowvector	_symeigenvalues(*A*)

where inputs are

    A: *numeric matrix*
  rcond: *real scalar* (whether *rcond* desired)
nobalance: *real scalar* (whether to suppress balancing)

and outputs are

    X: *numeric matrix* of eigenvectors
    L: *numeric vector* of eigenvalues
  rcond: *real vector* of reciprocal condition numbers

The columns of *X* will contain the eigenvectors except when using _lefteigensystem(), in which case the rows of *X* contain the eigenvectors.

The following routines are used in implementing the above routines:

    real scalar _eigen_la(*real scalar todo*, *numeric matrix A*, *X*, *L*, *real scalar rcond*, *real scalar nobalance*)

    real scalar _symeigen_la(*real scalar todo*, *numeric matrix A*, *X*, *L*)

Description

These routines calculate eigenvalues and eigenvectors of square matrix A.

eigensystem(A, X, L, *rcond*, *nobalance*) calculates eigenvalues and eigenvectors of a general, real or complex, square matrix A. Eigenvectors are returned in X and eigenvalues in L. The remaining arguments are optional:

1. If *rcond* is not specified, then reciprocal condition numbers are not returned in *rcond*.

    If *rcond* is specified and contains a value other than 0 or missing—*rcond*=1 is suggested—in *rcond* will be placed a vector of the reciprocals of the condition numbers for the eigenvalues. Each element of the new *rcond* measures the accuracy to which the corresponding eigenvalue has been calculated; large numbers (numbers close to 1) are better and small numbers (numbers close to 0) indicate inaccuracy; see *Eigenvalue condition* below.

2. If *nobalance* is not specified, balancing is performed to obtain more accurate results.

    If *nobalance* is specified and is not zero nor missing, balancing is not used. Results are calculated more quickly, but perhaps a little less accurately; see *Balancing* below.

lefteigensystem(A, X, L, *rcond*, *nobalance*) mirrors eigensystem(), the difference being that lefteigensystem() solves for left eigenvectors solving $XA = \mathrm{diag}(L)*X$ instead of right eigenvectors solving $AX = X*\mathrm{diag}(L)$.

eigenvalues(A, *rcond*, *nobalance*) returns the eigenvalues of square matrix A; the eigenvectors are not calculated. Arguments *rcond* and *nobalance* are optional.

symeigensystem(A, X, L) and symeigenvalues(A) mirror eigensystem() and eigenvalues(), the difference being that A is assumed to be symmetric (Hermitian). The eigenvalues returned are real. (Arguments *rcond* and *nobalance* are not allowed; *rcond* because symmetric matrices are inherently well conditioned; *nobalance* because it is unnecessary.)

The underscore routines mirror the routines of the same name without underscores, the difference being that A is damaged during the calculation and so the underscore routines use less memory.

_eigen_la() and _symeigen_la() are the interfaces into the [M-1] **LAPACK** routines used to implement the above functions. Their direct use is not recommended.

Remarks

Remarks are presented under the following headings:

> *Eigenvalues and eigenvectors*
> *Left eigenvectors*
> *Symmetric eigensystems*
> *Normalization and order*
> *Eigenvalue condition*
> *Balancing*
> *eigensystem( ) and eigenvalues( )*
> *lefteigensystem( )*
> *symeigensystem( ) and symeigenvalues( )*

Eigenvalues and eigenvectors

A scalar λ is said to be an eigenvalue of square matrix $A: n \times n$ if there is a nonzero column vector $x: n \times 1$ (called the eigenvector) such that

$$Ax = \lambda x \qquad (1)$$

(1) can also be written

$$(A - \lambda I)x = 0$$

where I is the $n \times n$ identity matrix. A nontrivial solution to this system of n linear homogeneous equations exists if and only if

$$\det(A - \lambda I) = 0 \qquad (2)$$

This nth degree polynomial in λ is called the characteristic polynomial or characteristic equation of A, and the eigenvalues λ are its roots, also known as the characteristic roots.

There are, in fact, n solutions (λ_i, x_i) that satisfy (1)—although some can be repeated—and we can compactly write the full set of solutions as

$$AX = X * \mathrm{diag}(L) \qquad (3)$$

where

$$X = (x_1, x_2, \ldots) \qquad (X : n \times n)$$
$$L = (\lambda_1, \lambda_2, \ldots) \qquad (L : 1 \times n)$$

```
: A = (1, 2 \ 9, 4)
: X = .
: L = .
: eigensystem(A, X, L)
: X
```

	1	2
1	-.316227766	-.554700196
2	-.948683298	.832050294

```
: L
```

	1	2
1	7	-2

The first eigenvalue is 7, and the corresponding eigenvector is $(-.316 \setminus -.949)$. The second eigenvalue is -2, and the corresponding eigenvector is $(-.555 \setminus .832)$.

In general, eigenvalues and vectors can be complex even if A is real.

Left eigenvectors

What we have defined above is properly known as the right-eigensystem problem:

$$Ax = \lambda x \qquad (1)$$

In the above, x is a column vector. The left-eigensystem problem is to find the row vector x satisfying

$$xA = \lambda x \qquad (1')$$

The eigenvalue λ is the same in (1) and (1'), but x can differ.

The n solutions (λ_i, x_i) that satisfy (1') can be compactly written

$$XA = \operatorname{diag}(L) * X \qquad (3')$$

where

$$X = \begin{bmatrix} x_1 \\ x_2 \\ \vdots \\ x_n \end{bmatrix}_{n \times n} \qquad L = \begin{bmatrix} \lambda_1 \\ \lambda_2 \\ \vdots \\ \lambda_n \end{bmatrix}_{n \times 1}$$

For instance,

```
: A = (1, 2 \ 9, 4)
: X = .
: L = .
: lefteigensystem(A, X, L)
: X
```

	1	2
1	-.832050294	-.554700196
2	-.948683298	.316227766

```
: L
```

	1
1	7
2	-2

The first eigenvalue is 7, and the corresponding eigenvector is $(-.832, -.555)$. The second eigenvalue is -2, and the corresponding eigenvector is $(-.949, .316)$.

The eigenvalues are the same as in the previous example; the eigenvectors are different.

Symmetric eigensystems

Below we use the term symmetric to encompass Hermitian matrices, even when we do not emphasize the fact.

Eigensystems of symmetric matrices are conceptually no different from general eigensystems, but symmetry introduces certain simplifications:

1. The eigenvalues associated with symmetric matrices are real, whereas those associated with general matrices may be real or complex.

2. The eigenvectors associated with symmetric matrices—which may be real or complex—are orthogonal.

3. The left and right eigenvectors of symmetric matrices are transposes of each other.

4. The eigenvalues and eigenvectors of symmetric matrices are more easily, and more accurately, computed.

For item 3, let us begin with the right-eigensystem problem:

$$AX = X * \mathrm{diag}(L)$$

Taking the transpose of both sides results in

$$X'A = \mathrm{diag}(L) * X'$$

because $A = A'$ if A is symmetric (Hermitian).

symeigensystem(A, X, L) calculates right eigenvectors. To obtain the left eigenvectors, you simply transpose X.

Normalization and order

If x is a solution to

$$Ax = \lambda x$$

then so is cx, c: 1×1, $c \neq 0$.

The eigenvectors returned by the above routines are scaled to have length (norm) 1.

The eigenvalues are combined and returned in a vector (L) and the eigenvectors in a matrix (X). The eigenvalues are ordered from largest to smallest in absolute value (or, if the eigenvalues are complex, in length). The eigenvectors are ordered to correspond to the eigenvalues.

Eigenvalue condition

Optional argument *rcond* may be specified as a value other than 0 or missing—*rcond* = 1 is suggested—and then *rcond* will be filled in with a vector containing the reciprocals of the condition numbers for the eigenvalues. Each element of *rcond* measures the accuracy with which the corresponding eigenvalue has been calculated; large numbers (numbers close to 1) are better and small numbers (numbers close to 0) indicate inaccuracy.

The reciprocal condition number is calculated as abs($y*x$), where y: $1 \times n$ is the left eigenvector and x: $n \times 1$ is the corresponding right eigenvector. Since y and x each have norm 1, abs($y*x$) = abs(cos(*theta*)), where *theta* is the angle made by the vectors. Thus $0 \leq$ abs($y*x$) ≤ 1. For symmetric matrices, $y*x$ will equal 1. It can be proved that abs($y*x$) is the reciprocal of the condition number for a simple eigenvalue, and so it turns out that the sensitivity of the eigenvalue to a perturbation is a function of how far the matrix is from symmetric on this scale.

Requesting that *rcond* be calculated increases the amount of computation considerably.

Balancing

By default, balancing is performed for general matrices. Optional argument *nobalance* allows you to suppress balancing.

Balancing is related to row-and-column equilibration; see [M-5] **_equilrc( )**. Here, however, a diagonal matrix D is found such that DAD^{-1} is better balanced, the eigenvalues and eigenvectors for DAD^{-1} are extracted, and then the eigenvalues and eigenvectors are adjusted by D so that they reflect those for the original A matrix.

There is no gain from these machinations when A is symmetric, so the symmetric routines do not have a *nobalance* argument.

eigensystem( ) and eigenvalues( )

1. Use $L =$ eigenvalues(A) and eigensystem(A, X, L) for general matrices A.

2. Use $L =$ eigenvalues(A) when you do not need the eigenvectors; it will save both time and memory.

3. The eigenvalues returned by $L =$ eigenvalues(A) and by eigensystem(A, X, L) are of storage type complex even if the eigenvalues are real (i.e., even if Im(L)==0). If the eigenvalues are known to be real, you can save computer memory by subsequently coding

    L = Re(L)

    If you wish to test whether the eigenvalues are real, examine mreldifre(L); see [M-5] **reldif( )**.

4. The eigenvectors returned by eigensystem(A, X, L) are of storage type complex even if the eigenvectors are real (i.e., even if Im(X)==0). If the eigenvectors are known to be real, you can save computer memory by subsequently coding

    X = Re(X)

    If you wish to test whether the eigenvectors are real, examine mreldifre(X); see [M-5] **reldif( )**.

5. If you are using eigensystem(A, X, L) interactively (outside a program), X and L must be predefined. Type

    : eigensystem(A, X=., L=.)

lefteigensystem( )

What was just said about eigensystem( ) applies equally well to lefteigensystem( ).

If you need only the eigenvalues, use $L =$ eigenvalues(A). The eigenvalues are the same for both left and right systems.

symeigensystem( ) and symeigenvalues( )

1. Use $L = \text{symeigenvalues}(A)$ and $\text{symeigensystem}(A, X, L)$ for symmetric or Hermitian matrices A.

2. Use $L = \text{symeigenvalues}(A)$ when you do not need the eigenvectors; it will save both time and memory.

3. The eigenvalues returned by $L = \text{symeigenvalues}(A)$ and by $\text{symeigensystem}(A, X, L)$ are of storage type real. Eigenvalues of symmetric and Hermitian matrices are always real.

4. The eigenvectors returned by $\text{symeigensystem}(A, X, L)$ are of storage type real when A is of storage type real and of storage type complex when A is of storage type complex.

5. If you are using $\text{symeigensystem}(A, X, L)$ interactively (outside a program), X and L must be predefined. Type

    : symeigensystem(A, X=., L=.)

Conformability

eigensystem(A, X, L, *rcond*, *nobalance*):

   input:

A:	$n \times n$	
rcond:	1×1	(optional, specify as 1 to obtain *rcond*)
nobalance:	1×1	(optional, specify as 1 to prevent balancing)

   output:

X:	$n \times n$	(columns contain eigenvectors)
L:	$1 \times n$	
rcond:	$1 \times n$	(optional)
result:	*void*	

lefteigensystem(A, X, L, *rcond*, *nobalance*):

   input:

A:	$n \times n$	
rcond:	1×1	(optional, specify as 1 to obtain *rcond*)
nobalance:	1×1	(optional, specify as 1 to prevent balancing)

   output:

X:	$n \times n$	(rows contain eigenvectors)
L:	$n \times 1$	
rcond:	$n \times 1$	(optional)
result:	*void*	

eigenvalues(A, *rcond*, *nobalance*):

   input:

A:	$n \times n$	
rcond:	1×1	(optional, specify as 1 to obtain *rcond*)
nobalance:	1×1	(optional, specify as 1 to prevent balancing)

   output:

rcond:	$1 \times n$	(optional)
result:	$1 \times n$	(contains eigenvalues)

[M-5] eigensystem( ) — Eigenvalues and eigenvectors

symeigensystem(*A*, *X*, *L*):
- *input*:
  - *A*: $n \times n$
- *output*:
  - *X*: $n \times n$ (columns contain eigenvectors)
  - *L*: $1 \times n$
  - *result*: void

symeigenvalues(*A*):
- *input*:
  - *A*: $n \times n$
- *output*:
  - *result*: $1 \times n$ (contains eigenvalues)

_eigensystem(*A*, *X*, *L*, *rcond*, *nobalance*):
- *input*:
  - *A*: $n \times n$
  - *rcond*: 1×1 (optional, specify as 1 to obtain *rcond*)
  - *nobalance*: 1×1 (optional, specify as 1 to prevent balancing)
- *output*:
  - *A*: 0×0
  - *X*: $n \times n$ (columns contain eigenvectors)
  - *L*: $1 \times n$
  - *rcond*: $1 \times n$ (optional)
  - *result*: void

_lefteigensystem(*A*, *X*, *L*, *rcond*, *nobalance*):
- *input*:
  - *A*: $n \times n$
  - *rcond*: 1×1 (optional, specify as 1 to obtain *rcond*)
  - *nobalance*: 1×1 (optional, specify as 1 to prevent balancing)
- *output*:
  - *A*: 0×0
  - *X*: $n \times n$ (rows contain eigenvectors)
  - *L*: $n \times 1$
  - *rcond*: $n \times 1$ (optional)
  - *result*: void

_eigenvalues(*A*, *rcond*, *nobalance*):
- *input*:
  - *A*: $n \times n$
  - *rcond*: 1×1 (optional, specify as 1 to obtain *rcond*)
  - *nobalance*: 1×1 (optional, specify as 1 to prevent balancing)
- *output*:
  - *A*: 0×0
  - *rcond*: $1 \times n$ (optional)
  - *result*: $1 \times n$ (contains eigenvalues)

_symeigensystem(A, X, L):
- *input*:
  - A: $n \times n$
- *output*:
  - A: 0×0
  - X: $n \times n$ (columns contain eigenvectors)
  - L: $1 \times n$
  - *result*: void

_symeigenvalues(A):
- *input*:
  - A: $n \times n$
- *output*:
  - A: 0×0
  - *result*: $1 \times n$ (contains eigenvalues)

_eigen_la(todo, A, X, L, rcond, nobalance):
- *input*:
  - todo: 1×1
  - A: $n \times n$
  - rcond: 1×1
  - nobalance: 1×1
- *output*:
  - A: 0×0
  - X: $n \times n$
  - L: $1 \times n$ or $n \times 1$
  - rcond: $1 \times n$ or $n \times 1$ (optional)
  - *result*: 1×1 (return code)

_symeigen_la(todo, A, X, L):
- *input*:
  - todo: 1×1
  - A: $n \times n$
- *output*:
  - A: 0×0
  - X: $n \times n$
  - L: $1 \times n$
  - *result*: 1×1 (return code)

Diagnostics

All functions return missing-value results if A has missing values.

symeigensystem(), symeigenvalues(), _symeigensystem(), and _symeigenvalues() use the lower triangle of A without checking for symmetry. When A is complex, only the real part of the diagonal is used.

Also See

[M-5] **matexpsym( )** — Exponentiation and logarithms of symmetric matrices

[M-5] **matpowersym( )** — Powers of a symmetric matrix

[M-4] **matrix** — Matrix functions

Title

[M-5] **eltype( )** — Element type and organizational type of object

Syntax

>   *string scalar*   eltype(*X*)
>
>   *string scalar*   orgtype(*X*)

Description

eltype() returns the current *eltype* of the argument.

orgtype() returns the current *orgtype* of the argument.

See [M-6] **glossary** for a definition of *eltype* and *orgtype*.

Remarks

If *X* is a matrix (syntax 1), returned is

eltype(*X*)	orgtype(*X*)
real	scalar
complex	rowvector
string	colvector
pointer	matrix
struct	

The returned value reflects the current contents of *X*. That is, *X* might be declared a transmorphic matrix, but at any instant, it contains something, and if that something were 5, returned would be "real" and "scalar".

For orgtype(), returned is "scalar" if the object is currently 1×1; "rowvector" if it is $1 \times k$, $k \neq 1$; "colvector" if it is $k \times 1$, $k \neq 1$; and "matrix" otherwise (it is $r \times c$, $r \neq 1$, $c \neq 1$).

X can be a function. Returned is

eltype(*(&func()))	orgtype(*(&func()))
transmorphic	matrix
numeric	vector
real	rowvector
complex	colvector
string	scalar
pointer	void
struct	
structdef	

[M-5] eltype( ) — Element type and organizational type of object

These types are obtained from the declaration of the function.

Aside: struct and structdef have to do with structures; see [M-2] **struct**. structdef indicates that the function not only returns a structure but is the routine that defines the structure as well.

Conformability

eltype(X), orgtype(X):
        X: $r \times c$
    result: 1×1

Diagnostics

None.

Also See

[M-5] **isreal( )** — Storage type of matrix

[M-5] **isview( )** — Whether matrix is view

[M-4] **utility** — Matrix utility functions

Title

[M-5] **epsilon( )** — Unit roundoff error (machine precision)

Syntax

>   *real scalar*   epsilon(*real scalar x*)

Description

epsilon(*x*) returns the unit roundoff error in quantities of size abs(*x*).

Remarks

On all computers on which Stata and Mata are currently implemented—which are computers following IEEE standards—epsilon(1) is 1.0X−34, or about 2.22045e−16. This is the smallest amount by which a real number can differ from 1.

epsilon(*x*) is abs(*x*)*epsilon(1). This is an approximation of the smallest amount by which a real number can differ from *x*. The approximation is exact at integer powers of 2.

Conformability

epsilon(*x*):
>   *x*: 1×1
>   *result*: 1×1

Diagnostics

epsilon(*x*) returns . if *x* is missing.

Also See

[M-5] **mindouble( )** — Minimum and maximum nonmissing value

[M-5] **edittozero( )** — Edit matrix for roundoff error (zeros)

[M-4] **utility** — Matrix utility functions

Title

[M-5] _equilrc( ) — Row and column equilibration

Syntax

void	_equilrc(*numeric matrix A*, *r*, *c*)
void	_equilr(*numeric matrix A*, *r*)
void	_equilc(*numeric matrix A*, *c*)
real scalar	_perhapsequilrc(*numeric matrix A*, *r*, *c*)
real scalar	_perhapsequilr(*numeric matrix A*, *r*)
real scalar	_perhapsequilc(*numeric matrix A*, *c*)
real colvector	rowscalefactors(*numeric matrix A*)
real rowvector	colscalefactors(*numeric matrix A*)

The types of *r* and *c* are irrelevant because they are overwritten.

Description

_equilrc(*A*, *r*, *c*) performs row and column equilibration (balancing) on matrix *A*, returning the equilibrated matrix in *A*, the row-scaling factors in *r*, and the column-scaling factors in *c*.

_equilr(*A*, *r*) performs row equilibration on matrix *A*, returning the row-equilibrated matrix in *A* and the row-scaling factors in *r*.

_equilc(*A*, *c*) performs column equilibration on matrix *A*, returning the column-equilibrated matrix in *A* and the column-scaling factors in *c*.

_perhapsequilrc(*A*, *r*, *c*) performs row and/or column equilibration on matrix *A*—as is necessary and which decision is made by _perhapsequilrc()—returning the equilibrated matrix in *A*, the row-scaling factors in *r*, the column-scaling factors in *c*, and returning 0 (no equilibration performed), 1 (row equilibration performed), 2 (column equilibration performed), or 3 (row and column equilibration performed).

_perhapsequilr(*A*, *r*) performs row equilibration on matrix *A*—if necessary and which decision is made by _perhapsequilr()—returning the equilibrated matrix in *A*, the row-scaling factors in *r*, and returning 0 (no equilibration performed) or 1 (row equilibration performed).

_perhapsequilc(*A*, *c*) performs column equilibration on matrix *A*—if necessary and which decision is made by _perhapsequilc()—returning the equilibrated matrix in *A*, the column-scaling factors in *c*, and returning 0 (no equilibration performed) or 1 (column equilibration performed).

rowscalefactors(*A*) returns the row-scaling factors of *A*.

colscalefactors(*A*) returns the column-scaling factors of *A*.

Remarks

Remarks are presented under the following headings:

> Introduction
> Is equilibration necessary?
> The _equil*( ) family of functions
> The _perhapsequil*( ) family of functions
> rowscalefactors( ) and colscalefactors( )

Introduction

Equilibration (also known as balancing) takes a matrix with poorly scaled rows and columns, such as

	1	2
1	1.00000e+10	5.00000e+10
2	2.00000e-10	8.00000e-10

and produces a related matrix, such as

	1	2
1	.2	1
2	.25	1

that will yield improved accuracy in, for instance, the solution to linear systems. The improved matrix above has been row equilibrated. All we did was find the maximum of each row of the original and then divide the row by its maximum. If we were to take the result and repeat the process on the columns—divide each column by the column's maximum—we would obtain

	1	2
1	.8	1
2	1	1

which is the row-and-column equilibrated form of the original matrix.

In terms of matrix notation, equilibration can be thought about in terms of multiplication by diagonal matrices. The row-equilibrated form of A is RA, where R contains the reciprocals of the row maximums on its diagonal. The column-equilibrated form of A is AC, where C contains the reciprocals of the column maximums on its diagonal. The row-and-column equilibrated form of A is RAC, where R contains the reciprocals of the row maximums of A on its diagonal, and C contains the reciprocals of the column maximums of RA on its diagonal.

Say we wished to find the solution x to

$$Ax = b$$

We could compute the solution by solving for y in the equilibrated system

$$(RAC)y = Rb$$

and then setting

$$x = Cy$$

Thus routines that perform equilibration need to return to you, in some fashion, R and C. The routines here do that by returning r and c, the reciprocals of the maximums in vector form. You could obtain R and C from them by coding

$$R = \text{diag}(r)$$
$$C = \text{diag}(c)$$

but that is not in general necessary, and it is wasteful of memory. In code, you will need to multiply by R and C, and you can do that using the :* operator with r and c:

$$RA \leftrightarrow r\!:\!*A$$
$$AC \leftrightarrow A\!:\!*c$$
$$RAC \leftrightarrow r\!:\!*A\!:\!*c$$

Is equilibration necessary?

Equilibration is not a panacea. Equilibration can reduce the condition number of some matrices and thereby improve the accuracy of the solution to linear systems, but equilibration is not guaranteed to reduce the condition number, and counterexamples exist in which equilibration actually decreases the accuracy of the solution. That said, you have to look long and hard to find such examples.

Equilibration is not especially computationally expensive, but neither is it cheap, especially when you consider the extra computational costs of using the equilibrated matrices. In statistical contexts, equilibration may buy you little because matrices are already nearly equilibrated. Data analysts know variables should be on roughly the same scale, and observations are assumed to be draws from an underlying distribution. The computational cost of equilibration is probably better spent somewhere else. For instance, consider obtaining regression estimates from $X'X$ and $X'y$. The gain from equilibrating $X'X$ and $X'y$, or even from equilibrating the original X matrix, is nowhere near that from the gain to be had in removal of the means before $X'X$ and $X'y$ are formed.

In the example in the previous section, we showed you a matrix that assuredly benefited from equilibration. Even so, after row equilibration, column equilibration was unnecessary. It is often the case that solely row or column equilibration is sufficient, and in those cases, although the extra equilibration will do no numerical harm, it will burn computer cycles. And, as we have already argued, some matrices do not need equilibration at all.

Thus programmers who want to use equilibration and obtain the best speed possible examine the matrix and on that basis perform (1) no equilibration, (2) row equilibration, (3) column equilibration, or (4) both. They then write four branches in their subsequent code to handle each case efficiently.

In terms of determining whether equilibration is necessary, measures of the row and column condition can be obtained from $\min(r)/\max(r)$ and $\min(c)/\max(c)$, where r and c are the scaling factors (reciprocals of the row and column maximums). If those measures are near 1 (LAPACK uses $\geq .1$), then equilibration can be skipped.

There is also the issue of the overall scale of the matrix. In theory, the overall scale should not matter, but many packages set tolerances in absolute rather than relative terms, and so overall scale does matter. In most of Mata's other functions, relative scales are used, but provisions are made so that you can specify tolerances in relative or absolute terms. In any case, LAPACK uses the rule that equilibration is necessary if the matrix is too small (its maximum value is less than epsilon(100), approximately 2.22045e−14) or too large (greater than 1/epsilon(100), approximately 4.504e+13).

To summarize,

1. In statistical applications, we believe that equilibration burns too much computer time and adds too much code complexity for the extra accuracy it delivers. This is a judgment call, and we would probably recommend the use of equilibration were it computationally free.

2. If you are going to use equilibration, there is nothing numerically wrong with simply equilibrating matrices in all cases, including those in which equilibration is not necessary. The advantages of this is that you will gain the precision to be had from equilibration while still keeping your code reasonably simple.

3. If you wish to minimize execution time, then you want to perform the minimum amount of equilibration possible and write code to deal efficiently with each case: (1) no equilibration, (2) row equilibration, (3) column equilibration, and (4) row and column equilibration. The defaults used by LAPACK and incorporated in Mata's _perhapsequil*() routines are

    a. Perform row equilibration if $\min(r)/\max(r) < .1$, or if $\min(\text{abs}(A)) < \text{epsilon}(100)$, or if $\min(\text{abs}(A)) > 1/\text{epsilon}(100)$.

    b. After performing row equilibration, perform column equilibration if $\min(c)/\max(c) < .1$, where c is calculated on $r*A$, the row-equilibrated A, if row equilibration was performed.

The _equil*() family of functions

The _equil*() family of functions performs equilibration as follows:

_equilrc(A, r, c) performs row equilibration followed by column equilibration; it returns in r and in c the row- and column-scaling factors, and it modifies A to be the fully equilibrated matrix $r{:}*A{:}*c$.

_equilr(A, r) performs row equilibration only; it returns in r the row-scaling factors, and it modifies A to be the row-equilibrated matrix $r{:}*A$.

_equilc(A, c) performs column equilibration only; it returns in c the row-scaling factors, and it modifies A to be the row-equilibrated matrix $A{:}*c$.

Here is code to solve $Ax = b$ using the fully equilibrated form, which damages A in the process:

```
_equilrc(A, r, c)
x = c:*lusolve(A, r:*b)
```

The _perhapsequil*() family of functions

The _perhapsequil*() family of functions mirrors _equil*(), except that these functions apply the rules mentioned above for whether equilibration is necessary.

Here is code to solve $Ax = b$, which may damage A in the process:

```
result = _perhapsequilrc(A, r, c)
if (result==0)        x = lusolve(A, b)
else if (result==1)   x = lusolve(A, r:*b)
else if (result==2)   x = c:*lusolve(A, b)
else if (result==3)   x = c:*lusolve(A, r:*b)
```

As a matter of fact, the _perhapsequil*() family returns a vector of 1s when equilibration is not performed, so you could code

```
(void) _perhapsequilrc(A, r, c)
x = c:*lusolve(A, r:*b)
```

but that would be inefficient.

rowscalefactors( ) and colscalefactors( )

rowscalefactors(A) and colscalefactors(A) return the scale factors (reciprocals of row and column maximums) to perform row and column equilibration. These functions are used by the other functions above and are provided for those who wish to write their own equilibration routines.

Conformability

_equilrc(A, r, c):
    input:
        A: $m \times n$
    output:
        A: $m \times n$
        r: $m \times 1$
        c: $1 \times n$

_equilr(A, r):
    input:
        A: $m \times n$
    output:
        A: $m \times n$
        r: $m \times 1$

_equilc(A, c):
    input:
        A: $m \times n$
    output:
        A: $m \times n$
        c: $1 \times n$

_perhapsequilrc(A, r, c):
    input:
        A: $m \times n$
    output:
        A: $m \times n$ (unmodified if *result* = 0)
        r: $m \times 1$
        c: $1 \times n$
        result: 1×1

_perhapsequilr(A, r):
   input:
        A: $m \times n$
   output:
        A: $m \times n$ (unmodified if *result* = 0)
        r: $m \times 1$
    result: 1×1

_perhapsequilc(A, c):
   input:
        A: $m \times n$
   output:
        A: $m \times n$ (unmodified if *result* = 0)
        c: $1 \times n$
    result: 1×1

rowscalefactors(A):
        A: $m \times n$
    result: $m \times 1$

colscalefactors(A):
        A: $m \times n$
    result: $1 \times n$

Diagnostics

Scale factors used and returned by all functions are calculated by rowscalefactors(A) and colscalefactors(A). The functions are defined as 1:/rowmaxabs(A) and 1:/colmaxabs(A), with missing values changed to 1. Thus rows or columns that contain missing or are entirely zero are defined to have scale factors of 1.

Equilibration functions do not equilibrate rows or columns that contain missing or all zeros.

The _equil*() functions convert A to an array if A was a view. The Stata dataset is not changed.

The _perhapsequil*() functions convert A to an array if A was a view and returned is nonzero. The Stata dataset is not changed.

Also See

[M-4] **matrix** — Matrix functions

Title

[M-5] **error( )** — Issue error message

Syntax

>   *real scalar*   error(*real scalar rc*)
>
>   *void*   _error(*real scalar errnum*)
>
>   *void*   _error(*string scalar errtxt*)
>
>   *void*   _error(*real scalar errnum, string scalar errtxt*)

Description

error(*rc*) displays the standard Stata error message associated with return code *rc* and returns *rc*; see [P] **error** for a listing of return codes. error() does not abort execution; standard usage is exit(error(*rc*)).

_error() aborts execution and produces a traceback log.

_error(*errnum*) produces a traceback log with standard Mata error message *errnum*; see [M-2] **errors** for a listing of the standard Mata error codes.

_error(*errtxt*) produces a traceback log with error number 3498 and custom text *errtext*.

_error(*errnum, errtxt*) produces a traceback log with error number *errnum* and custom text *errtext*.

If *errtxt* is specified, it should contain straight text; SMCL codes are not interpreted.

Remarks

Remarks are presented under the following headings:

>   Use of _error( )
>   Use of error( )

Use of _error( )

_error() aborts execution and produces a traceback log:

```
        : myfunction(A,B)
                    mysub():  3200  conformability error
               myfunction():     -  function returned error
                    <istmt>:     -  function returned error
        r(3200);
```

The above output was created because function mysub() contained the line

>   _error(3200)

and 3200 is the error number associated with the standard message "conformability error"; see
[M-2] **errors**. Possibly, the code read

```
if (rows(A)!=rows(B) | cols(A)!=cols(B)) {
        _error(3200)
}
```

Another kind of mistake might produce

```
: myfunction(A,B)
           mysub():  3498   zeros on diagonal not allowed
      myfunction():    -    function returned error
           <istmt>:    -    function returned error
r(3498);
```

and that could be produced by the code

```
if (diag0cnt(A)>0) {
        _error("zeros on diagonal not allowed")
}
```

If we wanted to produce the same text but change the error number to 3300, we could have coded

```
if (diag0cnt(A)>0) {
        _error(3300, "zeros on diagonal not allowed")
}
```

Coding _error() is not always necessary. In our conformability-error example, imagine that more of the code read

```
...
if (rows(A)!=rows(B) | cols(A)!=cols(B)) {
        _error(3200)
}
C = A + B
...
```

If we simplified the code to read

```
...
C = A + B
...
```

the conformability error would still be detected because + requires p-conformability:

```
: myfunction(A,B)
                +:  3200   conformability error
           mysub():    -    conformability error
      myfunction():    -    function returned error
           <istmt>:    -    function returned error
r(3200);
```

Sometimes, however, you must detect the error yourself. For instance,

```
        ...
        if (rows(A)!=rows(B) | cols(A)!=cols(B) | rows(A)!=2*cols(A)) {
                _error(3200)
        }
        C = A + B
        ...
```

We assume we have some good reason to require that A has twice as many rows as columns. +, however, will not require that, and perhaps no other calculation we will make will require that, either. Or perhaps it will be subsequently detected, but in a way that leads to a confusing error message for the caller.

Use of error( )

error(*rc*) does not cause the program to terminate. Standard usage is

```
        exit(error(rc))
```

such as

```
        exit(error(503))
```

In any case, error( ) does not produce a traceback log:

```
        : myfunction(A,B)
        conformability error
        r(503);
```

error( ) is intended to be used in functions that are subroutines of ado-files:

── top of example.ado ────────
```
        program example
                version 10
                ...
                mata: myfunction("`mat1'", "`mat2'")
                ...
        end
        version 10
        mata:
        void myfunction(string scalar matname1, string scalar matname2)
        {
                ...
                A = st_matrix(matname1)
                B = st_matrix(matname2)
                if (rows(A)!=rows(B) | cols(A)!=cols(B)) {
                        exit(error(503))
                }
                C = A + B
                ...
        }
        end
```
── end of example.ado ────────

This way, when the user uses our example command incorrectly, he or she will see

```
        . example ...
        conformability error
        r(503);
```

rather than the traceback log that would have been produced had we omitted the test and
exit(error(503)):

```
          . example ...
                        +:  3200  conformability error
            myfunction():    -    function returned error
                  <istmt>:   -    function returned error
          r(3200);
```

Conformability

error(rc):
        rc: 1 × 1
   result: 1 × 1

_error(errnum):
  errnum: 1 × 1
   result: void

_error(errtxt):
   errtxt: 1 × 1
   result: void

_error(errnum, errtxt):
  errnum: 1 × 1
   errtxt: 1 × 1
   result: void

Diagnostics

error(*rc*) does not abort execution; code exit(error(*rc*)) if that is your desire; see [M-5] **exit( )**.

The code error(*rc*) returns can differ from *rc* if *rc* is not a standard code or if there is a better code associated with it.

error(*rc*) with *rc* = 0 produces no output and returns 0.

_error(*errnum*), _error(*errtxt*), and _error(*errnum*, *errtxt*) always abort with error. _error() will abort with error because you called it wrong if you specify an *errnum* less than 1 or greater than 2,147,483,647 or if you specify an *errtxt* longer than 100 characters. If you specify an *errnum* that is not a standard code, the text of the error messages will read "Stata returned error".

Also See

[M-2] **errors** — Error codes

[M-5] **exit( )** — Terminate execution

[M-4] **programming** — Programming functions

Title

[M-5] **errprintf( )** — Format output and display as error message

Syntax

> *void* errprintf(*string scalar fmt*, r_1, r_2, ... , r_N)

Description

errprintf() is a convenience tool for displaying an error message.

errprintf(...) is equivalent to printf(...) except that it executes displayas("error") before the printf() is executed; see [M-5] **printf( )** and [M-5] **displayas( )**.

Remarks

You have written a program. At one point in the code, you have variable fn that should contain the name of an existing file:

```
if (!fileexists(fn)) {
    // you wish to display the error message
    // file ____ not found
    exit(601)
}
```

One solution is

```
if (!fileexists(fn)) {
    displayas("error")
    printf("file %s not found\n", fn)
    exit(601)
}
```

Equivalent is

```
if (!fileexists(fn)) {
    errprintf("file %s not found\n", fn)
    exit(601)
}
```

It is important that you either displayas("error") before using printf() or that you use errprintf(), to ensure that your error message is displayed (is not suppressed by a quietly) and that it is displayed in red; see [M-5] **displayas( )**.

Conformability

errprintf(*fmt*, r_1, r_2, ..., r_N)
        fmt: 1×1
        r_1: 1×1
        r_2: 1×1
        ...
        r_N: 1×1
     result: *void*

Diagnostics

errprintf() aborts with error if a %*fmt* is misspecified, if a numeric %*fmt* corresponds to a string result or a string %*fmt* corresponds to a numeric result, or there are too few or too many %*fmts* in *fmt* relative to the number of *results* specified.

Also See

[M-5] **printf( )** — Format output

[M-5] **displayas( )** — Set display level

[M-4] **io** — I/O functions

Title

[M-5] **exit( )** — Terminate execution

Syntax

>exit(*real scalar rc*)

>exit()

Description

exit(*rc*) terminates execution and sets the overall return code to *rc*.

exit() with no argument specified is equivalent to exit(0).

Remarks

Do not confuse exit() and return. return stops execution of the current function and returns to the caller, whereupon execution continues. exit() terminates execution. For instance, consider

```
function first()
{
        "begin execution"
        second()
        "this message will never be seen"
}
function second()
{
        "hello from second()"
        exit(0)
}
```

The result of running this would be

```
: first
  begin execution
  hello from second()
```

If we changed the exit(0) to be exit(198) in second(), the result would be

```
: first
  begin execution
  hello from second()
r(198);
```

No error message is presented. If you want to present an error message and exit, you should code exit(error(198)); see [M-5] **error( )**.

Conformability

exit(*rc*):
    rc: 1×1 (optional)

Diagnostics

exit(*rc*) and exit() do not return.

Also See

[M-5] **error( )** — Issue error message

[M-4] **programming** — Programming functions

Title

[M-5] exp( ) — Exponentiation and logarithms

Syntax

>   *numeric matrix*   exp(*numeric matrix Z*)

>   *numeric matrix*   ln(*numeric matrix Z*)

>   *numeric matrix*   log(*numeric matrix Z*)

>   *numeric matrix*   log10(*numeric matrix Z*)

Description

exp(*Z*) returns the elementwise exponentiation of *Z*. exp() returns real if *Z* is real and complex if *Z* is complex.

ln(*Z*) and log(*Z*) return the elementwise natural logarithm of *Z*. The functions are synonyms for each other. ln() and log() return real if *Z* is real and complex if *Z* is complex.

log10(*Z*) returns the elementwise log base 10 of *Z*. log10() returns real if *Z* is real and complex if *Z* is complex.

Conformability

exp(*Z*), ln(*Z*), log(*Z*), log10(*Z*):
>   *Z*: $r \times c$
>   *result*: $r \times c$

Diagnostics

exp(*Z*) returns missing when Re(*Z*) > 709.

ln(*Z*), log(*Z*), and log10(*Z*) return missing when *Z* is real and $Z \leq 0$.

Also See

[M-4] **scalar** — Scalar mathematical functions

Title

[M-5] factorial( ) — Factorial and gamma function

Syntax

real matrix	factorial(*real matrix R*)
real matrix	lnfactorial(*real matrix R*)
numeric matrix	lngamma(*numeric matrix Z*)
numeric matrix	gamma(*numeric matrix Z*)
real matrix	digamma(*real matrix R*)
real matrix	trigamma(*real matrix R*)

Description

factorial(*R*) returns the elementwise factorial of *R*.

lnfactorial(*R*) returns the elementwise ln(factorial(*R*)), calculated differently. Very large values of *R* may be evaluated.

lngamma(*Z*), for *Z* real, returns the elementwise real result ln(abs(gamma(*Z*))), but calculated differently. lngamma(*Z*), for *Z* complex, returns the elementwise ln(gamma(*Z*)), calculated differently. Thus, lngamma(-2.5) = $-.056244$, whereas lngamma(-2.5+0i) = $-.056244 + 3.1416i$. In both cases, very large values of *Z* may be evaluated.

gamma(*Z*) returns exp(lngamma(*Z*)) for complex arguments and Re(exp(lngamma(C(*Z*)))) for real arguments. Thus gamma() can correctly calculate, say, gamma(-2.5) even for real arguments.

digamma(*R*) returns the derivative of lngamma() for $R > 0$, sometimes called the psi function. digamma() requires a real argument.

trigamma(*R*) returns the second derivative of lngamma() for $R > 0$. trigamma() requires a real argument.

Conformability

All functions return a matrix of the same dimension as input, containing element-by-element calculated results.

Diagnostics

factorial() returns missing for noninteger arguments, negative arguments, and arguments > 167.

lnfactorial() returns missing for noninteger arguments, negative arguments, and arguments $> 2{,}147{,}483{,}647$.

lngamma() returns missing for 0, negative integer arguments, negative arguments $< -2{,}147{,}483{,}647$, and arguments $> 1\text{e}{+}307$.

gamma() returns missing for real arguments > 171, for negative integer arguments, and for arguments $< -2{,}147{,}483{,}647$.

digamma() returns missing for 0 and negative integer arguments, and for arguments $< -10{,}000{,}000$.

trigamma() returns missing for 0 and negative integer arguments, and for arguments $< -10{,}000{,}000$.

Also See

[M-4] **scalar** — Scalar mathematical functions

[M-4] **statistical** — Statistical functions

Title

[M-5] **favorspeed()** — Whether speed or space is to be favored

Syntax

    real scalar favorspeed()

Description

favorspeed() returns 1 if the user has mata set matafavor speed and 0 if the user has mata set matafavor space or has not set matafavor at all; see [M-3] **mata set**.

Remarks

Sometimes in programming you can choose between writing code that runs faster but consumes more memory or writing code that conserves memory at the cost of execution speed. favorspeed() tells you the user's preference:

```
if (favorspeed()) {
    /* code structured for speed over memory */
}
else {
    /* code structured for memory over speed */
}
```

Conformability

favorspeed():
    result: 1×1

Diagnostics

None.

Also See

[M-3] **mata set** — Set and display Mata system parameters

[M-4] **programming** — Programming functions

Title

> [M-5] **ferrortext( )** — Text and return code of file error code

Syntax

> *string scalar* ferrortext(*real scalar ec*)
>
> *real scalar* freturncode(*real scalar ec*)

Description

ferrortext(*ec*) returns the text associated with a file error code returned by, for instance, _fopen(), _fwrite(), fstatus(), or any other file-processing functions that return an error code. See [M-5] **fopen( )**.

freturncode(*ec*) returns the Stata return code associated with the file error code.

Remarks

Most file-processing functions abort with error if something goes wrong. You attempt to read a nonexisting file, or attempt to write a read-only file, and the file functions you use to do that, usually documented in [M-5] **fopen( )**, abort with error. Abort with error means not only that the file function you called stops when the error arises but also that the calling function is aborted. The user is presented with a traceback log documenting what went wrong.

Sometimes you will want to write code to handle such errors for itself. For instance, you are writing a subroutine for a command to be used in Stata and, if the file does not exist, you want the subroutine to present an informative error message and exit without a traceback log but with a nonzero return code. Or in another application, if the file does not exist, that is not an error at all; your code will create the file.

Most file-processing functions have a corresponding underscore function that, rather than aborting, returns an error code when things go wrong. fopen() opens a file or aborts with error. _fopen() opens a file or returns an error code. The error code is sufficient for your program to take the appropriate action. One uses the underscore functions when the calling program will deal with any errors that might arise.

Let's take the example of simply avoiding traceback logs. If you code

```
fh = fopen(filename, "r")
```

and the file does not exist, execution aborts and you see a traceback log. If you code

```
if ((fh = _fopen(filename, "r"))<0) {
        errprintf("%s\n", ferrortext(fh))
        exit(freturncode(fh))
}
```

execution still stops if the file does not exist, but this time, it stops because you coded exit(). You still see an error message, but this time, you see the message because you coded errprintf(). No traceback log is produced because you did not insert code to produce one. You could have coded _exit() if you wanted one.

[M-5] ferrortext( ) — Text and return code of file error code

The file error codes and the messages associated with them are

Negative code	Meaning
0	all is well
−1	end of file *(this code is usually not an error)*
−601	file not found
−602	file already exists
−603	file could not be opened
−608	file is read-only
−610	file not Stata format
−612	unexpected end of file
−630	web files not supported in this version of Stata
−631	host not found
−632	web file not allowed in this context
−633	may not write to web file
−639	file transmission error — checksums do not match
−660	proxy host not found
−662	proxy server refused request to send
−663	remote connection to proxy failed
−665	could not set socket nonblocking
−667	wrong version of `winsock.dll`
−668	could not find valid `winsock.dll` or `astsys0.lib`
−669	invalid URL
−670	invalid network port number
−671	unknown network protocol
−672	server refused to send file
−673	authorization required by server
−674	unexpected response from server
−675	server reported error
−676	server refused request to send
−677	remote connection failed — see `r(677)` for troubleshooting
−678	could not open local network socket
−679	unexpected web error
−680	could not find valid `odbc32.dll`
−681	too many open files
−682	could not connect to ODBC data source name
−683	could not fetch variable in ODBC table
−684	could not find valid `dlxabi32.dll`
−691	I/O error
−699	insufficient disk space

−3601	invalid file handle
−3602	invalid filename
−3611	too many open files
−3621	attempt to write read-only file
−3622	attempt to read write-only file
−3623	attempt to seek append-only file
−3698	file seek error

File error codes are usually negative, but neither ferrortext(ec) nor freturncode(ec) cares whether ec is of the positive or negative variety.

Conformability

ferrortext(ec), freturncode(ec):
        cn: 1×1
    result: 1×1

Diagnostics

ferrortext(ec) and freturncode(ec) treat $ec = -1$ (end of file) the same as $ec = 612$ (unexpected end of file). Code -1 usually is not an error; it just denotes end of file. It is assumed that you will not call ferrortext() and freturncode() in such cases. If you do call the functions here, it is assumed that you mean that the end of file was unexpected.

Also See

[M-5] **fopen( )** — File I/O

[M-4] **io** — I/O functions

Title

[M-5] **fft( )** — Fourier transform

Syntax

    complex vector   fft(*numeric vector h*)

    numeric vector   invfft(*numeric vector H*)

    void   _fft(*complex vector h*)

    void   _invfft(*complex vector H*)

    numeric vector   convolve(*numeric vector r, numeric vector s*)

    numeric vector   deconvolve(*numeric vector r, numeric vector sm*)

    numeric vector   Corr(*numeric vector g, numeric vector h, real scalar k*)

    real vector   ftperiodogram(*numeric vector H*)

    numeric vector   ftpad(*numeric vector h*)

    numeric vector   ftwrap(*numeric vector r, real scalar n*)

    numeric vector   ftunwrap(*numeric vector H*)

    numeric vector   ftretime(*numeric vector r, numeric vector s*)

    real vector   ftfreqs(*numeric vector H, real scalar delta*)

Description

H=fft(*h*) and h=invfft(*H*) calculate the Fourier transform and inverse Fourier transform. The length of *h* (*H*) must be a power of 2.

_fft(*h*) and _invfft(*H*) do the same thing, but they perform the calculation in place, replacing the contents of *h* and *H*.

convolve(*r, s*) returns the convolution of the signal *s* with the response function *r*. deconvolve(*r, sm*) deconvolves the smeared signal *sm* with the response function *r* and is thus the inverse of convolve().

Corr(*g, h, k*) returns a $2k + 1$ element vector containing the correlations of *g* and *h* for lags and leads as large as *k*.

ftperiodogram(*H*) returns a real vector containing the one-sided periodogram of *H*.

ftpad(*h*) returns *h* padded with 0s to have a length that is a power of 2.

ftwrap(*r, n*) converts the symmetrically stored response function *r* into wraparound format of length *n*, $n \geq$ rows(*r*)*cols(*r*) and rows(*r*)*cols(*r*) odd.

ftunwrap(*H*) unwraps frequency-wraparound order such as returned by fft(). You may find this useful when graphing or listing results, but it is otherwise unnecessary.

ftretime(*r*, *s*) retimes the signal *s* to be on the same time scale as convolve(*r*, *s*). This is useful in graphing data and listing results but is otherwise not required.

ftfreqs(*H*, *delta*) returns a vector containing the frequencies associated with the elements of *H*; *delta* is the sampling interval and is often specified as 1.

Remarks

Remarks are presented under the following headings:

> *Definitions, notation, and conventions*
> *Fourier transform*
> *Convolution and deconvolution*
> *Correlation*
> *Utility routines*
> *Warnings*

Definitions, notation, and conventions

A signal *h* is a row or column vector containing real or complex elements. The length of the signal is defined as the number of elements of the vector. It is occasionally necessary to pad a signal to a given length. This is done by forming a new vector equal to the original and with zeros added to the end.

The Fourier transform of a signal *h*, typically denoted by capital letter *H* of *h*, is stored in frequency-wraparound order. That is, if there are n elements in *H*:

$H[1]$	frequency 0
$H[2]$	frequency 1
$H[3]$	frequency 2
$\vdots$	
$H[n/2]$	frequency $n/2-1$
$H[n/2+1]$	frequency $n/2$ ($-n/2$, aliased)
$H[n/2+2]$	frequency $-(n/2-1)$
$\vdots$	
$H[n-1]$	frequency -2
$H[n]$	frequency -1

All routines expect and use this order, but see ftunwrap() below.

A response function *r* is a row or column vector containing $m = 2k + 1$ real or complex elements. m is called the duration of the response function. Response functions are generally stored symmetrically, although the response function itself need not be symmetric. The response vector contains

$r[1]$	response at lag $-k$
$r[2]$	response at lag $-k+1$
$\vdots$	
$r[k]$	response at lag -1
$r[k+1]$	contemporaneous response
$r[k+2]$	response at lead 1
$r[k+3]$	response at lead 2
$\vdots$	
$r[2k+1]$	response at lead k

Response functions always have odd lengths. Response vectors are never padded.

You may occasionally find it convenient to store a response vector in "wraparound" order (similar to frequency-wraparound order), although none of the routines here require this. In wraparound order:

wrap[1]	contemporaneous response
wrap[2]	response at lead 1
wrap[3]	response at lead 2
$\vdots$	
wrap[$k+1$]	response at lead k
wrap[$k+2$]	response at lag $-k$
wrap[$k+3$]	response at lag $-k+1$
$\vdots$	
wrap[$2k+1$]	response at lag -1

Response vectors stored in wraparound order may be internally padded (as opposed to merely padded) to a given length by the insertion of zeros between wrap[$k+1$] and wrap[$k+2$].

Fourier transform

fft(*h*) returns the discrete Fourier transform of *h*. *h* may be either real or complex, but its length must be a power of 2, so one typically codes fft(ftpad(*h*)); see ftpad(), below. The returned result is p-conformable with *h*. The calculation is performed by _fft().

invfft(*H*) returns the discrete inverse Fourier transform of *H*. *H* may be either real or complex, but its length must be a power of 2. The returned result is p-conformable with *H*. The calculation is performed by _invfft().

invfft(*H*) may return a real or complex. This should be viewed as a feature, but if you wish to ensure the complex interpretation, code C(invfft(*H*)).

_fft(*h*) is the built-in procedure that performs the fast Fourier transform in place. *h* must be complex, and its length must be a power of 2.

_invfft(*H*) is the built-in procedure that performs the inverse fast Fourier transform in place. *H* must be complex, and its length must be a power of 2.

Convolution and deconvolution

convolve(*r*, *s*) returns the convolution of the signal *s* with the response function *r*. Calculation is performed by taking the fft() of the elements, multiplying, and using invfft() to transform the results back. Nevertheless, it is not necessary that the length of *s* be a power of 2. convolve() handles all paddings necessary, including paddings to *s* required to prevent the result from being contaminated by erroneous wrapping around of *s*. Although one thinks of the convolution operator as being commutative, convolve() is not commutative since required zero-padding of the response and signal differ.

If n is the length of the signal and $2k + 1$ is the length of the response function, the returned result has length $n + 2k$. The first k elements are the convoluted signal before the true signal begins, and the last k elements are the convoluted signal after the true signal ends. See ftretime(), below. In any case, you may be interested only in the elements convolve()[|k+1\n-k|], the part contemporaneous with *s*.

The returned vector is a row vector if *s* is a row vector and a column vector otherwise. The result is guaranteed to be real if both *r* and *s* are real; the result may be complex or real, otherwise.

It is not required that the response function be shorter than the signal, although this will typically be the case.

deconvolve(*r*, *sm*) deconvolves the smeared signal *sm* with the response function *r* and is thus the inverse of convolve(). In particular,

$$\text{deconvolve}(r, \text{convolve}(r,s)) = s \qquad \text{(up to roundoff error)}$$

Everything said about convolve() applies equally to deconvolve().

Correlation

Here we refer to correlation in the signal-processing sense, not the statistical sense.

Corr(*g*, *h*, *k*) returns a $2k + 1$ element vector containing the correlations of *g* and *h* for lags and leads as large as *k*. For instance, Corr(*g*, *h*, 2) returns a five-element vector, the first element of which contains the correlation for lag 2, the second element lag 1, the third (middle) element the contemporaneous correlation, the fourth element lead 1, and the fifth element lead 2. *k* must be greater than or equal to 1. The returned vector is a row or column vector depending on whether *g* is a row or column vector. *g* and *h* must have the same number of elements but need not be p-conformable.

The result is obtained by padding with zeros to avoid contamination, taking the Fourier transform, multiplying $G \times \text{conj}(H)$, and rearranging the inverse transformed result. Nevertheless, it is not required that the number of elements of *g* and *h* be powers of 2 because the program pads internally.

Utility routines

ftpad(*h*) returns *h* padded with 0s to have a length that is a power of 2. For instance,

```
: h = (1,2,3,4,5)
: ftpad(h)
```

	1	2	3	4	5	6	7	8
1	1	2	3	4	5	0	0	0

If *h* is a row vector, a row vector is returned. If *h* is a column vector, a column vector is returned.

ftwrap(*r*, *n*) converts the symmetrically stored response function *r* into wraparound format of length *n*, $n \geq \text{rows}(r)*\text{cols}(r)$ and $\text{rows}(r)*\text{cols}(r)$ odd. A symmetrically stored response function is a vector of odd length, for example:

(.1, .5, 1, .2, .4)

The middle element of the vector represents the response function at lag 0. Elements before the middle represent lags while elements after the middle represent leads. Here .1 is the response for lag 2 and .5 for lag 1, 1 the contemporaneous response, .2 the response for lead 1, and .4 the response for lead 2. The wraparound format of a response function records the response at times 0, 1, and so on in the first positions, has some number of zeros, and then records the most negative time value of the response function, and so on.

For instance,

: r

	1	2	3	4	5
1	.1	.5	1	.2	.4

: ftwrap(r, 5)

	1	2	3	4	5
1	1	.2	.4	.1	.5

: ftwrap(r, 6)

	1	2	3	4	5	6
1	1	.2	.4	0	.1	.5

: ftwrap(r, 8)

	1	2	3	4	5	6	7	8
1	1	.2	.4	0	0	0	.1	.5

ftunwrap(*H*) unwraps frequency-wraparound order such as returned by fft(). You may find this useful when graphing or listing results, but it is otherwise unnecessary. Frequency-unwrapped order is defined as

unwrap[1]	frequency $-(n/2) + 1$
unwrap[2]	frequency $-(n/2) + 2$
$\vdots$	
unwrap[$n/2 - 1$]	frequency -1
unwrap[$n/2$]	frequency 0
unwrap[$n/2 + 1$]	frequency 1
$\vdots$	
unwrap[$n - 1$]	frequency $n/2 - 1$
unwrap[n]	frequency $n/2$

Here we assume that *n* is even, as will usually be true. The aliased (highest) frequency is assigned the positive sign.

Also see `ftperiodogram()`, below.

`ftretime(r, s)` retimes the signal s to be on the same time scale as `convolve(r, s)`. This is useful in graphing and listing results but is otherwise not required. `ftretime()` uses only the length of r, and not its contents, to perform the retiming. If the response vector is of length $2k + 1$, a vector containing k zeros, s, and k more zeros is returned. Thus the result of `ftretime(r, s)` is p-conformable with `convolve(r, s)`.

`ftfreqs(H, delta)` returns a p-conformable-with-H vector containing the frequencies associated with the elements of H. $delta$ is the sampling interval and is often specified as 1.

`ftperiodogram(H)` returns a real vector of length $n/2$ containing the one-sided periodogram of H (length n), calculated as

$$|H(f)|^2 + |H(-f)|^2$$

excluding frequency 0. Thus `ftperiodogram(H)[1]` corresponds to frequency 1 (-1), `ftperiodogram(H)[2]` to frequency 2 (-2), and so on.

Warnings

`invfft(H)` will cast the result down to real if possible. Code `C(invfft(H))` if you want to be assured of the result being stored as complex.

`convolve(r, s)` is not the same as `convolve(s, r)`.

`convolve(r, s)` will cast the result down to real if possible. Code `C(convolve(r, s))` if you want to be assured of the result being stored as complex.

For `convolve(r, s)`, the response function r must have odd length.

Conformability

```
fft(h):
            h:      1 × n       or     n × 1,  n a power of 2
       result:      1 × n       or     n × 1
invfft(H):
            H:      1 × n       or     n × 1,  n a power of 2
       result:      1 × n       or     n × 1
_fft(h):
            h:      1 × n       or     n × 1,  n a power of 2
       result:      void
_invfft(H):
            h:      1 × n       or     n × 1,  n a power of 2
       result:      void
convolve(r, s):
            r:      1 × n            or                n × 1,      n > 0, n odd
            s:      1 × 2k + 1       or                2k + 1 × 1, i.e., s of odd length
       result:      1 × 2k + n       or                2k + n × 1
```

deconvolve(r, sm):

r:	$1 \times n$	or	$n \times 1$,	$n > 0$, n odd
sm:	$1 \times 2k + n$	or	$2k + n \times 1$	
result:	$1 \times 2k + 1$	or	$2k + 1 \times 1$	

Corr(g, h, k):

g:	$1 \times n$	or	$n \times 1$,	$n > 0$
h:	$1 \times n$	or	$n \times 1$	
k:	1×1	or	1×1,	$k > 0$
result:	$1 \times 2k + 1$	or	$2k + 1 \times 1$	

ftperiodogram(H):

H:	$1 \times n$	or	$n \times 1$,	n even
result:	$n/2 \times 1$	or	$1 \times n/2$	

ftpad(h):

h:	$1 \times n$	or	$n \times 1$	
result:	$1 \times N$	or	$N \times 1$,	$N = n$ rounded up to power of 2

ftwrap(r, n):

r:	$1 \times m$	or	$m \times 1$,	$m > 0$, m odd
n:	1×1	or	1×1,	$n \geq m$
result:	$1 \times n$	or	$n \times 1$	

ftunwrap(H):

H:	$1 \times n$	or	$n \times 1$
result:	$1 \times n$	or	$n \times 1$

ftretime(r, s):

r:	$1 \times n$	or	$n \times 1$,	$n > 0$, n odd
s:	$1 \times 2k + 1$	or	$2k + 1 \times 1$,	i.e., s of odd length
result:	$1 \times 2k + n$	or	$2k + n \times 1$	

ftfreqs(H, delta):

H:	$1 \times n$	or	$n \times 1$,	n even
result:	$1 \times n$	or	$n \times 1$	

Diagnostics

All functions abort with error if the conformability requirements are not met. This is always true, of course, but pay particular attention to the requirements outlined under *Conformability* directly above.

fft(h), _fft(h), invfft(H), _invfft(H), convolve(r, s), deconvolve(r, sm), and Corr(g, h, k) return missing results if any argument contains missing values.

ftwrap(r, n) aborts with error if n contains missing value.

Also See

[M-4] **mathematical** — Important mathematical functions

Title

[M-5] **fileexists( )** — Whether file exists

Syntax

>   *real scalar*   `fileexists(`*string scalar fn*`)`

Description

`fileexists(`*fn*`)` returns 1 if file *fn* exists and is readable and returns 0 otherwise.

Conformability

`fileexists(`*fn*`)`:
    fn: 1×1
  result: 1×1

Diagnostics

None.

Also See

[M-4] **io** — I/O functions

Title

[M-5] **_fillmissing( )** — Fill matrix with missing values

Syntax

>   *void*   `_fillmissing(`*transmorphic matrix A*`)`

Description

`_fillmissing(`*transmorphic matrix A*`)` changes the contents of *A* to missing values.

Remarks

The definition of missing depends on the storage type of *A*:

Storage type	Contents
real	.
complex	C(.)
string	""
pointer	NULL

Conformability

`_fillmissing(`*A*`)`:
>   *input*:
>   >   *A*: $r \times c$
>
>   *output*:
>   >   *A*: $r \times c$

Diagnostics

None.

Also See

[M-4] **manipulation** — Matrix manipulation

Title

[M-5] **findexternal( )** — Find, create, and remove external globals

Syntax

>   *pointer()*   *scalar*   findexternal(*string scalar name*)
>
>   *pointer()*   *scalar*   crexternal(*string scalar name*)
>
>   *void*   rmexternal(*string scalar name*)
>
>   *string*   *scalar*   nameexternal(*pointer() scalar p*)

Description

findexternal(*name*) returns a pointer (see [M-2] **pointers**) to the external global matrix, vector, or scalar whose name is specified by *name*, or to the external global function if the contents of *name* end in (). findexternal() returns NULL if the external global is not found.

crexternal(*name*) creates a new external global 0×0 real matrix with the specified name and returns a pointer to it; it returns NULL if an external global of that name already exists.

rmexternal(*name*) removes (deletes) the specified external global or does nothing if no such external global exists.

nameexternal(*p*) returns the name of **p*.

Remarks

Remarks are presented under the following headings:

>   Definition of a global
>   Use of globals

Also see *Linking to external globals* in [M-2] **declarations**.

Definition of a global

When you use Mata interactively, any variables you create are known, equivalently, as externals, globals, or external globals.

```
: myvar = x
```

Such variables can be used by subsequent functions that you run, and there are two ways that can happen:

```
function example1(...)
{
        external real myvar

        ... myvar ...
}
```

and

```
function example2(...)
{
        pointer(real) p

        p = findexternal("myvar")
        ... *p ...
}
```

Using the first method, you must know the name of the global at the time you write the source code, and when you run your program, if the global does not exist, it will refuse to run (abort with myvar not found). With the second method, the name of the global can be specified at run time and what is to happen when the global is not found is up to you.

In the second example, although we declared p as a pointer to a real, myvar will not be required to contain a real. After p = findexternal("myvar"), if p!=NULL, p will point to whatever myvar contains, whether it be real, complex, string, or another pointer. (You can diagnose the contents of *p using eltype(*p) and orgtype(*p).)

Use of globals

Globals are useful when a function must remember something from one call to the next:

```
function example3(real scalar x)
{
        pointer() scalar p

        if ( (p = findexternal("myprivatevar")) == NULL) {
                printf("you haven't called me previously")
                p = crexternal("myprivatevar")
        }
        else {
                printf("last time, you said "%g", *p)
        }
        *p = x
}
```

```
: example3(2)
you haven't called me previously
: example3(31)
last time, you said 2
: example3(9)
last time, you said 31
```

Note our use of the name myprivatevar. It actually is not a private variable; it is global, and you would see the variable listed if you described the contents of Mata's memory. Because global variables are so exposed, it is best that you give them long and unlikely names.

In general, programs do not need global variables. The exception is when a program must remember something from one invocation to the next, and especially if that something must be remembered from one invocation of Mata to the next.

[M-5] findexternal( ) — Find, create, and remove external globals

When you do need globals, you probably will have more than one thing you will need to recall. There are two ways to proceed. One way is simply to create separate global variables for each thing you need to remember. The other way is to create one global pointer vector and store everything in that. In the following example, we remember one scalar and one matrix:

```
function example4()
{
        pointer(pointer() vector) scalar    p
        scalar                              s
        real matrix                         X
        pointer() scalar                    ps, pX

        if ( (p = findexternal("mycollection")) == NULL) {
                ... calculate scalar s and X from nothing ...
                ... and save them:
                p = crexternal("mycollection")
                *p = (&s, &X)
        }
        else {
                ps = (*p)[1]
                pX = (*p)[2]
                ... calculate using *ps and *pX ...
        }
}
```

In the above example, even though crexternal() created a 0×0 real global, we morphed it into a 1×2 pointer vector:

```
p = crexternal("mycollection")       *p is 0 × 0 real
*p = (&s, &X)                        *p is 1 × 2 vector
```

just as we could with any nonpointer object.

In the else part of our program, where we use the previous values, we do not use variables s and X, but ps and pX. Actually, we did not really need them, we could just as well have used *((*p)[1]) and *((*p)[2]), but the code is more easily understood by introducing *ps and *pX.

Actually, we could have used the variables s and X by changing the else part of our program to read

```
else {
        s = *(*p)[1]
        X = *(*p)[2]
        ... calculate using s and X ...
        *p = (&s, &X)            ← remember to put them back
}
```

Doing that is inefficient because s and X contain copies of the global values. Obviously, the amount of inefficiency depends on the sizes of the elements being copied. For s, there is really no inefficiency at all because s is just a scalar. For X, the amount of inefficiency depends on the dimensions of X. Making a copy of a small X matrix would introduce just a little inefficiency.

The best balance between efficiency and readability is achieved by introducing a subroutine:

```
function example5()
{
        pointer(pointer() vector) scalar    p
        scalar                              s
        real matrix                         X

        if ( (p = findexternal("mycollection")) == NULL) {
                example5_sub(1, s=., X=J(0,0,.))
                p = crexternal("mycollection")
                *p = (&s, &X)
        }
        else {
                example5_sub(0, (*p)[1], (*p)[2])
        }
}

function example5_sub(scalar firstcall, scalar x, matrix X)
{
        ...
}
```

The last two lines in the not-found case

```
p = crexternal("mycollection")
*p = (&s, &X)
```

could also be coded

```
*crexternal("mycollection") = (&s, &X)
```

Conformability

findexternal(*name*), crexternal(*name*):
    name: 1×1
    result: 1×1

rmexternal(*name*):
    name: 1×1
    result: *void*

nameexternal(*p*):
    p: 1×1
    result: 1×1

Diagnostics

findexternal(*name*), crexternal(*name*), and rmexternal(*name*) abort with error if *name* contains an invalid name.

findexternal(*name*) returns NULL if *name* does not exist.

180 [M-5] findexternal( ) — Find, create, and remove external globals

crexternal(*name*) returns NULL if *name* already exists.

nameexternal(*p*) returns "" if p = NULL. Also, nameexternal() may be used not just with pointers to globals but pointers to locals as well. For example, you can code nameexternal(&myx), where myx is declared in the same program or a calling program. nameexternal() will usually return the expected local name, such as "myx". In such cases, however, it is also possible that "" will be returned. That can occur because, in the compilation/optimization process, the identity of local variables can be lost.

Also See

[M-5] **valofexternal( )** — Obtain value of external global

[M-4] **programming** — Programming functions

Title

[M-5] **findfile( )** — Find file

Syntax

>   *string scalar*   findfile(*string scalar fn, string scalar dirlist*)

>   *string scalar*   findfile(*string scalar fn*)

Description

findfile(*fn, dirlist*) looks for file *fn* along the semicolon-separated list of directories *dirlist* and returns the fully qualified path and filename if *fn* is found. findfile() returns "" if the file is not found.

findfile(*fn*) is equivalent to findfile(*fn*, c("adopath")). findfile() with one argument looks along the official Stata ado-path for file *fn*.

Remarks

For instance,

>   findfile("kappa.ado")

might return C:\Program Files\Stata10\ado\updates\k\kappa.ado.

Conformability

findfile(*fn, dirlist*):
>   *fn*: 1×1
>   *dirlist*: 1×1 (optional)
>   *result*: 1×1

Diagnostics

findfile(*fn, dirlist*) and findfile(*fn*) return "" if the file is not found. If the file is found, the returned fully qualified path and filename is guaranteed to exist and be readable at the instant findfile() concluded.

Also See

[M-4] **io** — I/O functions

Title

[M-5] **floatround( )** — Round to float precision

Syntax

> *real matrix* floatround(*real matrix x*)

Description

floatround(*x*) returns *x* rounded to IEEE 4-byte real (float) precision. floatround() is the element-by-element equivalent of Stata's float() function. The Mata function could not be named float() because the word float is reserved in Mata.

Remarks

```
: printf("   %21x\n", .1)
   +1.999999999999aX-004
: printf("   %21x\n", floatround(.1))
   +1.99999a0000000X-004
```

Conformability

floatround(*x*):
    x: $r \times c$
  result: $r \times c$

Diagnostics

floatround(*x*) returns missing (.) if $x < -1.\text{fffffeX}+7e$ (approximately $-1.70141173319e+38$) or $x > 1.\text{fffffeX}+7e$ (approximately $1.70141173319e+38$).

By contrast with most functions, floatround(*x*) returns the same kind of missing value if *x* contains missing; . if $x == .$, .a if $x == .a$, .b if $x == .b$, ..., and .z if $x == .z$.

Also See

[M-4] **utility** — Matrix utility functions

Title

> **[M-5] fmtwidth( )** — Width of %fmt

Syntax

>   *real matrix* fmtwidth(*string matrix f*)

Description

fmtwidth(*f*) returns the width of the %*fmt* contained in *f*.

Remarks

fmtwidth("%9.2f") returns 9.

fmtwidth("%20s") returns 20.

fmtwidth("%tc") returns 18.

fmtwidth("%tcDay_Mon_DD_hh:mm:ss_!C!D!T_CCYY") returns 28.

fmtwidth("not a format") returns . (missing).

Conformability

fmtwidth(*f*):
    f: $r \times c$
  result: $r \times c$

Diagnostics

fmtwidth(*f*) returns . (missing) when *f* does not contain a valid Stata format.

Also See

[M-5] **strlen( )** — Length of string

[M-4] **string** — String manipulation functions

Title

[M-5] fopen( ) — File I/O

Syntax

>*real scalar*   fopen(*string scalar fn, mode*)

>*real scalar*   fopen(*string scalar fn, mode, public*)

>*real scalar*   _fopen(*string scalar fn, mode*)

>*real scalar*   _fopen(*string scalar fn, mode, public*)

where

>*mode*: *string scalar* containing "r", "w", "rw", or "a"
>
>*public*: optional *real scalar* containing zero or nonzero

In what follows, *fh* is the value returned by fopen( ) or _fopen( ):

>*void*   fclose(*fh*)
>
>*real scalar*   _fclose(*fh*)

>*string scalar*   fget(*fh*)
>
>*string scalar*   _fget(*fh*)
>
>*string scalar*   fgetnl(*fh*)
>
>*string scalar*   _fgetnl(*fh*)

>*string scalar*   fread(*fh, real scalar len*)
>
>*string scalar*   _fread(*fh, real scalar len*)

>*void*   fput(*fh, string scalar s*)
>
>*real scalar*   _fput(*fh, string scalar s*)

>*void*   fwrite(*fh, string scalar s*)
>
>*real scalar*   _fwrite(*fh, string scalar s*)

>*matrix*   fgetmatrix(*fh*)
>
>*matrix*   _fgetmatrix(*fh*)
>
>*void*   fputmatrix(*fh, transmorphic matrix X*)
>
>*real scalar*   _fputmatrix(*fh, transmorphic matrix X*)

real scalar	`fstatus(`*fh*`)`
real scalar	`ftell(`*fh*`)`
real scalar	`_ftell(`*fh*`)`
void	`fseek(`*fh*, *real scalar offset*, *real scalar whence*`)`
real scalar	`_fseek(`*fh*, *real scalar offset*, *real scalar whence*`)`
	(*whence* is coded −1, 0, or 1, meaning from start of file, from current position, or from end of file; *offset* is signed: positive values mean after *whence* and negative values mean before)
void	`ftruncate(`*fh*`)`
real scalar	`_ftruncate(`*fh*`)`

Description

These functions read and write files. First, open the file and get back a file handle (*fh*). The file handle, which is nothing more than an integer, is how you refer to the file in the calls to other file I/O functions. When you are finished, close the file.

Most file I/O functions come in two varieties: without and with an underscore in front of the name, such as `fopen()` and `_fopen()`, and `fwrite()` and `_fwrite()`.

In functions without a leading underscore, errors cause execution to be aborted. For instance, you attempt to open a file for read and the file does not exist. Execution stops. Or, having successfully opened a file, you attempt to write into it and the disk is full. Execution stops. When execution stops, the appropriate error message is presented.

In functions with the leading underscore, execution continues and no error message is displayed; it is your responsibility (1) to verify that things went well and (2) to take the appropriate action if they did not. Concerning (1), some underscore functions return a status value; others require that you call `fstatus()` to obtain the status information.

You can mix and match use of underscore and nonunderscore functions, using, say, `_fopen()` to open a file and `fread()` to read it, or `fopen()` to open and `_fwrite()` to write.

Remarks

Remarks are presented under the following headings:

> *Opening and closing files*
> *Reading from a file*
> *Writing to a file*
> *Reading and writing in the same file*
> *Reading and writing matrices*
> *Repositioning in a file*
> *Truncating a file*
> *Error codes*

Opening and closing files

Functions

>   fopen(*string scalar fn, string scalar mode*)
>
>   _fopen(*string scalar fn, string scalar mode*)
>
>   fopen(*string scalar fn, string scalar mode, real scalar public*)
>
>   _fopen(*string scalar fn, string scalar mode, real scalar public*)

open a file. The file may be on a local disk, a network disk, or even on the web (such as http://www.stata.com/index.html). *fn* specifies the filename, and *mode* specifies how the file is to opened:

mode	meaning
"r"	Open for reading; file must exist and be readable. File may be "http://..." file. File will be positioned at the beginning.
"w"	Open for writing; file must not exist and the directory be writable. File may not be "http://..." file. File will be positioned at the beginning.
"rw"	Open for reading and writing; file must either exist and be writable or not exist and directory be writable. File may not be "http://..." file. File will be positioned at the beginning (new file) or at the end (existing file).
"a"	Open for appending; file must either exist and be writable or not exist and directory be writable. File may not be "http://..." file. File will be positioned at the end.

Other values for *mode* cause fopen() and _fopen() to abort with an invalid-mode error.

Optional third argument *public* specifies whether the file, if it is being created, should be given permissions so that everyone can read it, or if it instead should be given the normal permissions. Not specifying *public*, or specifying *public* as 0, gives the file the normal permissions. Specifying *public* as nonzero makes the file publicly readable. *public* is relevant only when the file is being created, i.e., is being opened "w", or being opened "rw" and not previously existing.

fopen() returns a file handle; the file is opened or execution is aborted.

_fopen() returns a file handle or returns a negative number. If a negative number is returned, the file is not open, and the number indicates the reason. For _fopen(), there are a few likely possibilities

negative value	meaning
−601	file not found
−602	file already exists
−603	file could not be opened
−608	file is read-only
−691	I/O error

and there are many other possibilities. For instance, perhaps you attempted to open a file on the web (say, *http://www.newurl.org/upinfo.doc*) and the URL was not found, or the server refused to send back the file, etc. See *Error codes* below for a complete list of codes.

After opening the file, you use the other file I/O commands to read and write it, and then you close the file with fclose() or _fclose(). fclose() returns nothing; if the file cannot be closed, execution is aborted. _fclose returns 0 if successful, or a negative number otherwise. For _fclose(), the likely possibilities are

negative value	meaning
−691	filesystem I/O error

Reading from a file

You may read from a file opened "r" or "rw". The commands to read are

    fget(*fh*)

    fgetnl(*fh*)

    fread(*fh*, *real scalar len*)

and, of course,

    _fget(*fh*)

    _fgetnl(*fh*)

    _fread(*fh*, *real scalar len*)

All functions, with or without an underscore, require a file handle be specified, and all the functions return a string scalar or they return J(0,0,""), a 0 × 0 string matrix. They return J(0,0,"") on end of file and, for the underscore functions, when the read was not successful for other reasons. When using the underscore functions, you use fstatus() to obtain the status code; see *Error codes* below. The underscore read functions are rarely used because the only reason a read can fail is I/O error, and there is not much that can be done about that except abort, which is exactly what the nonunderscore functions do.

fget(*fh*) is for reading ASCII files; the next line from the file is returned, without end-of-line characters. (If the line is longer then 32,768 characters, the first 32,768 characters are returned.)

fgetnl(*fh*) is much the same as fget(), except that the new-line characters are not removed from the returned result. (If the line is longer then 32,768 characters, the first 32,768 characters are returned.)

fread(*fh*, *len*) is usually used for reading binary files and returns the next *len* characters (bytes) from the file or, if there are fewer than that remaining to be read, however many remain. (*len* may not exceed 2,147,483,647 on 32-bit computers and 9,007,199,254,740,991 [*sic*] on 64-bit computers; memory shortages for storing the result will arise long before these limits are reached on most computers.)

The following code reads and displays a file:

```
fh = fopen(filename, "r")
while ((line=fget(fh))!=J(0,0,"")) {
    printf("%s\n", line)
}
fclose(fh)
```

Writing to a file

You may write to a file opened "w", "rw", or "a". The functions are

fput(*fh*, *string scalar s*)

fwrite(*fh*, *string scalar s*)

and, of course,

_fput(*fh*, *string scalar s*)

_fwrite(*fh*, *string scalar s*)

fh specifies the file handle, and *s* specifies the string to be written. fput() writes *s* followed by the new-line characters appropriate for your operating system. fwrite() writes *s* alone.

fput() and fwrite() return nothing; _fput() and _fwrite() return a real scalar equal to 0 if all went well or a negative error code; see *Error codes* below.

The following code copies text from one file to another:

```
fh_in  = fopen(inputname, "r")
fh_out = fopen(outputname, "w")
while ((line=fget(fh_in))!=J(0,0,"")) {
    fput(fh_out, line)
}
fclose(fh_out)
fclose(fh_in)
```

The following code reads a file (binary or text) and changes every occurrence of "a" to "b":

```
fh_in  = fopen(inputname, "r")
fh_out = fopen(outputname, "w")
while ((c=fread(fh_in, 1))!=J(0,0,"")) {
      fwrite(fh_out, (c=="a" ? "b" : c))
}
fclose(fh_out)
fclose(fh_in)
```

Reading and writing in the same file

You may read and write from a file opened "rw", using any of the read or write functions described above. When reading and writing in the same file, one often uses file repositioning functions, too; see *Repositioning in a file* below.

Reading and writing matrices

Functions

>fputmatrix(*fh, transmorphic matrix X*)

and

>_fputmatrix(*fh, transmorphic matrix X*)

will write a matrix to a file. In the usual fashion, fputmatrix() returns nothing (it aborts if there is an I/O error) and _fputmatrix() returns a scalar equal to 0 if all went well and a negative error code otherwise.

Functions

>fgetmatrix(*fh*)

and

>_fgetmatrix(*fh*)

will read a matrix written by fputmatrix() or _fputmatrix(). Both functions return the matrix read or return J(0,0,.) on end of file (both functions) or error (_fgetmatrix() only). Since J(0,0,.) could be the matrix that was written, distinguishing between that and end of file requires subsequent use of fstatus(). fstatus() will return 0 if fgetmatrix() or _fgetmatrix() returned a written matrix, −1 if end of file, or (after _fgetmatrix()) a negative error code.

fputmatrix() writes matrices in a compact, efficient, and portable format; a matrix written on a Windows computer can be read back on a Macintosh or Unix computer and vice versa.

The following code creates a file containing three matrices

```
fh = fopen(filename, "w")
fputmatrix(fh, a)
fputmatrix(fh, b)
fputmatrix(fh, c)
fclose(fh)
```

and the following code reads them back:

```
fh = fopen(filename, "r")
a = fgetmatrix(fh)
b = fgetmatrix(fh)
c = fgetmatrix(fh)
fclose(fh)
```

❑ Technical Note

You may even write pointer matrices

```
mats = (&a, &b, &c, NULL)
fh = fopen(filename,"w")
fputmatrix(fh, mats)
fclose(fh)
```

and read them back:

```
fh = fopen(filename, "r")
mats = fgetmatrix(fh)
fclose(fh)
```

When writing pointer matrices, fputmatrix() writes NULL for any pointer-to-function value. It is also recommended that you do not write self-referential matrices (matrices that point to themselves, either directly or indirectly), although the elements of the matrix may be cross linked and even recursive themselves. If you are writing pointer matrix p, no element of p, *p, **p, etc., should contain &p. That one address cannot be preserved because in the assignment associated with reading back the matrix (the "*result=*" part of *result*=fgetmatrix(*fh*), a new matrix with a different address is associated with the contents.

❑

Repositioning in a file

The function

ftell(*fh*)

returns a real scalar reporting where you are in a file, and function

fseek(*fh*, *real scalar offset*, *real scalar whence*)

changes where you are in the file to be *offset* bytes from the beginning of the file (*whence* = −1), *offset* bytes from the current position (*whence* = 0), or *offset* bytes from the end of the file (*whence* = 1).

Functions _ftell() and _fseek() do the same thing as ftell() and fseek(), the difference being that, rather than aborting on error, the underscore functions return negative error codes. _ftell() is pretty well useless as the only error that can arise is I/O error, and what else are you going to do other than abort? _fseek(), however, has a use, because it allows you to try out a repositioning and check whether it was successful. With fseek(), if the repositioning is not successful, execution is aborted.

Say you open a file for read:

    fh = fopen(filename, "r")

After opening the file in mode r, you are positioned at the beginning of the file or, in the jargon of file processing, at position 0. Now say that you read 10 bytes from the file:

    part1 = fread(fh, 10)

Assuming that was successful, you are now at position 10 of the file. Say that you next read a line from the file

    line = fget(fh)

and assume that fget() returns "abc". You are now at position 14 or 15. (No, not 13: fget() read the line and the new-line characters and returned the line. abc was followed by carriage return and line feed (two characters) if the file was written under Windows and by a carriage return or line feed alone (one character) if the file was written under Macintosh or Unix).

ftell(*fh*) and _ftell(*fh*) tell you where you are in the file. Coding

    pos = ftell(fh)

would store 14 or 15 in pos. Later in your code, after reading more of the file, you could return to this position by coding

    fseek(fh, pos, -1)

You could return to the beginning of the file by coding

    fseek(fh, 0, -1)

and you could move to the end of the file by coding

    fseek(fh, 0, 1)

ftell(*fh*) is equivalent to _fseek(*fh*, 0, 0).

Repositioning functions cannot be used when the file has been opened "a".

Truncating a file

Truncation refers to making a longer file shorter. If a file was opened "w" or "rw", you may truncate it at its current position by using

    ftruncate(*fh*)

or

    _ftruncate(*fh*)

ftruncate() returns nothing; _ftruncate() returns 0 on success and otherwise returns a negative error code.

The following code shortens a file to its first 100 bytes:

    fh = fopen(filename, "rw")
    fseek(fh, 100, -1)
    ftruncate(fh)
    fclose(fh)

Error codes

If you use the underscore I/O functions, if there is an error, they will return a negative code. Those codes are

negative code	meaning
0	all is well
−1	end of file
−601	file not found
−602	file already exists
−603	file could not be opened
−608	file is read-only
−610	file format error
−630	web files not supported in this version of Stata
−631	host not found
−632	web file not allowed in this context
−633	web file not allowed in this context
−660	proxy host not found
−662	proxy server refused request to send
−663	remote connection to proxy failed
−665	could not set socket nonblocking
−669	invalid URL
−670	invalid network port number
−671	unknown network protocol
−672	server refused to send file
−673	authorization required by server
−674	unexpected response from server
−678	could not open local network socket
−679	unexpected web error
−691	I/O error
−699	insufficient disk space
−3601	invalid file handle
−3602	invalid filename
−3611	too many open files
−3621	attempt to write read-only file
−3622	attempt to read write-only file
−3623	attempt to seek append-only file
−3698	file seek error

Other codes in the −600 to −699 range are possible. The codes in this range correspond to the negative of the corresponding Stata return code; see [P] **error**.

Underscore functions that return a real scalar will return one of these codes if there is an error.

If an underscore function does not return a real scalar, then you obtain the outcome status using
fstatus(). For instance, the read-string functions return a string scalar or J(0,0,"") on end of
file. The underscore variants do the same, and they also return J(0,0,"") on error, meaning error
looks like end of file. You can determine the error code using the function

    fstatus(*fh*)

fstatus() returns 0 (no previous error) or one of the negative codes above.

fstatus() may be used after any underscore I/O command to obtain the current error status.

fstatus() may also be used after the nonunderscore I/O commands; then fstatus() will return
−1 or 0 because all other problems would have stopped execution.

Conformability

fopen(*fn, mode, public*), _fopen(*fn, mode, public*):
    fn: 1×1
    mode: 1×1
    public: 1×1 (optional)
    result: 1×1

fclose(*fh*):
    fh: 1×1
    result: *void*

_fclose(*fh*):
    fh: 1×1
    result: 1×1

fget(*fh*), _fget(*fh*), fgetnl(*fh*), _fgetnl(*fh*):
    fh: 1×1
    result: 1×1 or 0×0 if end of file

fread(*fh, len*), _fread(*fh, len*):
    fh: 1×1
    len: 1×1
    result: 1×1 or 0×0 if end of file

fput(*fh, s*), fwrite(*fh, s*):
    fh: 1×1
    s: 1×1
    result: *void*

_fput(*fh, s*), _fwrite(*fh, s*):
    fh: 1×1
    s: 1×1
    result: 1×1

fgetmatrix(*fh*), _fgetmatrix(*fh*):
    fh: 1×1
    result: $r \times c$ or 0×0 if end of file

fputmatrix(*fh*, *X*):
    fh: 1×1
    X: $r \times c$
    result: *void*

_fputmatrix(*fh*, *X*):
    fh: 1×1
    X: $r \times c$
    result: 1×1

fstatus(*fh*):
    fh: 1×1
    result: 1×1

ftell(*fh*), _ftell(*fh*):
    fh: 1×1
    result: 1×1

fseek(*fh*, *offset*, *whence*):
    fh: 1×1
    offset: 1×1
    whence: 1×1
    result: *void*

_fseek(*fh*, *offset*, *whence*):
    fh: 1×1
    offset: 1×1
    whence: 1×1
    result: 1×1

ftruncate(*fh*):
    fh: 1×1
    result: *void*

_ftruncate(*fh*):
    fh: 1×1
    result: 1×1

Diagnostics

fopen(*fn*, *mode*) aborts with error if *mode* is invalid or if *fn* cannot be opened or if an attempt is made to open too many files simultaneously.

_fopen(*fn*, *mode*) aborts with error if *mode* is invalid or if an attempt is made to open too many files simultaneously. _fopen() returns the appropriate negative error code if *fn* cannot be opened.

All remaining I/O functions—even functions with leading underscore—abort with error if *fh* is not a handle to a currently open file.

Also, the functions that do not begin with an underscore abort with error when a file was opened read-only and a request requiring write access is made, when a file is opened write-only and a request requiring read access is made, etc. See *Error codes* above; all problems except code -1 (end of file) cause the nonunderscore functions to abort with error.

Finally, the following functions will also abort with error for the following specific reasons:

fseek(*fh*, *offset*, *whence*) and _fseek(*fh*, *offset*, *whence*) abort with error if *offset* is outside the range ±2,147,483,647 on 32-bit computers; if *offset* is outside the range ±9,007,199,254,740,991 on 64-bit computers; or, on all computers, if *whence* is not −1, 0, or 1.

Also See

[M-5] **cat()** — Load file into string matrix sprintf() in

[M-5] **printf()** — Format output

[M-5] **bufio()** — Buffered (binary) I/O

[M-4] **io** — I/O functions

Title

[M-5] **fullsvd( )** — Full singular value decomposition

Syntax

void	fullsvd(*numeric matrix A, U, s, Vt*)
numeric matrix	fullsdiag(*numeric colvector s, real scalar k*)
void	_fullsvd(*numeric matrix A, U, s, Vt*)
real scalar	_svd_la(*numeric matrix A, U, s, Vt*)

Description

fullsvd(*A, U, s, Vt*) calculates the singular value decomposition of $m \times n$ matrix A, returning the result in U, s, and Vt. Singular values in s are sorted from largest to smallest.

fullsdiag(*s, k*) converts column vector s returned by fullsvd() into matrix S. In all cases, the appropriate call for this function is

  S = fullsdiag(*s*, rows(*A*)-cols(*A*))

_fullsvd(*A, U, s, Vt*) does the same as fullsvd(), except that, in the process, it destroys A. Use of _fullsvd() in place of fullsvd() conserves memory.

_svd_la() is the interface into the [M-1] **LAPACK** SVD routines and is used in the implementation of the previous functions. There is no reason you should want to use it. _svd_la() is similar to _fullsvd(). It differs in that it returns a real scalar equal to 1 if the numerical routines fail to converge, and it returns 0 otherwise. The previous SVD routines set s to contain missing values in this unlikely case.

Remarks

Remarks are presented under the following headings:

> *Introduction*
> *Relationship between the full and thin SVDs*
> *The contents of s*
> *Possibility of convergence problems*

Documented here is the full SVD, appropriate in all cases, but of interest mainly when A: $m \times n$, $m < n$. There is a thin SVD that conserves memory when $m \geq n$; see [M-5] **svd( )**. The relationship between the two is discussed in *Relationship between the full and thin SVDs* below.

Introduction

The SVD is used to compute accurate solutions to linear systems and least-squares problems, to compute the 2-norm, and to determine the numerical rank of a matrix.

The singular value decomposition (SVD) of A: $m \times n$ is given by

$$A = USV'$$

where
- U: $m \times m$ and orthogonal (unitary)
- S: $m \times n$ and diagonal
- V: $n \times n$ and orthogonal (unitary)

When A is complex, the transpose operator $'$ is understood to mean the conjugate transpose operator.

Diagonal matrix S contains the singular values and those singular values are real even when A is complex. It is usual (but not required) that S is arranged so that the largest singular value appears first, then the next largest, and so on. The SVD routines documented here do this.

The full SVD routines return U and $Vt = V'$. S is returned as a column vector s, and S can be obtained by

$$S = \text{fullsdiag}(s, \text{rows}(A)-\text{cols}(A))$$

so we will write the SVD as

$$A = U * \text{fullsdiag}(s, \text{rows}(A)-\text{cols}(A)) * Vt$$

Function `fullsvd(A, U, s, Vt)` returns the U, s, and Vt corresponding to A.

Relationship between the full and thin SVDs

A popular variant of the SVD is known as the thin SVD and is suitable for use when $m \geq n$. Both SVDs have the same formula,

$$A = USV'$$

but U and S have reduced dimensions in the thin version:

Matrix	Full SVD	Thin SVD
U:	$m \times m$	$m \times n$
S:	$m \times n$	$n \times n$
V:	$n \times n$	$n \times n$

When $m = n$, the two variants are identical.

The thin SVD is of use when $m > n$, because then only the first m diagonal elements of S are nonzero, and therefore only the first m columns of U are relevant in $A = USV'$. There are considerable memory savings to be had in calculating the thin SVD when $m \gg n$.

As a result, many people call the thin SVD the SVD and ignore the full SVD altogether. If the matrices you deal with have $m \geq n$, you will want to do the same. To obtain the thin SVD, see [M-5] **svd()**.

Regardless of the dimension of your matrix, you may wish to obtain only the singular values. In this case, see `svdsv()` documented in [M-5] **svd()**. That function is appropriate in all cases.

The contents of s

Given A: $m \times n$, the singular values are returned in s: $\min(m,n) \times 1$.

Let's consider the $m = n$ case first. A is $m \times m$ and the m singular values are returned in s, an $m \times 1$ column vector. If A were 3×3, perhaps we would get back

```
: s
                   1
         1    13.47
         2     5.8
         3     2.63
```

If we needed it, we could obtain S from s simply by creating a diagonal matrix from s

```
: S = diag(s)
: S
[symmetric]
              1       2       3
         1  13.47
         2    0      5.8
         3    0       0     2.63
```

although the official way we are supposed to do this is

```
: S = fullsdiag(s, rows(A)-cols(A))
```

and that will return the same result.

Now let's consider $m < n$. Let's pretend that A is 3×4. The singular values will be returned in 3×1 vector s. For instance, s might still contain

```
: s
                   1
         1    13.47
         2     5.8
         3     2.63
```

The S matrix here needs to be 3×4, and `fullsdiag()` will form it:

```
: fullsdiag(s, rows(A)-cols(A))
              1       2       3       4
         1  13.47     0       0       0
         2    0      5.8      0       0
         3    0       0     2.63      0
```

The final case is $m > n$. We will pretend that A is 4×3. The s vector we get back will look the same

          : s
                   1
              ┌─────────┐
            1 │ 13.47 │
            2 │   5.8   │
            3 │   2.63 │
              └─────────┘

but this time, we need a 4 × 3 rather than a 3 × 4 matrix formed from it.

          : fullsdiag(s, rows(A)-cols(A))
                    1   2   3
              ┌─────────────────────────┐
            1 │ 13.47   0   0   │
            2 │   0   5.8   0   │
            3 │   0   0   2.63 │
            4 │   0   0   0   │
              └─────────────────────────┘

Possibility of convergence problems

See *Possibility of convergence problems* in [M-5] **svd( )**; what is said there applies equally here.

Conformability

fullsvd(*A*, *U*, *s*, *Vt*):

    input:

               A: $m \times n$

    output:

               U: $m \times m$
              s: $\min(m, n) \times 1$
              Vt: $n \times n$
          result: void

fullsdiag(*s*, *k*):

    input:

               s: $r \times 1$
              k: 1×1

    output:

          result: $r + k \times r$,   if k $\geq$ 0
                           $r \times r - k$,   otherwise

_fullsvd(*A*, *U*, *s*, *Vt*):

    input:

               A: $m \times n$

    output:

               A: 0×0
              U: $m \times m$
              s: $\min(m, n) \times 1$
              Vt: $n \times n$
          result: void

_svd_la(*A*, *U*, *s*, *Vt*):
   input:
                    A: $m \times n$
   output:
                    A: $m \times n$, but contents changed
                    U: $m \times m$
                    s: $\min(m,n) \times 1$
                    Vt: $n \times n$
           result: 1×1

Diagnostics

fullsvd(*A*, *U*, *s*, *Vt*) and _fullsvd(*A*, *s*, *Vt*) return missing results if *A* contains missing. In all other cases, the routines should work, but there is the unlikely possibility of convergence problems, in which case missing results will also be returned; see *Possibility of convergence problems* in [M-5] **svd( )**.

_fullsvd( ) aborts with error if *A* is a view.

Direct use of _svd_la( ) is not recommended.

Also See

[M-5] **svd( )** — Singular value decomposition

[M-5] **svsolve( )** — Solve AX=B for X using singular value decomposition

[M-5] **pinv( )** — Moore–Penrose pseudoinverse

[M-5] **norm( )** — Matrix and vector norms

[M-5] **rank( )** — Rank of matrix

[M-4] **matrix** — Matrix functions

Title

[M-5] **ghk( )** — Geweke–Hajivassiliou–Keane (GHK) multivariate normal simulator

Syntax

> *real scalar* ghk(*real vector x, real matrix V, real vector opt, real scalar rank*)
>
> *real scalar* ghk(*real vector x, real matrix V, real vector opt, real scalar rank,
> real rowvector dfdx, real rowvector dfdv*)

Description

ghk(*x, V, opt, rank*) returns a real scalar containing the simulated value of the multivariate normal distribution with variance–covariance *V* at the point *x*. The vector *opt*, containing options for the simulation, is of length 2–4.

opt	meaning
$opt[1] = 1$	Halton sequence
$= 2$	Hammersley sequence
$= 3$	uniform pseudorandom sequence
$opt[2] > 0$	number of points to use in the simulation
$opt[3] > 0$	index of the first draw (optional for Halton or Hammersley sets)
$opt[4] \neq 0$	use antithetic draws (optional)

On return, *rank* contains the rank of *V*.

ghk(*x, V, opt, rank, dfdx, dfdv*) does the same thing but also returns the first-order derivatives of the simulated probability with respect to *x* in *dfdx* and the simulated probability derivatives with respect to vech(*V*) in *dfdv*. See vech( ) in [M-5] **vec( )** for details of the half-vectorized operator.

Remarks

Halton and Hammersley point sets are composed of deterministic sequences on [0,1] and, for sets of dimension less than 10, generally have better coverage than that of the uniform pseudorandom sequences.

Antithetic draws effectively double the number of points and reduce the variability of the simulated probability. For draw *u*, the antithetic draw is $1 - u$.

If you are using ghk( ) in a likelihood evaluator for ml, be sure to use the same sequence with each call to the likelihood evaluator. For a uniform pseudorandom sequence ($opt[1] = 3$), set the seed of the uniform random-number generator—see uniformseed( ) in [M-5] **uniform( )**—to the same value with each call to the likelihood evaluator. If you are using the Halton or Hammersley sets, you will want to keep the sequences going with each call to ghk( ) within one likelihood evaluation. First, initialize $opt[3] = 1$ on entering the ml likelihood evaluator and compute the increment $opt[3] = opt[3] + opt[2]$ after each call to ghk( ).

Conformability

ghk(x, V, opt, $rank$):
   input:
            x: $1 \times m$ or $m \times 1$
            V: $m \times m$ (symmetric, positive definite)
           opt: 1×4 or 4×1
   output:
           $result$: 1×1
           $rank$: 1×1

ghk(x, V, opt, $rank$, $dfdx$, $dfdv$):
   input:
            x: $1 \times m$ or $m \times 1$
            V: $m \times m$ (symmetric, positive definite)
            opt: 1×4 or 4×1
   output:
           $result$: 1×1
           $rank$: 1×1
           $dfdx$: $1 \times m$
           $dfdv$: $1 \times m(m+1)/2$

Diagnostics

The maximum dimension, m, is 20.

V must be symmetric and preferably positive definite. ghk() will not terminate if V is not positive definite, although the returned value of *rank* will be less than rows(V). The routine uses a Cholesky routine that pivots out the rows and columns of V (as well as elements of x) that make V indefinite. The corresponding elements of *dfdx* and *dfdv* are zero. Although the multivariate normal distribution is not defined for indefinite variance–covariance matrices, an indefinite V can occur early in an ml optimization. This flexible behavior may allow the optimization process to continue. You might decide to check for *rank* $<$ rows(V) and exit with an error message when it occurs.

Also See

[M-5] **ghkfast()** — GHK multivariate normal simulator using pregenerated points

[M-5] **halton()** — Generate a Halton or Hammersley set

[M-4] **statistical** — Statistical functions

Title

[M-5] **ghkfast( )** — GHK multivariate normal simulator using pregenerated points

Syntax

P = ghkfastsetup(*real scalar n*, *real scalar m*, *real scalar d*, *string scalar type*)

real colvector ghkfast(*P*, *real matrix X*, *real matrix V*, *rank*
  [, *real scalar i*
    [, *real scalar anti*
      [, *dfdx*, *dfdv*
  ]]])

where *P*, if it is declared, should be declared

  transmorphic *P*

and where *rank*, *dfdx*, and *dfdv* are outputs

  real scalar rank
  real matrix dfdx
  real matrix dfdv

Description

Please see [M-5] **ghk( )**. The routines documented here do the same thing. They are a little more difficult to use, but they are faster. Rather than simply calling ghk() to obtain the simulated value, you first code P = ghkfastsetup(...) and then use ghkfast(P, ...) to obtain the simulated values. There is a time savings because one can do the setup once and then obtain simulated values repeatedly. Also, ghkfast() can generate simulated probabilities from the generalized Halton sequence; see [M-5] **halton( )**.

ghkfastsetup(*n*, *m*, *d*, *type*) calculates the points to be used by ghkfast(). Inputs *n*, *m*, and *d* are the number of observations, the number of repetitions for the simulation, and the dimension of the multivariate normal (MVN) distribution, respectively. Input *type* specifies the type of points to generate and may be one of "halton", "hammersley", "random", or "ghalton".

ghkfast(*P*, *X*, *V*, *rank*) returns an $n \times 1$ real column vector containing the simulated values of the MVN distribution with $d \times d$ variance–covariance matrix *V* at the points stored in the rows of the $n \times d$ matrix *X*. On return, *rank* will contain the rank of *V*.

ghkfast(*P*, *X*, *V*, *rank*, *i*) does the same thing but evaluates the simulated probability only at $X[i,]$, using the *i*th set of simulation points set by ghkfastsetup(), $1 \leq i \leq n$, where *n* is the *n* specified in ghkfastsetup(). Since only $X[i]$ is used, *X* may be specified as a row vector. ghkfast(*P*, *X*, *V*, *rank*, *i*) returns a scalar.

ghkfast(*P*, *X*, *V*, *rank*, *i*, *anti*) does the same thing but also uses antithetic points when evaluating the simulated probability. Set *i* to zero if you wish to evaluate all the rows in *X*.

ghkfast(*P*, *X*, *V*, *rank*, *i*, *anti*, *dfdx*, *dfdv*) does the same thing but also returns the first-order derivatives of the simulated probability with respect to the rows of *X* in *dfdx* and the simulated probability derivatives with respect to vech(*V*) in *dfdv*. See vech() in [M-5] **vec( )** for details of the half-vectorized operator.

Remarks

For problems where repetitive calls to the GHK algorithm are required, ghkfast() might be a preferred alternative to ghk(). Generating the points once at the outset of a program produces a speed increase. For problems with many observations and/or many simulation points per observation, ghkfast() will be faster than ghk() at the cost of requiring more memory.

In addition to the sequences "halton", "hammersley", and "random", ghkfast() will use sequence "ghalton". "ghalton" sequences are like "random" sequences in that you should set the random-number seed before using them; see [M-5] **uniform( )**.

Conformability

ghkfastsetup(*n*, *m*, *d*, *type*):

n:	1×1
m:	1×1
d:	1×1
type:	1×1
result:	*transmorphic*

ghkfast(*P*, *X*, *V*, *rank*):

input:

P:	*transmorphic*
X:	$n \times d$
V:	$d \times d$ (symmetric, positive definite)

output:

rank:	1×1
result:	$n \times 1$

ghkfast(*P*, *X*, *V*, *rank*, *i*, *anti*, *dfdx*, *dfdv*):

input:

P:	*transmorphic*		
X:	$n \times d$	or	$1 \times d$
V:	$d \times d$ (symmetric, positive definite)		
i:	1×1		
anti:	1×1		

output:

result:	$n \times 1$	or	1×1
rank:	1×1		
dfdx:	$n \times d$	or	$1 \times d$
dfdv:	$n \times d(d+1)/2$	or	$1 \times d(d+1)/2$

Diagnostics

ghkfastsetup(n, m, d, $type$) aborts with error if the dimension d is greater than 20.

ghkfast(P, X, V, ...) requires that V be symmetric but does not check. V should preferably be positive definite. If it is not, the returned value in *rank* will be less than rows(V).

Also See

[M-5] **ghk( )** — Geweke–Hajivassiliou–Keane (GHK) multivariate normal simulator

[M-5] **halton( )** — Generate a Halton or Hammersley set

[M-4] **statistical** — Statistical functions

Title

[M-5] halton( ) — Generate a Halton or Hammersley set

Syntax

real matrix	halton(*real scalar n, real scalar d*)	
real matrix	halton(*real scalar n, real scalar d, real scalar start*)	
real matrix	halton(*real scalar n, real scalar d, real scalar start, real scalar hammersley*)	
void	_halton(*real matrix x*)	
void	_halton(*real matrix x, real scalar start*)	
void	_halton(*real matrix x, real scalar start, real scalar hammersley*)	
real colvector	ghalton(*real scalar n, real scalar base, real scalar u*)	

Description

halton(*n, d*) returns an $n \times d$ matrix containing a Halton set of length *n* and dimension *d*.

halton(*n, d, start*) does the same thing, but the first row of the returned matrix contains the sequences starting at index *start*. The default is *start* = 1.

halton(*n, d, start, hammersley*), with *hammersley* $\neq$ 0, returns a Hammersley set of length *n* and dimension *d* with the first row of the returned matrix containing the sequences starting at index *start*.

_halton(*x*) modifies the $n \times d$ matrix *x* so that it contains a Halton set of dimension *d* of length *n*.

_halton(*x, start*) does the same thing, but the first row of the returned matrix contains the sequences starting at index *start*. The default is *start* = 1.

_halton(*x, start, hammersley*), with *hammersley* $\neq$ 0, returns a Hammersley set of length *n* and dimension *d* with the first row of the returned matrix containing the sequences starting at index *start*.

ghalton(*n, base, u*) returns an $n \times 1$ vector containing a (generalized) Halton sequence using base *base* and starting from scalar $0 \leq u < 1$. For $u = 0$, the standard Halton sequence is generated. If *u* is uniform $(0, 1)$, a randomized Halton sequence is generated.

Remarks

The Halton sequences are generated from the first *d* primes and generally have more uniform coverage over the unit cube of dimension *d* than that of sequences generated from pseudouniform random numbers. However, Halton sequences based on large primes ($d > 10$) can be highly correlated, and their coverage can be worse than that of the pseudorandom uniform sequences.

The Hammersley set contains the sequence $(2*i - 1)/(2*n)$, $i = 1, \ldots, n$, in the first dimension and Halton sequences for dimensions $2, \ldots, d$.

_halton() modifies *x* and can be used when repeated calls are made to generate long sequences in blocks. Here update the *start* index between calls by using *start* = *start* + rows(*x*).

ghalton() uses the base *base*, preferably a prime, and generates a Halton sequence using $0 \leq u < 1$ as a starting value. If *u* is uniform $(0, 1)$, the sequence is a randomized Halton sequence. For $u = 0$, the sequence is the standard Halton sequence. Blocks of sequences can be generated by ghalton() by using the last value in the vector returned from a previous call as *u*. For example,

```
x = J(n,1,0)
for (i=1; i<=k; i++) {
    x[.] = ghalton(n, base, x[n])
    ...
}
```

Conformability

halton(*n*, *d*, *start*, *hammersley*):

> *input*:
>
> | | | |
> |---|---|---|
> | *n*: | 1×1 | |
> | *d*: | 1×1 | |
> | *start*: | 1×1 | (optional) |
> | *hammersley*: | 1×1 | (optional) |
>
> *output*:
>
> | | |
> |---|---|
> | *result*: | $n \times d$ |

_halton(*x*, *start*, *hammersley*):

> *input*:
>
> | | | |
> |---|---|---|
> | *x*: | $n \times d$ | |
> | *start*: | 1×1 | (optional) |
> | *hammersley*: | 1×1 | (optional) |
>
> *output*:
>
> | | |
> |---|---|
> | *x*: | $n \times d$ |

ghalton(*n*, *base*, *u*):

> *input*:
>
> | | |
> |---|---|
> | *n*: | 1×1 |
> | *base*: | 1×1 |
> | *u*: | 1×1 |
>
> *output*:
>
> | | |
> |---|---|
> | *result*: | $n \times 1$ |

Diagnostics

The maximum dimension, *d*, is 20. The scalar index *start* must be a positive integer, and the scalar *u* must be such that $0 \leq u < 1$.

Also See

[M-4] **mathematical** — Important mathematical functions

Title

> [M-5] **Hilbert( )** — Hilbert matrices

Syntax

> *real matrix*   Hilbert(*real scalar n*)
>
> *real matrix*   invHilbert(*real scalar n*)

Description

Hilbert(*n*) returns the $n \times n$ Hilbert matrix, defined as matrix H with elements $H_{ij} = 1/(i+j-1)$.

invHilbert(*n*) returns the inverse of the $n \times n$ Hilbert matrix, defined as the matrix with elements $-1^{i+j}(i+j-1) \times \text{comb}(n+i-1, n-j) \times \text{comb}(n+j-1, n-i) \times \text{comb}(i+j-2, i-1)^2$.

Remarks

Hilbert(*n*) and invHilbert(*n*) are used in testing Mata. Hilbert matrices are notoriously ill conditioned. The determinants of the first five Hilbert matrices are 1, 1/12, 1/2,160, 1/6,048,000, and 1/266,716,800,000.

Conformability

Hilbert(*n*), invHilbert(*n*):
    n: 1×1
   result: $\text{trunc}(n) \times \text{trunc}(n)$

Diagnostics

None.

David Hilbert (1862–1943) was born near Königsberg, Prussia (now Kaliningrad, Russia), and studied mathematics at the university there. He joined the staff from 1886 to 1895, when he moved to Göttingen, where he stayed despite tempting offers to move. Hilbert was one of the outstanding mathematicians of his time, producing major work in several fields, including invariant theory, algebraic number theory, the foundations of geometry, functional analysis, integral equations, and the calculus of variations. In 1900 he identified 23 key problems in an address to the Second International Congress of Mathematicians in Paris that continues to influence directions in research (Hilbert 1902). Hilbert's name is perhaps best remembered through the idea of Hilbert space. His work on what are now known as Hilbert matrices was published in 1894.

References

Choi, M.-D. 1983. Tricks or treats with the Hilbert matrix. *American Mathematical Monthly* 90: 301–312.

Hilbert, D. 1894. Ein Beitrag zur Theorie des Legendreschen Polynoms. *Acta Mathematica* 18: 155–159.

———. 1902. Mathematical problems. *Bulletin of the American Mathematical Society* 8: 437–479. (reprinted in same journal, 2000, 37: 407–436).

Reid, C. 1970. *Hilbert*. Berlin: Springer.

Also See

[M-4] **standard** — Functions to create standard matrices

Title

[M-5] I() — Identity matrix

Syntax

> *real matrix*   I(*real scalar n*)
>
> *real matrix*   I(*real scalar m*, *real scalar n*)

Description

I(*n*) returns the $n \times n$ identity matrix.

I(*m*, *n*) returns an $m \times n$ matrix with 1s down its principal diagonal and 0s elsewhere.

Remarks

I() must be typed in uppercase.

Conformability

I(*n*):
> *n*: 1×1
> *result*: $n \times n$

I(*m*, *n*):
> *m*: 1×1
> *n*: 1×1
> *result*: $m \times n$

Diagnostics

I(*n*) aborts with error if *n* is less than 0 or is missing. *n* is interpreted as trunc(*n*).

I(*m*, *n*) aborts with error if *m* or *n* are less than 0 or if they are missing. *m* and *n* are interpreted as trunc(*m*) and trunc(*n*).

Also See

[M-4] **standard** — Functions to create standard matrices

Title

[M-5] **inbase( )** — Base conversion

Syntax

> *string matrix*   inbase(*real scalar base*, *real matrix x* [, *real scalar fdigits* [, *err*]])
>
> *real matrix*   frombase(*real scalar base*, *string matrix s*)

Description

inbase(*base*, *x*) returns a string matrix containing the values of *x* in base *base*.

inbase(*base*, *x*, *fdigits*) does the same; *fdigits* specifies the maximum number of digits to the right of the base point to appear in the returned result when *x* has a fractional part. inbase(*base*, *x*) is equivalent to inbase(*base*, *x*, 8).

inbase(*base*, *x*, *fdigits*, *err*) is the same as inbase(*base*, *x*, *fdigits*), except that it returns in *err* the difference between *x* and the converted result.

x = frombase(*base*, *s*) is the inverse of *s* = inbase(*base*, *x*). It returns base *base* number *s* as a number. We are tempted to say, "as a number in base 10", but that is not exactly true. It returns the result as a *real*, that is, as an IEEE base-2 double-precision float that, when you display it, is displayed in base 10.

Remarks

Remarks are presented under the following headings:

> Positive integers
> Negative integers
> Numbers with nonzero fractional parts
> Use of the functions

Positive integers

inbase(2, 1691) is 11010011011; that is, 1691 base 10 equals 11010011011 base 2. frombase(2, "11010011011") is 1691.

inbase(3, 1691) is 2022122; that is, 1691 base 10 equals 2022122 base 3. frombase(3, "2022122") is 1691.

inbase(16, 1691) is 69b; that is, 1691 base 10 equals 1691 base 16. frombase(16, "69b") is 1691. (The base-16 digits are 0, 1, 2, 3, 4, 5, 6, 7, 8, 9, a, b, c, d, e, f.)

inbase(62, 1691) is rh; that is, 1691 base 10 equals rh base 62. frombase(62, "rh") is 1691. (The base-62 digits are 0, 1, 2, 3, 4, 5, 6, 7, 8, 9, a, b, ..., z, A, B, ..., Z.)

There is a one-to-one correspondence between the integers in different bases. The error of the conversion is always zero.

Negative integers

Negative integers are no different from positive integers. For instance, inbase(2, -1691) is -11010011011; that is, −1691 base 10 equals −11010011011 base 2. frombase(2, "-11010011011") is −1691.

The error of the conversion is always zero.

Numbers with nonzero fractional parts

inbase(2, 3.5) is 11.1; that is, 3.5 base 10 equals 11.1 base 2. frombase(2, "11.1") is 3.5.

inbase(3, 3.5) is 10.11111111.

inbase(3, 3.5, 20) is 10.11111111111111111111.

inbase(3, 3.5, 30) is 10.111111111111111111111111111111.

Therefore, 3.5 base 10 equals 1.1111... in base 3. There is no exact representation of one-half in base 3. The errors of the above three conversions are .0000762079, 1.433399e−10, and 2.45650e−15. Those are the values that would be returned in *err* if inbase(3, 3.5, *fdigits*, *err*) were coded.

frombase(3, "10.11111111") is 3.499923792.

frombase(3, "10.11111111111111111111") is 3.4999999998566.

frombase(3, "10.111111111111111111111111111111") is 3.49999999999999734.

inbase(16, 3.5) is 3.8; that is, 3.5 base 10 equals 3.8 base 16. The error is zero. frombase(16, "3.8") is 3.5.

inbase(62, 3.5) is 3.v; that is, 3.5 base 10 equals 3.v base 62. frombase(62, "3.v") is 3.5. The error is zero.

In inbase(*base*, *x*, *fdigits*), *fdigits* specifies the maximum number of digits to appear to the right of the base point. *fdigits* is required to be greater than or equal to 1. inbase(16, 3.5, *fdigits*) will be 3.8 regardless of the value of *fdigits* because more digits are unnecessary.

The error that is returned in inbase(*base*, *x*, *fdigits*, *err*) can be an understatement. For instance, inbase(16, .1, 14, *err*) is 0.1999999999999a and returned in *err* is 0 even though there is no finite-digit representation of 0.1 base 10 in base 16. That is because the .1 you specified in the call was not actually 0.1 base 10. The computer that you are using is binary, and it converted the .1 you typed to

0.0001100110011001100110011001100110011001100110011001100110011010 base 2

before inbase() was ever called. 0.1999999999999a base 16 is an exact representation of that number.

Use of the functions

These functions are used mainly for teaching, especially on the sources and avoidance of roundoff error; see Gould (2006).

The functions can have a use in data processing, however, when used with integer arguments. You have a dataset with 10-digit identification numbers. You wish to record the 10-digit number, but more densely. You could convert the number to base 62. The largest 10-digit ID number possible is 9999999999, or aUKYOz base 62. You can record the ID numbers in a six-character string by using inbase(). If you needed the original numbers back, you could use frombase().

In a similar way, Stata internally uses base 36 for naming temporary files, and that was important when filenames were limited to eight characters. Base 36 allows Stata to generate up to 2,821,109,907,455 filenames before wrapping of filenames occurs.

Conformability

inbase(*base*, *x*, *fdigits*, *err*):

	input:			
		base:	1×1	
		x:	$r \times c$	
		fdigits:	1×1	(optional)
	output:			
		err:	$r \times c$	(optional)
		result:	$r \times c$	

frombase(*base*, *s*):
    base: 1×1
    s: $r \times c$
    result: $r \times c$

Diagnostics

The digits used by inbase()/frombase() to encode/decode results are

0 0	10 a	20 k	30 u	40 E	50 O	60 Y
1 1	11 b	21 l	31 v	41 F	51 P	61 Z
2 2	12 c	22 m	32 w	42 G	52 Q	
3 3	13 d	23 n	33 x	43 H	53 R	
4 4	14 e	24 o	34 y	44 I	54 S	
5 5	15 f	25 p	35 z	45 J	55 T	
6 6	16 g	26 q	36 A	46 K	56 U	
7 7	17 h	27 r	37 B	47 L	57 V	
8 8	18 i	28 s	38 C	48 M	58 W	
9 9	19 j	29 t	39 D	49 N	59 X	

When *base* ≤ 36, frombase() treats A, B, C, ..., as if they were a, b, c, ...

inbase(*base*, *x*, *fdigits*, *err*) returns . (missing) if *base* < 2, *base* > 62, *base* is not an integer, or *x* is missing. If *fdigits* is less than 1 or *fdigits* is missing, results are as if *fdigits* = 8 were specified.

frombase(*base*, *s*) returns . (missing) if *base* < 2, *base* > 62, *base* is not an integer, or *s* is missing; if *s* is not a valid base *base* number; or if the converted value of *s* is greater than 8.988e+307 in absolute value.

Reference

Gould, W. 2006. Mata Matters: Precision. *Stata Journal* 6: 550–560.

Also See

[M-4] **mathematical** — Important mathematical functions

Title

[M-5] **indexnot( )** — Find character not in list

Syntax

> *real matrix*   indexnot(*string matrix* s_1, *string matrix* s_2)

Description

indexnot(s_1, s_2) returns the position of the first character of s_1 not found in s_2, or it returns 0 if all characters of s_1 are found in s_2.

Conformability

indexnot(s_1, s_2):
      s_1: $r_1 \times c_1$
      s_2: $r_2 \times c_2$, s_1 and s_2 r-conformable
   result: $\max(r_1,r_2) \times \max(c_1,c_2)$

Diagnostics

indexnot(s_1, s_2) returns 0 if all characters of s_1 are found in s_2.

Also See

[M-4] **string** — String manipulation functions

Title

[M-5] invorder( ) — Permutation vector manipulation

Syntax

> *real vector*   invorder(*real vector p*)

> *real vector*   revorder(*real vector p*)

where *p* is assumed to be a permutation vector.

Description

invorder(*p*) returns the permutation vector that undoes the permutation performed by *p*.

revorder(*p*) returns the permutation vector that is the reverse of the permutation performed by *p*.

Remarks

See [M-1] **permutation** for a description of permutation vectors. To summarize,

1. Permutation vectors *p* are used to permute the rows or columns of a matrix X: $r \times c$.

    If *p* is intended to permute the rows of X, the permuted X is obtained via $Y = X[p, .]$.

    If *p* is intended to permute the columns of X, the permuted X is obtained via $Y = X[., p]$.

2. If *p* is intended to permute the rows of X, it is called a row-permutation vector. Row-permutation vectors are $r \times 1$ column vectors.

3. If *p* is intended to permute the columns of X, it is called a column-permutation vector. Column-permutation vectors are $1 \times c$ row vectors.

4. Row-permutation vectors contain a permutation of the integers 1 to r.

5. Column-permutation vectors contain a permutation of the integers 1 to c.

Let us assume that *p* is a row-permutation vector, so that

$$Y = X[p, .]$$

invorder(*p*) returns the row-permutation vector that undoes *p*:

$$X = Y[\text{invorder}(p), .]$$

That is, using the matrix notation of [M-1] **permutation**,

$$Y = PX \quad \text{implies} \quad X = P^{-1}Y$$

If *p* is the permutation vector corresponding to permutation matrix P, invorder(*p*) is the permutation vector corresponding to permutation matrix P^{-1}.

revorder(*p*) returns the permutation vector that reverses the order of *p*. For instance, say that row-permutation vector *p* permutes the rows of *X* so that the diagonal elements are in ascending order. Then revorder(*p*) would permute the rows of *X* so that the diagonal elements would be in descending order.

Conformability

invorder(*p*), revorder(*p*):
        p: $r \times 1$ or $1 \times c$
  result: $r \times 1$ or $1 \times c$

Diagnostics

invorder(*p*) and revorder(*p*) can abort with error or can produce meaningless results when *p* is not a permutation vector.

Also See

[M-1] **permutation** — An aside on permutation matrices and vectors

[M-4] **manipulation** — Matrix manipulation

Title

[M-5] **invsym( )** — Symmetric real matrix inversion

Syntax

    real matrix   invsym(*real matrix A*)

    real matrix   invsym(*real matrix A*, *real vector order*)

    void   _invsym(*real matrix A*)

    void   _invsym(*real matrix A*, *real vector order*)

Description

invsym(*A*) returns a generalized inverse of real, symmetric, positive-definite matrix *A*.

invsym(*A*, *order*) does the same but allows you to specify which columns are to be swept first.

_invsym(*A*) and _invsym(*A*, *order*) do the same thing as invsym(*A*) and invsym(*A*, *order*) except that *A* is replaced with the generalized inverse result rather than the result being returned. _invsym( ) uses less memory than invsym( ).

invsym( ) and _invsym( ) are the routines Stata uses for calculating inverses of symmetric matrices.

Also see [M-5] **luinv( )**, [M-5] **qrinv( )**, and [M-5] **pinv( )** for general matrix inversion.

Remarks

Remarks are presented under the following headings:

    Definition of generalized inverse
    Specifying the order in which columns are dropped
    Determining the rank, or counting the number of dropped columns
    Extracting linear dependencies

Definition of generalized inverse

When the matrix is of full rank and positive definite, the generalized inverse equals the inverse, i.e., assuming *A* is $n \times n$,

$$\text{invsym}(A) * A = A * \text{invsym}(A) = I(n)$$

or, at least the above restriction is true up to roundoff error. When *A* is not full rank, the generalized inverse invsym( ) satisfies (ignoring roundoff error)

$$A * \text{invsym}(A) * A = A$$

$$\text{invsym}(A) * A * \text{invsym}(A) = \text{invsym}(A)$$

In the generalized case, there are an infinite number of inverse matrices that can satisfy the above restrictions. The one invsym() chooses is one that sets entire columns (and therefore rows) to 0, thus treating A as if it were of reduced dimension. Which columns (rows) are selected is determined on the basis of minimizing roundoff error.

In the above we talk as if determining whether a matrix is of full rank is an easy calculation. That is not true. Because of the roundoff error in the manufacturing and recording of A itself, columns that ought to be perfectly collinear will not be and yet you will still want invsym() to behave as if they were. invsym() tolerates a little deviation from collinearity in making the perfectly collinear determination.

Specifying the order in which columns are dropped

Left to make the decision itself, invsym() will choose which columns to drop (to set to 0) to minimize the overall roundoff error of the generalized inverse calculation. If column 1 and column 3 are collinear, then invsym() will choose to drop column 1 or column 3.

There are occasions, however, when you would like to ensure that a particular column or set of columns are not dropped. Perhaps column 1 corresponds to the intercept of a regression model and you would much rather, if one of columns 1 and 3 has to be dropped, that it be column 3.

Order allows you to specify the columns of the matrix that you would prefer not be dropped in the generalized inverse calculation. In the above example, to prevent column 1 from being dropped, you could code

> invsym(A, 1)

If you would like to keep columns 1, 5, and 10 from being dropped, you can code

> invsym(A, (1,5,10))

Specifying columns not to be dropped does not guarantee that they will not be dropped because they still might be collinear with each other or they might equal constants. However, if any other column can be dropped to satisfy your desire, it will be.

Determining the rank, or counting the number of dropped columns

If a column is dropped, 0 will appear on the corresponding diagonal entry. Hence, the rank of the original matrix can be extracted after inversion by invsym():

```
: Ainv = invsym(A)
: rank = rows(Ainv)-diag0cnt(Ainv)
```

See [M-5] **diag0cnt( )**.

Extracting linear dependencies

The linear dependencies can be read from the rows of $A*invsym(A)$:

```
: A*invsym(A)
```

	1	2	3
1	1	0	-5.20417e-17
2	-1	0	1
3	1.34441e-16	0	1

The above is interpreted to mean

$$x_1 = x_1$$
$$x_2 = -x_1 + x_3$$
$$x_3 = x_3$$

ignoring roundoff error.

Conformability

invsym(*A*), invsym(*A*, *order*):

      A: $n \times n$
   order: $1 \times k$ or $k \times 1$, $k \leq n$ (optional)
   result: $n \times n$

_invsym(*A*), _invsym(*A*, *order*):

      A: $n \times n$
   order: $1 \times k$ or $k \times 1$, $k \leq n$ (optional)
   returned:
      A: $n \times n$

Diagnostics

invsym(*A*), invsym(*A*, *order*), _invsym(*A*), and _invsym(*A*, *order*) assume that *A* is symmetric; they do not check. If *A* is nonsymmetric, they treat it as if it were symmetric and equal to its upper triangle.

invsym() and _invsym() return a result containing missing values if *A* contains missing values.

_invsym() aborts with error if *A* is a view. Both functions abort with argument-out-of-range error if *order* is specified and contains values less than 1, greater than rows(*A*), or the same value more than once.

invsym() and _invsym() return a matrix of zeros if *A* is not positive definite.

Also See

[M-5] **cholinv( )** — Symmetric, positive-definite matrix inversion

[M-5] **luinv( )** — Square matrix inversion

[M-5] **qrinv( )** — Generalized inverse of matrix via QR decomposition

[M-5] **pinv( )** — Moore–Penrose pseudoinverse

[M-5] **diag0cnt( )** — Count zeros on diagonal

[M-4] **matrix** — Matrix functions

[M-4] **solvers** — Functions to solve AX=B and to obtain A inverse

Title

[M-5] **invtokens( )** — Concatenate string rowvector into string scalar

Syntax

>*string scalar* invtokens(*string vector s*)

>*string scalar* invtokens(*string vector s*, *string scalar c*)

Description

invtokens(*s*) returns the elements of *s*, concatenated into a string scalar with the elements separated by spaces. invtokens(*s*) is equivalent to invtokens(*s*, " ").

invtokens(*s*, *c*) returns the elements of *s*, concatenated into a string scalar with the elements separated by *c*.

Remarks

invtokens(*s*) is the inverse of tokens() (see [M-5] **tokens( )**); invtokens() returns the string obtained by concatenating the elements of *s* into a space-separated list.

invtokens(*s*, *c*) places *c* between the elements of *s* even when the elements of *s* are equal to "". For instance,

```
: s = ("alpha", "", "gamma", "")
: invtokens(s, ";")
  alpha;;gamma;
```

To remove separators between empty elements, use select() (see [M-5] **select( )**) to remove the empty elements from *s* beforehand:

```
: s2 = select(s, strlen(s):>0)
: s2
           1        2
  1   alpha    gamma

: invtokens(s2, ";")
  alpha;gamma
```

Conformability

invtokens(*s*, *c*):
      s: $1 \times p$
      c: 1×1 (optional)
  result: 1×1

Diagnostics

If s is 1×0, invtokens(s,c) returns "".

Also See

[M-5] **tokens( )** — Obtain tokens from string

[M-4] **string** — String manipulation functions

Title

[M-5] **isdiagonal( )** — Whether matrix is diagonal

Syntax

>*real scalar* isdiagonal(*numeric matrix A*)

Description

isdiagonal(*A*) returns 1 if *A* has only zeros off the principal diagonal and returns 0 otherwise. isdiagonal( ) may be used with either real or complex matrices.

Remarks

See [M-5] **diag( )** for making diagonal matrices out of vectors or out of nondiagonal matrices; see [M-5] **diagonal( )** for extracting the diagonal of a matrix into a vector.

Conformability

isdiagonal(*A*):
>>*A*: $r \times c$
>*result*: 1×1

Diagnostics

isdiagonal(*A*) returns 1 if *A* is void.

Also See

[M-5] **diag( )** — Create diagonal matrix

[M-5] **diagonal( )** — Extract diagonal into column vector

[M-4] **utility** — Matrix utility functions

Title

[M-5] isfleeting( ) — Whether argument is temporary

Syntax

>   *real scalar*   `isfleeting`(*polymorphic matrix A*)

where *A* is an argument passed to your function.

Description

isfleeting(*A*) returns 1 if *A* was constructed for the sole purpose of passing to your function, and returns 0 otherwise. If an argument is fleeting, then you may change its contents and be assured that the caller will not care or even know.

Remarks

Let us assume that you have written function myfunc(*A*) that takes A: $r \times c$ and returns an $r \times c$ matrix. Just to fix ideas, we will pretend that the code for myfunc() reads

```
real matrix myfunc(real matrix A)
{
        real scalar     i
        real matrix     B

        B=A
        for (i=1; i<=rows(B); i++) B[i,i] = 1
        return(B)
}
```

Function myfunc(*A*) returns a matrix equal to *A*, but with ones along the diagonal. Now let's imagine myfunc() in use. A snippet of the code might read

```
...
C = A*myfunc(D)*C
...
```

Here D is passed to myfunc(), and the argument D is said not to be fleeting. Now consider another code snippet:

```
...
D = A*myfunc(D+E)*D
...
```

In this code snippet, the argument passed to myfunc() is D+E and that argument is fleeting. It is fleeting because it was constructed for the sole purpose of being passed to myfunc(), and once myfunc() concludes, the matrix containing D+E will be discarded.

Arguments that are fleeting can be reused to save memory.

Look carefully at the code for `myfunc()`. It makes a copy of the matrix that it was passed. It did that to avoid damaging the matrix that it was passed. Making that copy, however, is unnecessary if the argument received was fleeting, because damaging something that would have been discarded anyway does not matter. Had we not made the copy, we would have saved not only computer time but also memory. Function `myfunc()` could be recoded to read

```
real matrix myfunc(real matrix A)
{
        real scalar    i
        real matrix    B

        if (isfleeting(A)) {
                for (i=1; i<=rows(A); i++) A[i,i] = 1
                return(A)
        }
        B=A
        for (i=1; i<=rows(B); i++) B[i,i] = 1
        return(B)
}
```

Here we wrote separate code for the fleeting and nonfleeting cases. That is not always necessary. We could use a pointer here to combine the two code blocks:

```
real matrix myfunc(real matrix A)
{
        real scalar    i
        real matrix    B
        pointer scalar p

        if (isfleeting(A)) p = &A
        else {
                B = A
                p = &B
        }
        for (i=1; i<=rows(*p); i++) (*p)[i,i] = 1
        return(*p)
}
```

Many official library functions come in two varieties: $\_foo(A, \ldots)$, which replaces A with the calculated result, and $foo(A, \ldots)$, which returns the result leaving A unmodified. Invariably, the code for `foo()` reads

```
function foo(A, ...)
{
        matrix B

        if (isfleeting(A)) {
                _foo(A, ...)
                return(A)
        }
        _foo(B=A, ...)
        return(B)
}
```

This makes function foo() whoppingly efficient. If foo() is called with a temporary argument—an argument that could be modified without the caller being aware of it—then no extra copy of the matrix is ever made.

Conformability

```
isfleeting(A):
         A:     r × c
    result:    1 × 1
```

Diagnostics

isfleeting(A) returns 1 if A is fleeting and not a view. The value returned is indeterminate if A is not an argument of the function in which it appears, and therefore the value of isfleeting() is also indeterminate when used interactively.

Also See

[M-4] **programming** — Programming functions

Title

[M-5] **isreal( )** — Storage type of matrix

Syntax

>*real scalar*   `isreal(`*transmorphic matrix X*`)`
>
>*real scalar*   `iscomplex(`*transmorphic matrix X*`)`
>
>*real scalar*   `isstring(`*transmorphic matrix X*`)`
>
>*real scalar*   `ispointer(`*transmorphic matrix X*`)`

Description

`isreal(X)` returns 1 if X is a `real` and returns 0 otherwise.

`iscomplex(X)` returns 1 if X is a `complex` and returns 0 otherwise.

`isstring(X)` returns 1 if X is a `string` and returns 0 otherwise.

`ispointer(X)` returns 1 if X is a `pointer` and returns 0 otherwise.

Remarks

These functions base their results on storage type. `isreal()` is not the way to check whether a number is real, since it might be stored as a complex and yet still be a real number, such as $2 + 0i$. To determine whether x is real, you want to use `isrealvalues(X)`; see [M-5] **isrealvalues( )**.

Conformability

`isreal(X)`, `iscomplex(X)`, `isstring(X)`, `ispointer(X)`:
>   X: $r \times c$
>   result: 1×1

Diagnostics

These functions return 1 or 0; they cannot fail.

Also See

[M-5] **isrealvalues( )** — Whether matrix contains only real values

[M-5] **eltype( )** — Element type and organizational type of object

[M-4] **utility** — Matrix utility functions

Title

[M-5] **isrealvalues( )** — Whether matrix contains only real values

Syntax

>   *real scalar*   isrealvalues(*numeric matrix X*)

Description

isrealvalues(*X*) returns 1 if *X* is of storage type real or if *X* is of storage type complex and all elements are real or missing; 0 is returned otherwise.

Remarks

isrealvalues(*X*) is logically equivalent to *X*==Re(*X*) but is significantly faster and requires less memory.

Conformability

```
isrealvalues(X):
        X:      r × c
   result:      1 × 1
```

Diagnostics

isrealvalues(*X*) returns 1 if *X* is void and complex.

Also See

[M-5] **isreal( )** — Storage type of matrix

[M-4] **utility** — Matrix utility functions

Title

[M-5] **issymmetric( )** — Whether matrix is symmetric (Hermitian)

Syntax

>*real scalar*   issymmetric(*transmorphic matrix A*)

>*real scalar*   issymmetriconly(*transmorphic matrix A*)

Description

issymmetric(*A*) returns 1 if *A*==*A*′ and returns 0 otherwise. (Also see mreldifsym() in [M-5] **reldif( )**.)

issymmetriconly(*A*) returns 1 if *A*==transposeonly(*A*) and returns 0 otherwise.

Remarks

issymmetric(*A*) and issymmetriconly(*A*) return the same result except when *A* is complex.

In the complex case, issymmetric(*A*) returns 1 if *A* is equal to its conjugate transpose, i.e., if *A* is Hermitian, which is the complex analog of symmetric. *A* is symmetric (Hermitian) if its off-diagonal elements are conjugates of each other and its diagonal elements are real.

issymmetriconly(*A*), on the other hand, uses the mechanical definition of symmetry: *A* is symmetriconly [*sic*] if its off-diagonal elements are equal. issymmetriconly( ) is uninteresting, mathematically speaking, but can be useful in certain data-management programming situations.

Conformability

issymmetric(*A*), issymmetriconly(*A*):
>>*A*: $r \times c$
>>*result*: 1×1

Diagnostics

issymmetric(*A*) returns 0 if *A* is not square. If *A* is 0×0, it is symmetric.

issymmetriconly(*A*) returns 0 if *A* is not square. If *A* is 0×0, it is symmetriconly.

Charles Hermite (1822–1901) was born in Dieuze in eastern France and early showed an aptitude for mathematics, publishing two papers before entering university. He started studying at the Ecole Polytechnique but left after 1 year because of difficulties from a defect in his right foot. However, Hermite was soon appointed to a post at the same institution and later at the Sorbonne. He made outstanding contributions to number theory, algebra, and especially analysis, and he published the first proof that e is transcendental. Hermite's name carries on in Hermite polynomials, Hermite's differential equation, Hermite's formula of interpolation, and Hermitian matrices.

Reference

James, I. 2002. *Remarkable Mathematicians: From Euler to von Neumann.* Cambridge: Cambridge University Press.

Also See

[M-5] **makesymmetric( )** — Make square matrix symmetric (Hermitian)

[M-5] **reldif( )** — Relative/absolute difference

[M-4] **utility** — Matrix utility functions

Title

[M-5] **isview( )** — Whether matrix is view

Syntax

>   *real scalar*   isview(*transmorphic matrix X*)

Description

isview(*X*) returns 1 if *X* is a view and 0 otherwise.

See [M-6] **glossary** for a definition of a view.

Remarks

View matrices are created by st_view(); see [M-5] **st_view( )**.

Conformability

isview(*X*):
> *X*: $r \times c$
> *result*: 1×1

Diagnostics

None.

Also See

[M-5] **eltype( )** — Element type and organizational type of object

[M-4] **utility** — Matrix utility functions

Title

[M-5] J() — Matrix of constants

Syntax

>   *transmorphic matrix*   J(*real scalar r, real scalar c, scalar val*)

Description

J(*r, c, val*) returns an $r \times c$ matrix with each element equal to *val*.

Remarks

J() must be typed in uppercase.

J() can create any type of matrix:

function	returns
J(2, 3, 4)	2×3 real matrix, each element = 4
J(2, 3, 4+5i)	2×3 complex matrix, each element = 4 + 5i
J(2, 3, "hi")	2×3 string matrix, each element = "hi"
J(2, 3, &x)	2×3 pointer matrix, each element = address of x

Also, J() can create void matrices:

J(0, 0, .)	0×0 real
J(0, 1, .)	0×1 real
J(1, 0, .)	1×0 real
J(0, 0, 1i)	0×0 complex
J(0, 1, 1i)	0×1 complex
J(1, 0, 1i)	1×0 complex
J(0, 0, "")	0×0 string
J(0, 1, "")	0×1 string
J(1, 0, "")	1×0 string
J(0, 0, NULL)	0×0 pointer
J(0, 1, NULL)	0×1 pointer
J(1, 0, NULL)	1×0 pointer

See [M-2] **void**.

Conformability

J(*r*, *c*, *val*):
        r: 1×1
        c: 1×1
      val: 1×1
    result: $r \times c$

Diagnostics

J(*r*, *c*, *val*) aborts with error if $r < 0$ or $c < 0$, or if $r \geq$. or $c \geq$ .. Arguments *r* and *c* are interpreted as trunc(*r*) and trunc(*c*).

Also See

[M-5] **missingof( )** — Appropriate missing value

[M-4] **standard** — Functions to create standard matrices

Title

[M-5] **liststruct( )** — List structure's contents

Syntax

> *void* `liststruct`(*struct whatever matrix x*)

Description

`liststruct()` lists *x*'s contents, where *x* is an instance of structure *whatever*.

Remarks

`liststruct()` is often useful in debugging.

The dimension and type of all elements are listed, and the values of scalars are also shown.

Conformability

`liststruct(x)`:
    x: $r \times c$
    result: *void*

Diagnostics

None.

Also See

[M-2] **struct** — Structures

[M-4] **io** — I/O functions

Title

[M-5] **logit( )** — Log odds and complementary log-log

Syntax

real matrix logit(*real matrix X*)

real matrix invlogit(*real matrix X*)

real matrix cloglog(*real matrix X*)

real matrix invcloglog(*real matrix X*)

Description

logit(*X*) returns the log of the odds ratio of the elements of X, $\ln\{x/(1-x)\}$.

invlogit(*X*) returns the inverse of the logit() of the elements of X, $\exp(x)/\{1+\exp(x)\}$.

cloglog(*X*) returns the complementary log-log of the elements of X, $\ln\{-\ln(1-x)\}$.

invcloglog(*X*) returns the elementwise inverse of cloglog() of the elements of X, $1 - \exp\{-\exp(x)\}$.

Conformability

All functions return a matrix of the same dimension as input containing element-by-element calculated results.

Diagnostics

logit(*X*) and cloglog(*X*) return missing when $x \leq 0$ or $x \geq 1$.

Also See

[M-4] **statistical** — Statistical functions

Title

[M-5] **lowertriangle( )** — Extract lower or upper triangle

Syntax

> *numeric matrix*   lowertriangle(*numeric matrix A* [, *numeric scalar d*])
>
> *numeric matrix*   uppertriangle(*numeric matrix A* [, *numeric scalar d*])
>
> *void*   _lowertriangle(*numeric matrix A* [, *numeric scalar d*])
>
> *void*   _uppertriangle(*numeric matrix A* [, *numeric scalar d*])

where argument *d* is optional.

Description

lowertriangle() returns the lower triangle of *A*.

uppertriangle() returns the upper triangle of *A*.

_lowertriangle() replaces *A* with its lower triangle.

_uppertriangle() replaces *A* with its upper triangle.

Remarks

Remarks are presented under the following headings:

> *Optional argument d*
> *Nonsquare matrices*

Optional argument d

Optional argument *d* specifies the treatment of the diagonal. Specifying *d*>=., or not specifying *d* at all, means no special treatment; if

$$A = \begin{bmatrix} 1 & 2 & 3 \\ 4 & 5 & 6 \\ 7 & 8 & 9 \end{bmatrix}$$

then

$$\text{lowertriangle}(A, .) = \begin{bmatrix} 1 & 0 & 0 \\ 4 & 5 & 0 \\ 7 & 8 & 9 \end{bmatrix}$$

If a nonmissing value is specified for *d*, however, that value is substituted for each element of the diagonal, for example,

$$\texttt{lowertriangle}(A, 1) = \begin{bmatrix} 1 & 0 & 0 \\ 4 & 1 & 0 \\ 7 & 8 & 1 \end{bmatrix}$$

Nonsquare matrices

lowertriangle() and uppertriangle() may be used with nonsquare matrices. If

$$A = \begin{bmatrix} 1 & 2 & 3 & 4 \\ 5 & 6 & 7 & 8 \\ 9 & 10 & 11 & 12 \end{bmatrix}$$

then

$$\texttt{lowertriangle}(A) = \begin{bmatrix} 1 & 0 & 0 \\ 5 & 6 & 0 \\ 9 & 10 & 11 \end{bmatrix}$$

and

$$\texttt{uppertriangle}(A) = \begin{bmatrix} 1 & 2 & 3 & 4 \\ 0 & 6 & 7 & 8 \\ 0 & 0 & 11 & 12 \end{bmatrix}$$

_lowertriangle() and _uppertriangle(), however, may not be used with nonsquare matrices.

Conformability

lowertriangle(*A*, *d*):
    A: $r \times c$
    d: 1×1 (optional)
    result: $r \times \min(r, c)$

uppertriangle(*A*, *d*):
    A: $r \times c$
    d: 1×1 (optional)
    result: $\min(r, c) \times c$

_lowertriangle(*A*, *d*), _uppertriangle(*A*, *d*):

    input:
        A: $n \times n$
        d: 1×1 (optional)

    output:
        A: $n \times n$

Diagnostics

None.

Also See

[M-4] **manipulation** — Matrix manipulation

Title

[M-5] **lud( )** — LU decomposition

Syntax

>   *void* lud(*numeric matrix A, L, U, p*)

>   *void* _lud(*numeric matrix L, U, p*)

>   *void* _lud_la(*numeric matrix A, r*)

where

1. *A* may be real or complex and need not be square.
2. The types of *L*, *U*, and *p* are irrelevant; results are returned there.

Description

lud(*A, L, U, p*) returns the LU decomposition (with partial pivoting) of *A* in *L* and *U* along with a permutation vector *p*. The returned results are such that *A=L[p,.]*U* up to roundoff error.

_lud(*L, U, p*) is similar to lud(), except that it conserves memory. The matrix to be decomposed is passed in *L*, and the same storage location is overwritten with the calculated *L* matrix.

_lud_la(*A, r*) is the [M-1] **LAPACK** routine that the above functions use to calculate the LU decomposition. See *LAPACK routine* below.

Remarks

Remarks are presented under the following headings:

>   *LU decomposition*
>   *LAPACK routine*

LU decomposition

The LU decomposition of matrix *A* can be written

$$P'A = LU$$

where P' is a permutation matrix that permutes the rows of *A*. *L* is lower triangular and *U* is upper triangular. The decomposition can also be written

$$A = PLU$$

because, given that *P* is a permutation matrix, $P^{-1} = P'$.

Rather than returning P directly, returned is p corresponding to P. Lowercase p is a column vector that contains the subscripts of the rows in the desired order. That is,

$$PL = L[p,.]$$

The advantage of this is that p requires less memory than P and the reorganization, should it be desired, can be performed more quickly; see [M-1] **permutation**. In any case, the formula defining the LU decomposition can be written

$$A = L[p,.]*U$$

One can also write

$$B = LU, \text{ where } B[p,.] = A$$

LAPACK routine

_lud_la(A, r) is the interface into the [M-1] **LAPACK** routines that the above functions use to calculate the LU decomposition. You may use it directly if you wish.

Matrix A is decomposed, and the decomposition is placed back in A. U is stored in the upper triangle (including the diagonal) of A. L is stored in the lower triangle of A, and it is understood that L is supposed to have ones on its diagonal. r is a column vector recording the row swaps that account for the pivoting. This is the same information as stored in p, but in a different format.

r records the row swaps to be made. For instance, r = (1\2\2) means that (start at the end) the third row is to be swapped with the second row, then the second row is to stay where it is, and finally the first row is to stay where it is. r can be converted into p by the following logic:

```
p = 1::rows(r)
for (i=rows(r); i>=1; i--) {
        hold = p[i]
        p[i] = p[r[i]]
        p[r[i]] = hold
}
```

Conformability

lud(A, L, U, p):

    input:
        A: $r \times c$
    output:
        L: $r \times m$, $m = \min(r,c)$
        U: $m \times c$
        p: $r \times 1$

_lud(L, U, p):
   input:
                L: $r \times c$
   output:
                L: $r \times m$, $m = \min(r, c)$
                U: $m \times c$
                p: $r \times 1$

_lud_la(A, r):
   input:
                L: $r \times c$
   output:
                L: $r \times c$
                p: $r \times 1$

Diagnostics

lud(A, L, U, p) returns missing results if A contains missing values; L will have missing values below the diagonal, 1s on the diagonal, and 0s above the diagonal; U will have missing values on and above the diagonal and 0s below. Thus if there are missing values, U[1,1] will contain missing.

_lud(L, U, p) sets L and U as described above if A contains missing values.

_lud_la(A, r) aborts with error if A is a view.

Also See

[M-5] **det( )** — Determinant of matrix

[M-5] **lusolve( )** — Solve AX=B for X using LU decomposition

[M-4] **matrix** — Matrix functions

Title

[M-5] **luinv( )** — Square matrix inversion

Syntax

>   *numeric matrix*   `luinv(`*numeric matrix A*`)`
>
>   *numeric matrix*   `luinv(`*numeric matrix A, real scalar tol*`)`
>
>   *void*   `_luinv(`*numeric matrix A*`)`
>
>   *void*   `_luinv(`*numeric matrix A, real scalar tol*`)`
>
>   *real scalar*   `_luinv_la(`*numeric matrix A, real scalar tol*`)`

Description

`luinv(`*A*`)` and `luinv(`*A, tol*`)` return the inverse of real or complex, square matrix *A*.

`_luinv(`*A*`)` and `_luinv(`*A, tol*`)` do the same thing except that, rather than returning the inverse matrix, they overwrite the original matrix *A* with the inverse.

In all cases, optional argument *tol* specifies the tolerance for determining singularity; see *Remarks* below.

`_luinv_la(`*A, tol*`)` is the interface to the [M-1] **LAPACK** routines that do the work.

Remarks

These routines calculate the inverse of *A*. The inverse matrix A^{-1} of *A* satisfies the conditions

$$AA^{-1} = I$$
$$A^{-1}A = I$$

A is required to be square and of full rank. See [M-5] **qrinv( )** and [M-5] **pinv( )** for generalized inverses of nonsquare or rank-deficient matrices. See [M-5] **invsym( )** for inversion of real, symmetric matrices.

`luinv(`*A*`)` is logically equivalent to `lusolve(`*A*`, I(rows(`*A*`)))`; see [M-5] **lusolve( )** for details and for use of the optional *tol* argument.

Conformability

`luinv(`*A, tol*`)`:
>   *A*: $n \times n$
>   *tol*: 1×1 (optional)
>   *result*: $n \times n$

_luinv(A, tol):

  input:

	A:	$n \times n$	
	tol:	1×1	(optional)

  output:

	A:	$n \times n$

_luinv_la(A, tol):

  input:

	A:	$n \times n$
	tol:	1×1

  output:

	A:	$n \times n$
	result:	1×1

Diagnostics

The inverse returned by these functions is real if A is real and is complex if A is complex. If you use these functions with a singular matrix, returned will be a matrix of missing values. The determination of singularity is made relative to *tol*. See *Tolerance* under *Remarks* in [M-5] **lusolve( )** for details.

luinv(A) and _luinv(A) return a matrix containing missing if A contains missing values.

_luinv(A) aborts with error if A is a view.

_luinv_la(A, *tol*) should not be used directly; use _luinv( ).

See [M-5] **lusolve( )** and [M-1] **tolerance** for information on the optional *tol* argument.

Also See

[M-5] **invsym( )** — Symmetric real matrix inversion

[M-5] **cholinv( )** — Symmetric, positive-definite matrix inversion

[M-5] **qrinv( )** — Generalized inverse of matrix via QR decomposition

[M-5] **pinv( )** — Moore–Penrose pseudoinverse

[M-5] **lusolve( )** — Solve AX=B for X using LU decomposition

[M-5] **lud( )** — LU decomposition

[M-4] **matrix** — Matrix functions

[M-4] **solvers** — Functions to solve AX=B and to obtain A inverse

Title

[M-5] **lusolve( )** — Solve AX=B for X using LU decomposition

Syntax

numeric matrix	lusolve(*numeric matrix A*, *numeric matrix B*)
numeric matrix	lusolve(*numeric matrix A*, *numeric matrix B*, *real scalar tol*)
void	_lusolve(*numeric matrix A*, *numeric matrix B*)
void	_lusolve(*numeric matrix A*, *numeric matrix B*, *real scalar tol*)
real scalar	_lusolve_la(*numeric matrix A*, *numeric matrix B*)
real scalar	_lusolve_la(*numeric matrix A*, *numeric matrix B*, *real scalar tol*)

Description

lusolve(*A*, *B*) solves *AX=B* and returns *X*. lusolve( ) returns a matrix of missing values if *A* is singular.

lusolve(*A*, *B*, *tol*) does the same thing but allows you to specify the tolerance for declaring that *A* is singular; see *Tolerance* under *Remarks* below.

_lusolve(*A*, *B*) and _lusolve(*A*, *B*, *tol*) do the same thing except that, rather than returning the solution *X*, they overwrite *B* with the solution and, in the process of making the calculation, they destroy the contents of *A*.

_lusolve_la(*A*, *B*) and _lusolve_la(*A*, *B*, *tol*) are the interfaces to the [M-1] **LAPACK** routines that do the work. They solve *AX=B* for *X*, returning the solution in *B* and, in the process, using as workspace (overwriting) *A*. The routines return 1 if *A* was singular and 0 otherwise. If *A* was singular, *B* is overwritten with a matrix of missing values.

Remarks

The above functions solve *AX=B* via LU decomposition and are accurate. An alternative is qrsolve( ) (see [M-5] **qrsolve( )**), which uses QR decomposition. The difference between the two solutions is not, practically speaking, accuracy. When *A* is of full rank, both routines return equivalent results, and the LU approach is quicker, using approximately $O(2/3n^3)$ operations rather than $O(4/3n^3)$, where *A* is $n \times n$.

The difference arises when *A* is singular. Then the LU-based routines documented here return missing values. The QR-based routines documented in [M-5] **qrsolve( )** return a generalized (least squares) solution.

For more information on LU and QR decomposition, see [M-5] **lud( )** and see [M-5] **qrd( )**.

Remarks are presented under the following headings:

>*Derivation*
>*Relationship to inversion*
>*Tolerance*

Derivation

We wish to solve for X

$$AX = B \qquad (1)$$

Perform LU decomposition on A so that we have $A = PLU$. Then (1) can be written

$$PLUX = B$$

or, premultiplying by P' and remembering that $P'P = I$,

$$LUX = P'B \qquad (2)$$

Define

$$Z = UX \qquad (3)$$

Then (2) can be rewritten

$$LZ = P'B \qquad (4)$$

It is easy to solve (4) for Z because L is a lower-triangular matrix. Once Z is known, it is easy to solve (3) for X because U is upper triangular.

Relationship to inversion

Another way to solve

$$AX = B$$

is to obtain A^{-1} and then calculate

$$X = A^{-1}B$$

It is, however, better to solve $AX = B$ directly because fewer numerical operations are required, and the result is therefore more accurate and obtained in less computer time.

Indeed, rather than thinking about how solving a system of equations can be implemented via inversion, it is more productive to think about how inversion can be implemented via solving a system of equations. Obtaining A^{-1} amounts to solving

$$AX = I$$

Thus lusolve() (or any other solve routine) can be used to obtain inverses. The inverse of A can be obtained by coding

        : Ainv = lusolve(A, I(rows(A)))

In fact, we provide luinv() (see [M-5] **luinv()**) for obtaining inverses via LU decomposition, but luinv() amounts to making the above calculation, although a little memory is saved because the matrix I is never constructed.

Hence, everything said about lusolve() applies equally to luinv().

Tolerance

The default tolerance used is

$$eta = (1e-13)*trace(abs(U))/n$$

where U is the upper-triangular matrix of the LU decomposition of A: $n \times n$. A is declared to be singular if any diagonal element of U is less than or equal to *eta*.

If you specify *tol* > 0, the value you specify is used to multiply *eta*. You may instead specify *tol* ≤ 0, and then the negative of the value you specify is used in place of *eta*; see [M-1] **tolerance**.

So why not specify *tol* $= 0$? You do not want to do that because, as matrices become close to being singular, results can become inaccurate. Here is an example:

```
: uniformseed(12345)
: A = lowertriangle(uniform(4,4))
: A[3,3] = 1e-15
: trux = uniform(4,1)
: b = A*trux
: /* the above created an Ax=b problem, and we have placed the true
>    value of x in trux.  We now obtain the solution via lusolve( )
>    and compare trux with the value obtained:
> */
: x = lusolve(A, b, 0)
: trux, x
```

1	.7997150919	.7997150919
2	.9102488109	.9102488109
3	.442547889	.3230683012
4	.756650276	.8673158447

← The discussed numerical instability can cause this output to vary a little across different computers

We would like to see the second column being nearly equal to the first—the estimated x being nearly equal to the true x—but there are substantial differences.

Even though the difference between x and xtrue is substantial, the difference between them is small in the prediction space:

```
: A*trux-b, A*x-b
```

	1	2
1	0	-2.77556e-17
2	0	0
3	0	-1.11022e-16
4	0	0

What made this problem so difficult was the line A[3,3] = 1e-15. Remove that and you would find that the maximum difference between x and trux would be 2.22045e−16.

The degree to which the residuals A*x-b are a reliable measure of the accuracy of x depends on the condition number of the matrix, which can be obtained by [M-5] **cond( )**, which for A, is 3.2947e+15. If the matrix is well conditioned, small residuals imply an accurate solution for x. If the matrix is ill conditioned, small residuals are not a reliable indicator of accuracy.

Another way to check the accuracy of x is to set *tol* $= 0$ and to see how well x could be obtained were $x = $ x:

```
: x  = lusolve(A, b, 0)
: x2 = lusolve(A, A*x, 0)
```

If x and x2 are virtually the same, then you can safely assume that x is the result of a numerically accurate calculation. You might compare x and x2 with mreldif(x2,x); see [M-5] **reldif**(). In our example, mreldif(x2,x) is .03, a large difference.

If A is ill conditioned, then small changes in A or B can lead to radical differences in the solution for X.

Conformability

lusolve(*A*, *B*, *tol*):

  input:
      A: $n \times n$
      B: $n \times k$
      tol: 1×1 (optional)

  output:
      result: $n \times k$

_lusolve(*A*, *B*, *tol*):

  input:
      A: $n \times n$
      B: $n \times k$
      tol: 1×1 (optional)

  output:
      A: 0×0
      B: $n \times k$

_lusolve_la(*A*, *B*, *tol*):

  input:
      A: $n \times n$
      B: $n \times k$
      tol: 1×1 (optional)

  output:
      A: 0×0
      B: $n \times k$
      result: 1×1

Diagnostics

lusolve(*A*, *B*, ...), _lusolve(*A*, *B*, ...), and _lusolve_la(*A*, *B*, ...) return a result containing missing if *A* or *B* contain missing values. The functions return a result containing all missing values if *A* is singular.

_lusolve(*A*, *B*, ...) and _lusolve_la(*A*, *B*, ...) abort with error if *A* or *B* is a view.

_lusolve_la(*A*, *B*, ...) should not be used directly; use _lusolve().

Also See

[M-5] **luinv( )** — Square matrix inversion

[M-5] **lud( )** — LU decomposition

[M-5] **solvelower( )** — Solve AX=B for X, A triangular

[M-5] **cholsolve( )** — Solve AX=B for X using Cholesky decomposition

[M-5] **qrsolve( )** — Solve AX=B for X using QR decomposition

[M-5] **svsolve( )** — Solve AX=B for X using singular value decomposition

[M-4] **matrix** — Matrix functions

[M-4] **solvers** — Functions to solve AX=B and to obtain A inverse

Title

> **[M-5] makesymmetric( )** — Make square matrix symmetric (Hermitian)

Syntax

>   *numeric matrix*   makesymmetric(*numeric matrix A*)
>
>   *void*   _makesymmetric(*numeric matrix A*)

Description

makesymmetric(*A*) returns *A* made into a symmetric (Hermitian) matrix by reflecting elements below the diagonal.

_makesymmetric(*A*) does the same thing but stores the result back in *A*.

Remarks

If *A* is real, elements below the diagonal are copied into their corresponding above-the-diagonal position.

If *A* is complex, the conjugate of the elements below the diagonal are copied into their corresponding above-the-diagonal positions, and the imaginary part of the diagonal is set to zero.

Whether *A* is real or complex, roundoff error can make matrix calculations that are supposed to produce symmetric matrices produce matrices that vary a little from symmetry, and makesymmetric() can be used to correct the situation.

Conformability

    makesymmetric(A):
            A: n × n
        result: n × n
    _makesymmetric(A):
            A: n × n

Diagnostics

makesymmetric(*A*) and _makesymmetric(*A*) abort with error if *A* is not square. Also, _makesymmetric() aborts with error if *A* is a view.

Also See

[M-5] **issymmetric( )** — Whether matrix is symmetric (Hermitian)

[M-4] **manipulation** — Matrix manipulation

Title

[M-5] **matexpsym( )** — Exponentiation and logarithms of symmetric matrices

Syntax

    numeric matrix   matexpsym(*numeric matrix A*)

    numeric matrix   matlogsym(*numeric matrix A*)

    void   _matexpsym(*numeric matrix A*)

    void   _matlogsym(*numeric matrix A*)

Description

matexpsym(*A*) returns the matrix exponential of the symmetric (Hermitian) matrix *A*.

matlogsym(*A*) returns the matrix natural logarithm of the symmetric (Hermitian) matrix *A*.

_matexpsym(*A*) and _matlogsym(*A*) do the same thing as matexpsym() and matlogsym(), but instead of returning the result, they store the result in *A*.

Remarks

Do not confuse matexpsym(*A*) with exp(*A*), nor matlogsym(*A*) with log(*A*).

matexpsym(2*matlogsym(*A*)) produces the same result as *A**A. exp() and log() return elementwise results.

Exponentiated results and logarithms are obtained by extracting the eigenvalues and eigenvectors of *A*, performing the operation on the eigenvalues, and then rebuilding the matrix. That is, first X and L are found such that

$$AX = X * \mathrm{diag}(L) \tag{1}$$

For symmetric (Hermitian) matrix *A*, X is orthogonal, meaning $X'X = XX' = I$. Thus

$$A = X * \mathrm{diag}(L) * X' \tag{2}$$

matexpsym(*A*) is then defined

$$A = X * \mathrm{diag}(\exp(L)) * X' \tag{3}$$

and matlogsym(*A*) is defined

$$A = X * \mathrm{diag}(\log(L)) * X' \tag{4}$$

(1) is obtained via symeigensystem(); see [M-5] **eigensystem( )**.

Conformability

matexpsym(*A*), matlogsym(*A*):
   A: $n \times n$
 result: $n \times n$

_matexpsym(*A*), _matlogsym(*A*):
 input:
   A: $n \times n$
 output:
   A: $n \times n$

Diagnostics

matexpsym(*A*), matlogsym(*A*), _matexpsym(*A*), and _matlogsym(*A*) return missing results if *A* contains missing values.

Also:

1. These functions do not check that *A* is symmetric or Hermitian. If *A* is a real matrix, only the lower triangle, including the diagonal, is used. If *A* is a complex matrix, only the lower triangle and the real parts of the diagonal elements are used.

2. These functions return a matrix of the same storage type as *A*.

   For symatlog(*A*), this means that if *A* is real and the result cannot be expressed as a real, a matrix of missing values is returned. If you want the generalized solution, code matlogsym(C(*A*)). This is the same rule as with scalars: log(-1) evaluates to missing, but log(C(-1)) is 3.14159265i.

3. These functions are guaranteed to return a matrix that is numerically symmetric, Hermitian, or symmetriconly if theory states that the matrix should be symmetric, Hermitian, or symmetriconly. See [M-5] **matpowersym()** for a discussion of this issue.

   For the functions here, real function exp(*x*) is defined for all real values of *x* (ignoring overflow), and thus the matrix returned by matexpsym() will be symmetric (Hermitian).

   The same is not true for matlogsym(). log(*x*) is not defined for $x = 0$, so if any of the eigenvalues of *A* are 0 or very small, a matrix of missing values will result. Also, log(*x*) is complex for $x < 0$, and thus if any of the eigenvalues are negative, the resulting matrix will be (1) missing if *A* is real stored as real, (2) symmetriconly if *A* contains reals stored as complex, and (3) general if *A* is complex.

Also See

[M-5] **matpowersym()** — Powers of a symmetric matrix

[M-5] **eigensystem()** — Eigenvalues and eigenvectors

[M-4] **matrix** — Matrix functions

Title

[M-5] **matpowersym( )** — Powers of a symmetric matrix

Syntax

> *numeric matrix*   matpowersym(*numeric matrix A, real scalar p*)
>
> *void*   _matpowersym(*numeric matrix A, real scalar p*)

Description

matpowersym(*A*, *p*) returns A^p for symmetric matrix or Hermitian matrix *A*. The matrix returned is real if *A* is real and complex is *A* is complex.

_matpowersym(*A*, *p*) does the same thing, but instead of returning the result, it stores the result in *A*.

Remarks

Do not confuse matpowersym(*A*, *p*) and *A*:^*p*. If *p*==2, the first returns *A*∗*A* and the second returns *A* with each element squared.

Powers can be positive, negative, integer, or noninteger. Thus matpowersym(*A*, .5) is a way to find the square-root matrix *R* such that *R*∗*R*==*A*, and matpowersym(*A*, -1) is a way to find the inverse. For inversion, you could obtain the result more quickly using other routines.

Powers are obtained by extracting the eigenvalues and eigenvectors of *A*, raising the eigenvalues to the specified power, and then rebuilding the matrix. That is, first *X* and *L* are found such that

$$AX = X * \mathrm{diag}(L) \qquad (1)$$

For symmetric (Hermitian) matrix *A*, *X* is orthogonal, meaning $X'X = XX' = I$. Thus

$$A = X * \mathrm{diag}(L) X' \qquad (2)$$

A^p is then defined

$$A = X * \mathrm{diag}(L\text{:}\hat{\ }p) * X' \qquad (3)$$

(1) is obtained via symeigensystem(); see [M-5] **eigensystem( )**.

Conformability

matpowersym(*A*, *p*):

      A: $n \times n$
      p: 1×1
 result: $n \times n$

_matpowersym(*A*, *p*):

  input:
        A: $n \times n$
        p: 1×1

  output:
        A: $n \times n$

Diagnostics

matpowersym(*A*, *p*) and _matpowersym(*A*, *p*) return missing results if *A* contains missing values.

Also:

1. These functions do not check that *A* is symmetric or Hermitian. If *A* is a real matrix, only the lower triangle, including the diagonal, is used. If *A* is a complex matrix, only the lower triangle and the real parts of the diagonal elements are used.

2. These functions return a matrix of the same storage type as *A*. That means that if *A* is real and A^p cannot be expressed as a real, a matrix of missing values is returned. If you want the generalized solution, code matpowersym(C(*A*), *p*). This is the same rule as with scalars: (-1)^.5 is missing, but C(-1)^.5 is 1i.

3. These functions are guaranteed to return a matrix that is numerically symmetric, Hermitian, or symmetriconly if theory states that the matrix should be symmetric, Hermitian, or symmetriconly.

   Concerning theory, the returned result is not necessarily symmetric (Hermitian). The eigenvalues *L* of a symmetric (Hermitian) matrix are real. If L:^p are real, then the returned matrix will be symmetric (Hermitian), but otherwise, it will not. Think of a negative eigenvalue and *p* = .5: this results in a complex eigenvalue for A^p. Then if the original matrix was real (the eigenvectors were real), the resulting matrix will be symmetriconly. If the original matrix was complex (the eigenvectors were complex), the resulting matrix will have no special structure.

Also See

[M-5] **matexpsym( )** — Exponentiation and logarithms of symmetric matrices

[M-5] **eigensystem( )** — Eigenvalues and eigenvectors

[M-4] **matrix** — Matrix functions

Title

> [M-5] **mean( )** — Means, variances, and correlations

Syntax

>   *real rowvector*   mean(*X*[, *w*])
>
>   *real matrix*   variance(*X*[, *w*])
>   *real matrix*   quadvariance(*X*[, *w*])
>
>   *real matrix*   meanvariance(*X*[, *w*])
>   *real matrix*   quadmeanvariance(*X*[, *w*])
>
>   *real matrix*   correlation(*X*[, *w*])
>   *real matrix*   quadcorrelation(*X*[, *w*])

where

>   *X*:   *real matrix X* (rows are observations, columns are variables)
>   *w*:   *real colvector w* and is optional

Description

mean(*X*, *w*) returns the weighted-or-unweighted column means of data matrix *X*. mean() uses quad precision in forming sums and so is very accurate.

variance(*X*, *w*) returns the weighted-or-unweighted variance matrix of *X*. In the calculation, means are removed and those means are calculated in quad precision, but quad precision is not otherwise used.

quadvariance(*X*, *w*) returns the weighted-or-unweighted variance matrix of *X*. Calculation is highly accurate; quad precision is used in forming all sums.

meanvariance(*X*, *w*) returns mean(*X*,*w*)\variance(*X*,*w*).

quadmeanvariance(*X*, *w*) returns mean(*X*,*w*)\quadvariance(*X*,*w*).

correlation(*X*, *w*) returns the weighted-or-unweighted correlation matrix of *X*. correlation() obtains the variance matrix from variance().

quadcorrelation(*X*, *w*) returns the weighted-or-unweighted correlation matrix of *X*. quadcorrelation() obtains the variance matrix from quadvariance().

In all cases, *w* specifies the weight. Omit *w*, or specify *w* as 1 to obtain unweighted means.

In all cases, rows of *X* or *w* that contain missing values are omitted from the calculation, which amounts to casewise deletion.

Remarks

1. There is no quadmean() function because mean(), in fact, is quadmean(). The fact that mean() defaults to the quad-precision calculation reflects our judgment that the extra computational cost in computing means in quad precision is typically justified.

2. The fact that variance() and correlation() do not default to using quad precision for their calculations reflects our judgment that the extra computational cost is typically not justified. The emphasis on this last sentence is on the word typically.

   It is easier to justify means in part because the extra computational cost is less: there are only k means but $k(k+1)/2$ variances and covariances.

3. If you need just the mean or just the variance matrix, call mean() or variance() (or quadvariance()). If you need both, there is a CPU-time savings to be had by calling meanvariance() instead of the two functions separately (or quadmeanvariance() instead of calling mean() and quadvariance()).

   The savings is not great—one mean() calculation is saved—but the greater rows(X), the greater the savings.

   Upon getting back the combined result, it can be efficiently broken into its components via

          : var   = meanvariance(X)
          : means = var[1,.]
          : var   = var[|2,1 .,.|]

Conformability

mean(X, w):
       X: $n \times k$
       w: $n \times 1$ or 1×1 (optional, $w=1$ assumed)
       result: $1 \times k$

variance(X, w), quadvariance(X, w), correlation(X, w), quadcorrelation(X, w):
       X: $n \times k$
       w: $n \times 1$ or 1×1 (optional, $w=1$ assumed)
       result: $k \times k$

meanvariance(X, w), quadmeanvariance(X, w):
       X: $n \times k$
       w: $n \times 1$ or 1×1 (optional, $w=1$ assumed)
       result: $(k+1) \times k$

Diagnostics

All functions omit from the calculation rows that contain missing values unless all rows contain missing values. Then the returned result contains all missing values.

Also See

[M-4] **stata** — Stata interface functions

[M-4] **statistical** — Statistical functions

Title

[M-5] **mindouble( )** — Minimum and maximum nonmissing value

Syntax

>   *real scalar*   mindouble()
>
>   *real scalar*   maxdouble()
>
>   *real scalar*   smallestdouble()

Description

mindouble() returns the largest negative, nonmissing value.

maxdouble() returns the largest positive, nonmissing value.

smallestdouble() returns the smallest value of e, $e > 0$. The largest value of e, $e < 0$ is $-$smallestdouble().

Remarks

All nonmissing values x fulfill mindouble() $\leq x \leq$ maxdouble().

All missing values m fulfill $m >$ maxdouble()

Missing values also fulfill $m \geq$.

On all computers on which Stata and Mata are currently implemented, which are computers following IEEE standards:

function	exact hexadecimal value	approximate decimal value
mindouble()	-1.ffffffffffffX+3ff	$-1.7977e+308$
smallestdouble()	+1.ffffffffffffX-3fe	$4.4501e-308$
epsilon(1)	+1.0000000000000X-034	$2.2205e-016$
maxdouble()	+1.ffffffffffffX+3fe	$8.9885e+307$

The smallest missing value (. < .a < $\cdots$ < .z) is +1.0000000000000X+3ff.

Do not confuse smallestdouble() with the more interesting value epsilon(1). smallestdouble() is the smallest value of e, $e > 0$. epsilon(1) is the smallest value of e, $e + 1 > 1$; see [M-5] **epsilon( )**.

Conformability

mindouble(), maxdouble(), smallestdouble():
    result: 1×1

Diagnostics

None.

Also See

[M-5] **epsilon( )** — Unit roundoff error (machine precision)

[M-4] **utility** — Matrix utility functions

Title

[M-5] **minindex( )** — Indices of minimums and maximums

Syntax

> *void* minindex(*real vector v, real scalar k, i, w*)
>
> *void* maxindex(*real vector v, real scalar k, i, w*)

Results are returned in *i* and *w*.

> *i* will be a *real colvector*.
>
> *w* will be a $K \times 2$ *real matrix*, $K \leq |k|$.

Description

minindex(*v, k, i, w*) returns in *i* and *w* the indices of the *k* minimums of *v*.

maxindex(*v, k, i, w*) does the same, except that it returns the indices of the *k* maximums.

minindex( ) may be called with $k < 0$; it is then equivalent to maxindex( ).

maxindex( ) may be called with $k < 0$; it is then equivalent to minindex( ).

Remarks

Remarks are presented under the following headings:

> Use of functions when v has all unique values
> Use of functions when v has repeated (tied) values
> Summary

Remarks are cast in terms of minindex( ) but apply equally to maxindex( ).

Use of functions when v has all unique values

Consider $v = (3, 1, 5, 7, 6)$.

1. minindex(v, 1, i, w) returns i = 2, which means that v[2] is the minimum value in v.

2. minindex(v, 2, i, w) returns i = (2, 1)', which means that v[2] is the minimum value of v and that v[1] is the second minimum.

   ...

5. minindex(v, 5, i, w) returns i = (2, 1, 3, 5, 4)', which means that the ordered values in v are v[2], v[1], v[3], v[5], and v[4].

6. minindex(v, 6, i, w), minindex(v, 7, i, w), and so on, return the same as (5), because there are only five minimums in a five-element vector.

When v has unique values, the values returned in w are irrelevant.

- In (1), w will be $(1, 1)$.
- In (2), w will be $(1, 1\backslash 2, 1)$.
- ...
- In (5), w will be $(1, 1\backslash 2, 1\backslash 3, 1\backslash 4, 1\backslash 5, 1)$.

The second column of *w* records the number of tied values. Since the values in *v* are unique, the second column of *w* will be ones. If you have a problem where you are uncertain whether the values in *v* are unique, code

```
if (!allof(w[,2], 1)) {
        /* uniqueness assumption false */
}
```

Use of functions when v has repeated (tied) values

Consider $v = (3, 2, 3, 2, 3, 3)$.

1. minindex(v, 1, i, w) returns i = (2, 4)', which means that there is one minimum value and that it is repeated in two elements of v, namely, v[2] and v[4].

   Here, w will be (1, 2), but you can ignore that. There are two values in i corresponding to the same minimum.

   When k==1, rows(i) equals the number of observations in v corresponding to the minimum, as does w[1,2].

2. minindex(v, 2, i, w) returns $i = (2, 4, 1, 3, 5, 6)'$ and $w = (1, 2\backslash 3, 4)$.

   Begin with w. The first row of w is (1, 2), which states that the indices of the first minimums of v start at i[1] and consist of two elements. Thus the indices of the first minimums are i[1] and i[2] (the minimums are v[i[1]] and v[i[2]], which of course are equal).

   The second row of w is (3, 4), which states that the indices of the second minimums of v start at i[3] and consist of four elements: i[3], i[4], i[5], and i[6] (which are 1, 3, 5, and 6).

   In summary, rows(w) records the number of minimums returned. w[m,1] records where in i the mth minimum begins (it begins at i[w[m,1]]). w[m,2] records the total number of tied values. Thus one could step across the minimums and the tied values by coding

   ```
   minindex(v, k, i, w)
   for (m=1; m<=rows(w); m++) {
           for (j=w[m,1]; j<w[m,1]+w[m,2]; j++) {
                   /* i[j] is the index in v of an mth minimum */
           }
   }
   ```

3. minindex(v, 3, i, w), minindex(v, 4, i, w), and so on, return the same as (2) because, with v = (3, 2, 3, 2, 3, 3), there are only two minimums.

Summary

Consider minindex(v, k, i, w). Returned will be

$$w = \begin{bmatrix} i1 & n1 \\ i2 & n2 \\ . & . \\ . & . \end{bmatrix} \quad w : K \times 2, \quad K \leq |k|$$

$$i = \begin{bmatrix} j1 \\ j2 \\ j3 \\ j4 \\ . \\ . \\ . \end{bmatrix} \begin{matrix} \leftarrow i[i1] \text{ is start of first minimums} \\ \\ \\ \leftarrow i[i2] \text{ is start of second minimums} \\ \\ \text{etc.} \\ i : 1 \times m, \quad m = n1 + n2 + \ldots \end{matrix} \begin{matrix} \} \text{ has } n1 \text{ values} \\ \\ \} \text{ has } n2 \text{ values} \end{matrix}$$

$j1, j2, \ldots,$ are indices into v.

Conformability

minindex(v, k, i, w), maxindex(v, k, i, w):

   input:
         v: $n \times 1$ or $1 \times n$
         k: 1×1
   output:
         i: $L \times 1, \quad L \geq K$
         w: $K \times 2, \quad K \leq |k|$

Diagnostics

minindex(v, k, i, w) and maxindex(v, k, i, w) abort with error if *i* or *w* is a view.

In minindex(v, k, i, w) and maxindex(v, k, i, w), missing values in *v* are ignored in obtaining minimums and maximums.

In the examples above, we have shown input vector *v* as a row vector. It can also be a column vector; it makes no difference.

In minindex(v, k, i, w), input argument *k* specifies the number of minimums to be obtained. *k* may be zero. If *k* is negative, $-k$ maximums are obtained.

Similarly, in maxindex(v, k, i, w), input argument *k* specifies the number of maximums to be obtained. *k* may be zero. If *k* is negative, $-k$ minimums are obtained.

minindex() and maxindex() are designed for use when *k* is small relative to length(v); otherwise, see order() in [M-5] **sort( )**.

Also See

[M-5] **minmax( )** — Minimums and maximums

[M-4] **utility** — Matrix utility functions

Title

[M-5] **minmax( )** — Minimums and maximums

Syntax

real colvector	`rowmin(`*real matrix X*`)`
real rowvector	`colmin(`*real matrix X*`)`
real scalar	`min(`*real matrix X*`)`
real colvector	`rowmax(`*real matrix X*`)`
real rowvector	`colmax(`*real matrix X*`)`
real scalar	`max(`*real matrix X*`)`
real matrix	`rowminmax(`*real matrix X*`)`
real matrix	`colminmax(`*real matrix X*`)`
real rowvector	`minmax(`*real matrix X*`)`
real matrix	`rowminmax(`*real matrix X*, *real scalar usemiss*`)`
real matrix	`colminmax(`*real matrix X*, *real scalar usemiss*`)`
real rowvector	`minmax(`*real matrix X*, *real scalar usemiss*`)`
real colvector	`rowmaxabs(`*numeric matrix A*`)`
real rowvector	`colmaxabs(`*numeric matrix A*`)`

Description

These functions return the indicated minimums and maximums of *X*.

`rowmin(`*X*`)` returns the minimum of each row of *X*, `colmin(`*X*`)` returns the minimum of each column, and `min(`*X*`)` returns the overall minimum. Elements of *X* that contain missing are ignored.

`rowmax(`*X*`)` returns the maximum of each row of *X*, `colmax(`*X*`)` returns the maximum of each column, and `max(`*X*`)` returns the overall maximum. Elements of *X* that contain missing are ignored.

`rowminmax(`*X*`)` returns the minimum and maximum of each row of *X* in an $r \times 2$ matrix; `colminmax(`*X*`)` returns the minimum and maximum of each column in a $2 \times c$ matrix, and `minmax(`*X*`)` returns the overall minimum and maximum. Elements of *X* that contain missing are ignored.

The two-argument versions of `rowminmax()`, `colminmax()`, and `minmax()` allow you to specify how missing values are to be treated. Specifying a second argument with value 0 is the same as using the single-argument versions of the functions. In the two-argument versions, if the second argument is not zero, missing values are treated like all other values in determining the minimums and maximums: *nonmissing* $<$. $<$ .a $<$ .b $< \cdots <$ .z.

rowmaxabs(*A*) and colmaxabs(*A*) return the same result as rowmax(abs(*A*)) and colmax(abs(*A*)). The advantage is that matrix abs(*A*) is never formed or stored, and so these functions use less memory.

Conformability

rowmin(*X*), rowmax(*X*):
        X: $r \times c$
    result: $r \times 1$

colmin(*X*), colmax(*X*):
        X: $r \times c$
    result: $1 \times c$

min(*X*), max(*X*):
        X: $r \times c$
    result: 1×1

rowminmax(*X*, *usemiss*):
        X: $r \times c$
   usemiss: 1×1
    result: $r \times 2$

colminmax(*X*, *usemiss*)
        X: $r \times c$
   usemiss: 1×1
    result: $2 \times c$

minmax(*X*, *usemiss*)
        X: $r \times c$
   usemiss: 1×1
    result: 1×2

rowmaxabs(*A*):
        A: $r \times c$
    result: $r \times 1$

colmaxabs(*A*):
        A: $r \times c$
    result: $1 \times c$

Diagnostics

row*() functions return missing value for the corresponding minimum or maximum when the entire row contains missing.

col*() functions return missing value for the corresponding minimum or maximum when the entire column contains missing.

min() and max() return missing value when the entire matrix contains missing.

Also See

[M-5] **minindex( )** — Indices of minimums and maximums

[M-4] **mathematical** — Important mathematical functions

[M-4] **utility** — Matrix utility functions

Title

[M-5] **missing( )** — Count missing and nonmissing values

Syntax

    real rowvector   colmissing(*numeric matrix X*)

    real colvector   rowmissing(*numeric matrix X*)

    real scalar   missing(*numeric matrix X*)

    real rowvector   colnonmissing(*numeric matrix X*)

    real colvector   rownonmissing(*numeric matrix X*)

    real scalar   nonmissing(*numeric matrix X*)

Description

These functions return the indicated count of missing or nonmissing values.

colmissing(X) returns the count of missing values of each column of X, rowmissing(X) returns the count of missing values of each row, and missing(X) returns the overall count.

colnonmissing(X) returns the count of nonmissing values of each column of X, rownonmissing(X) returns the count of nonmissing values of each row, and nonmissing(X) returns the overall count.

Remarks

    colnonmissing(X) = rows(X) :- colmissing(X)
    rownonmissing(X) = cols(X) :- rowmissing(X)
    nonmissing(X) = rows(X)*cols(X) - missing(X)

Conformability

colmissing(X), colnonmissing(X):
    X: $r \times c$
    result: $1 \times c$

rowmissing(X), rownonmissing(X):
    X: $r \times c$
    result: $r \times 1$

missing(X), nonmissing(X):
    X: $r \times c$
    result: 1×1

Diagnostics

None.

Also See

[M-5] **editmissing( )** — Edit matrix for missing values

[M-4] **utility** — Matrix utility functions

Title

[M-5] **missingof( )** — Appropriate missing value

Syntax

> *transmorphic scalar* missingof(*transmorphic matrix A*)

Description

missingof(*A*) returns a missing of the same element type as *A*:

- if *A* is real, a real missing is returned;
- if *A* is complex, a complex missing is returned;
- if *A* is pointer, NULL is returned;
- if *A* is string, "" is returned.

Remarks

missingof( ) is useful when creating empty matrices of the same type as another matrix; e.g.,

> newmat = J(rows(x), cols(x), missingof(x))

Conformability

missingof(*A*)
        A: $r \times c$
  result: 1×1

Diagnostics

None.

Also See

[M-4] **utility** — Matrix utility functions

Title

[M-5] mod( ) — Modulus

Syntax

> *real matrix* mod(*real matrix x, real matrix y*)

Description

mod(x, y) returns the elementwise modulus of x with respect to y. mod() is defined

$$\mathrm{mod}(x, y) = x - y * \mathrm{trunc}(x/y)$$

Conformability

mod(x, y):
    x: $r_1 \times c_1$
    y: $r_2 \times c_2$, x and y r-conformable
    result: $\max(r_1, r_2) \times \max(c_1, c_2)$ (elementwise calculation)

Diagnostics

mod(x, y) returns missing when either argument is missing or when $y = 0$.

Also See

[M-4] **scalar** — Scalar mathematical functions

Title

[M-5] **more( )** — Create –more– condition

Syntax

>   *void*   more()
>
>   *real scalar* setmore()
>   *void*   setmore(*real scalar onoff*)
>
>   *void*   setmoreonexit(*real scalar onoff*)

Description

more() displays --more-- and waits for a key to be pressed. That is, more() does that if more is on, which it usually is, and it does nothing otherwise. more can be turned on and off by Stata's set more command or by the functions below.

setmore() returns whether more is on or off, encoded as 1 and 0.

setmore(*onoff*) sets more on if *onoff* $\neq 0$ and sets more off otherwise.

setmoreonexit(*onoff*) sets more on or off when the current execution ends. It has no effect on the current setting. The specified setting will take effect when control is passed back to the Mata prompt or to the calling ado-file or to Stata itself, and it will take effect regardless of whether execution ended because of a return, exit(), error, or abort. Only the first call to setmoreonexit() has that effect. Later calls have no effect whatsoever.

Remarks

setmoreonexit() is used to ensure that the more setting is restored if a program wants to temporarily reset it:

>   setmoreonexit(setmore())
>   setmore(0)

Only the first invocation of setmoreonexit() has any effect. This way, a subroutine that is used in various contexts might also contain

>   setmoreonexit(setmore())
>   setmore(0)

and that will not cause the wrong more setting to be restored if an earlier routine had already done that and yet still cause the right setting to be restored if the subroutine is the first to issue setmoreonexit().

Conformability

more() takes no arguments and returns *void*.

setmore():
    result: 1 × 1

setmore(*onoff*), setmoreonexit(*onoff*):
    onoff: 1 × 1
    result: *void*

Diagnostics

None.

Also See

[P] **more** — Pause until key is pressed

[M-4] **io** — I/O functions

Title

[M-5] **norm( )** — Matrix and vector norms

Syntax

> *real scalar* norm(*numeric matrix A*)
>
> *real scalar* norm(*numeric matrix A, real scalar p*)

Description

norm(*A*) returns norm(*A*, 2).

norm(*A, p*) returns the value of the norm of *A* for the specified *p*. The possible values and the meaning of *p* depend on whether *A* is a vector or a matrix.

When *A* is a vector, norm(*A, p*) returns

> sum(abs(*A*):^p) ^ (1/p)   if $1 \leq p <$.
>
> max(abs(*A*))   if $p \geq$.

When *A* is a matrix, returned is

p	norm(*A, p*)
0	sqrt(trace(conj(*A*)'*A*))
1	max(colsum(abs(*A*)))
2	max(svdsv(*A*))
.	max(rowsum(abs(*A*)))

Remarks

norm(*A*) and norm(*A, p*) calculate vector norms and matrix norms. *A* may be real or complex and need not be square when it is a matrix.

The formulas presented above are not the actual ones used in calculation. In the vector-norm case when $1 \leq p <$., the formula is applied to *A*:/max(abs(*A*)) and the result then multiplied by max(abs(*A*)). This prevents numerical overflow. A similar technique is used in calculating the matrix norm for $p = 0$, and that technique also avoids storage of conj(*A*)'*A*.

Conformability

norm(*A*)
>> *A*: $r \times c$
>> *result*: 1×1

norm(*A, p*)
>> *A*: $r \times c$
>> *p*: 1×1
>> *result*: 1×1

Diagnostics

The norm( ) is defined to return 0 if A is void and missing if any element of A is missing.

norm(A, p) aborts with error if p is out of range. When A is a vector, p must be greater than or equal to 1. When A is a matrix, p must be 0, 1, 2, or . (missing).

norm(A) and norm(A, p) return missing if the 2-norm is requested and the singular value decomposition does not converge, an event not expected to occur; see [M-5] **svd**( ).

Also See

[M-4] **matrix** — Matrix functions

Title

[M-5] **normal( )** — Cumulatives, reverse cumulatives, and densities

Syntax

Gaussian normal

$f = \text{normalden}(z)$

$f = \text{normalden}(x, sd)$

$f = \text{normalden}(x, mean, sd)$

$p = \text{normal}(z)$

$z = \text{invnormal}(p)$

$ln(f) = \text{lnnormalden}(z)$

$ln(f) = \text{lnnormalden}(x, sd)$

$ln(f) = \text{lnnormalden}(x, mean, sd)$

$ln(p) = \text{lnnormal}(z)$

Binormal

$p = \text{binormal}(z_1, z_2, rho)$

Chi-squared

$p = \text{chi2}(n, x)$

$q = \text{chi2tail}(n, x)$

$x = \text{invchi2}(n, p)$

$x = \text{invchi2tail}(n, q)$

Noncentral chi-squared

$p = \text{nchi2}(n, L, x)$

$x = \text{invnchi2}(n, L, p)$

$L = \text{npnchi2}(n, x, p)$

Student's t

$f = \text{tden}(n, t)$

$q = \text{ttail}(n, t)$

$t = \text{invttail}(n, q)$

F

$f = \text{Fden}(n_1, n_2, Fstat)$

$p = \text{F}(n_1, n_2, Fstat)$

$q = \text{Ftail}(n_1, n_2, Fstat)$

$Fstat = \text{invF}(n_1, n_2, p)$

$Fstat = \text{invFtail}(n_1, n_2, q)$

Noncentral F

$$f = \text{nFden}(n_1, n_2, L, F)$$
$$q = \text{nFtail}(n_1, n_2, L, F)$$
$$F = \text{invnFtail}(n_1, n_2, L, q)$$

Binomial

$$p = \text{binomial}(n, k, pi)$$
$$q = \text{binomialtail}(n, k, pi)$$
$$pi = \text{invbinomial}(n, k, p)$$
$$pi = \text{invbinomialtail}(n, k, q)$$

Beta

$$f = \text{betaden}(a, b, x)$$
$$p = \text{ibeta}(a, b, x)$$
$$q = \text{ibetatail}(a, b, x)$$
$$x = \text{invibeta}(a, b, p)$$
$$x = \text{invibetatail}(a, b, q)$$

Noncentral Beta

$$f = \text{nbetaden}(a, b, L, x)$$
$$p = \text{nibeta}(a, b, L, x)$$
$$x = \text{invnibeta}(a, b, L, p)$$

Gamma

$$f = \text{gammaden}(a, b, g, x)$$
$$p = \text{gammap}(a, x)$$
$$q = \text{gammaptail}(a, x)$$
$$x = \text{invgammap}(a, p)$$
$$x = \text{invgammaptail}(a, q)$$
$$dg/da = \text{dgammapda}(a, x)$$
$$dg/dx = \text{dgammapdx}(a, x)$$
$$d2g/da2 = \text{dgammapdada}(a, x)$$
$$d2g/dadx = \text{dgammapdadx}(a, x)$$
$$d2g/dx2 = \text{dgammapdxdx}(a, x)$$

where

1. All functions return real and all arguments are real.

2. The left-hand-side notation is used to assist in interpreting the meaning of the returned value:

$$f = \text{density value}$$
$$p = \text{left cumulative} = \Pr(-\infty < \textit{statistic} \leq x)$$
$$q = \text{right cumulative} = 1 - p$$

Description

The above functions return density values, cumulatives, reverse cumulatives, and in one case, derivatives of the indicated probability density function. These functions mirror the Stata functions of the same name and in fact are the Stata functions.

See [D] **functions** for function details. In the syntax diagram above, some arguments have been renamed in hope of aiding understanding, but the function arguments match one to one with the underlying Stata functions.

Remarks

Remarks are presented under the following headings:

> R-conformability
> A note concerning invbinomial( ) and invbinomialtail( )
> A note concerning ibeta( )
> A note concerning gammap( )

R-conformability

The above functions are usually used with scalar arguments and then return a scalar result:

```
: x = chi2(10, 12)
: x
  .7149434997
```

The arguments may, however, be vectors or matrices. For instance,

```
: x = chi2((10,11,12), 12)
: x
                1              2              3
    1   .7149434997    .6363567795    .5543203586

: x = chi2(10, (12,12.5,13))
: x
                1              2              3
    1   .7149434997    .7470146767    .7763281832

: x = chi2((10,11,12), (12,12.5,13))
: x
                1              2              3
    1   .7149434997    .6727441644    .6309593164
```

In the last example, the numbers correspond to chi2(10,12), chi2(11,12.5), and chi2(12,13).

Arguments are required to be r-conformable (see [M-6] **glossary**), and thus,

```
: x = chi2((10\11\12), (12,12.5,13))
: x
                  1              2              3
      ┌─────────────────────────────────────────────┐
    1 │  .7149434997   .7470146767   .7763281832    │
    2 │  .6363567795   .6727441644   .7066745906    │
    3 │  .5543203586   .593595966    .6309593164    │
      └─────────────────────────────────────────────┘
```

which corresponds to

```
                  1              2              3
      ┌─────────────────────────────────────────────┐
    1 │  chi2(10,12)   chi2(10,12.5)   chi2(10,13)  │
    2 │  chi2(11,12)   chi2(11,12.5)   chi2(11,13)  │
    3 │  chi2(12,12)   chi2(12,12.5)   chi2(12,13)  │
      └─────────────────────────────────────────────┘
```

A note concerning invbinomial( ) and invbinomialtail( )

invbinomial(n, k, pi) invbinomialtail(n, k, q) are useful for calculating confidence intervals for *pi*, the probability of a success. invbinomial() returns the probability *pi* such that the probability of observing k or fewer successes in n trials is *p*. invbinomialtail() returns the probability *pi* such that the probability of observing k or more successes in n trials is *q*.

A note concerning ibeta( )

ibeta(a, b, x) is known as the cumulative beta distribution, and it is known as the incomplete beta function $I_x(a, b)$.

A note concerning gammap( )

gammap(a, x) is known as the cumulative gamma distribution, and it is known as the incomplete gamma function $P(a, x)$.

Conformability

All functions require that arguments be r-conformable; see *R-conformability* above. Returned is matrix of max(argument rows) rows and max(argument columns) columns containing element-by-element calculated results.

Diagnostics

All functions return missing when arguments are out of range.

Also See

[M-4] **statistical** — Statistical functions

Title

[M-5] **optimize( )** — Function optimization

Syntax

S = optimize_init()

(varies)	optimize_init_which(S [, { "max" | "min" }])
(varies)	optimize_init_evaluator(S [, &*function*()])
(varies)	optimize_init_evaluatortype(S [, *evaluatortype*])
(varies)	optimize_init_params(S [, *real rowvector initialvalues*])
(varies)	optimize_init_nmsimplexdeltas(S [, *real rowvector delta*])
(varies)	optimize_init_argument(S, *real scalar k* [, *X*])
(varies)	optimize_init_narguments(S [, *real scalar K*])
(varies)	optimize_init_technique(S [, *technique*])
(varies)	optimize_init_singularHmethod(S [, *singularHmethod*])
(varies)	optimize_init_conv_maxiter(S [, *real scalar max*])
(varies)	optimize_init_conv_ptol(S [, *real scalar ptol*])
(varies)	optimize_init_conv_vtol(S [, *real scalar vtol*])
(varies)	optimize_init_conv_nrtol(S [, *real scalar nrtol*])
(varies)	optimize_init_valueid(S [, *string scalar id*])
(varies)	optimize_init_tracelevel(S [, *tracelevel*])
(varies)	optimize_init_constraints(S [, *real matrix Cc*])
(varies)	optimize_init_verbose(S [, *real scalar verbose*])
real rowvector	optimize(S)
real scalar	_optimize(S)
void	optimize_evaluate(S)
real scalar	_optimize_evaluate(S)

real rowvector	`optimize_result_params(`*S*`)`
real scalar	`optimize_result_value(`*S*`)`
real scalar	`optimize_result_value0(`*S*`)`
real rowvector	`optimize_result_gradient(`*S*`)`
real matrix	`optimize_result_scores(`*S*`)`
real matrix	`optimize_result_Hessian(`*S*`)`
real matrix	`optimize_result_V(`*S*`)`
string scalar	`optimize_result_Vtype(`*S*`)`
real matrix	`optimize_result_V_oim(`*S*`)`
real matrix	`optimize_result_V_opg(`*S*`)`
real matrix	`optimize_result_V_robust(`*S*`)`
real scalar	`optimize_result_iterations(`*S*`)`
real scalar	`optimize_result_converged(`*S*`)`
real colvector	`optimize_result_iterationlog(`*S*`)`
real scalar	`optimize_result_errorcode(`*S*`)`
string scalar	`optimize_result_errortext(`*S*`)`
real scalar	`optimize_result_returncode(`*S*`)`
void	`optimize_query(`*S*`)`

where *S*, if it is declared, should be declared

> `transmorphic` *S*

and where *evaluatortype* optionally specified in `optimize_init_evaluatortype()` is

evaluatortype	Description
`"d0"`	*function*() returns *scalar* value
`"d1"`	same as `"d0"` and returns gradient *rowvector*
`"d2"`	same as `"d1"` and returns Hessian *matrix*
`"d1debug"`	same as `"d1"` but checks gradient
`"d2debug"`	same as `"d2"` but checks gradient and Hessian
`"v0"`	*function*() returns *colvector* value
`"v1"`	same as `"v0"` and returns score *matrix*
`"v2"`	same as `"v1"` and returns Hessian *matrix*
`"v1debug"`	same as `"v1"` but checks gradient
`"v2debug"`	same as `"v2"` but checks gradient and Hessian

`"d0"` is default if not set.

and where *technique* optionally specified in `optimize_init_technique()` is

technique	Description
"nr"	modified Newton–Raphson
"dfp"	Davidon–Fletcher–Powell
"bfgs"	Broyden–Fletcher–Goldfarb–Shanno
"bhhh"	Berndt–Hall–Hall–Hausman
"nm"	Nelder–Mead

"nr" is default if not set.

and where *singularHmethod* optionally specified in optimize_init_singularHmethod() is

singularHmethod	Description
"m-marquardt"	modified Marquardt algorithm
"hybrid"	mixture of steepest descent and Newton

"m-marquardt" is default if not set;
"hybrid" is equivalent to ml's difficult option; see [R] **ml**.

and where *tracelevel* optionally specified in optimize_init_tracelevel() is

tracelevel	To be displayed each iteration
"none"	nothing
"value"	function value
"tolerance"	previous + convergence values
"params"	previous + parameter values
"step"	previous + stepping information
"gradient"	previous + gradient vector
"hessian"	previous + Hessian matrix

"value" is default if not set.

Description

These functions find parameter vector or scalar p such that function $f(p)$ is a maximum or a minimum.

optimize_init() begins the definition of a problem and returns S, a problem-description handle set to contain default values.

The optimize_init_*(S, ...) functions then allow you to modify those defaults. You use these functions to describe your particular problem: to set whether you wish maximization or minimization, to set the identity of function $f()$, to set initial values, and the like.

optimize(S) then performs the optimization. optimize() returns *real rowvector p* containing the values of the parameters that produce a maximum or minimum.

The optimize_result_*(S) functions can then be used to access other values associated with the solution.

Usually you would stop there. In other cases, you could restart optimization by using the resulting parameter vector as new initial values, change the optimization technique, and restart the optimization:

```
optimize_init_param(S, optimize_result_param(S))
optimize_init_technique(S, "dfp")
optimize(S)
```

Aside: The `optimize_init_*(S, ...)` functions have two modes of operation. Each has an optional argument that you specify to set the value and that you omit to query the value. For instance, the full syntax of `optimize_init_params()` is

> *void* `optimize_init_params`(*S, real rowvector initialvalues*)
>
> *real rowvector* `optimize_init_params`(*S*)

The first syntax sets the initial values and returns nothing. The second syntax returns the previously set (or default if not set) initial values.

All the `optimize_init_*(S, ...)` functions work the same way.

Remarks

Remarks are presented under the following headings:

> First example
> Notation
> Type d evaluators
> Example of d0, d1, and d2
> d1debug and d2debug
> Type v evaluators
> Example of v0, v1, and v2
> Functions
> > optimize_init()
> > optimize_init_which()
> > optimize_init_evaluator() and optimize_init_evaluatortype()
> > optimize_init_argument() and optimize_init_narguments()
> > optimize_init_params()
> > optimize_init_nmsimplexdeltas()
> > optimize_init_technique()
> > optimize_init_singularHmethod()
> > optimize_init_conv_maxiter()
> > optimize_init_conv_ptol(), ..._vtol(), ..._nrtol()
> > optimize_init_valueid()
> > optimize_init_tracelevel()
> > optimize_init_constraints()
> > optimize_init_verbose()
>
> > optimize()
> > _optimize()
> > optimize_evaluate()
> > _optimize_evaluate()
>
> > optimize_result_params()
> > optimize_result_value() and optimize_result_value0()
> > optimize_result_gradient()
> > optimize_result_scores()
> > optimize_result_Hessian()
> > optimize_result_V_oim(), ..._opg(), ..._robust()
> > optimize_result_V() and optimize_result_Vtype()
> > optimize_result_iterations()
> > optimize_result_converged()
> > optimize_result_iterationlog()
> > optimize_result_errorcode(), ..._errortext(), and ..._returncode()
> > optimize_query()

First example

The optimization functions may be used interactively.

Below we use the functions to find the value of x that maximizes $y = \exp(-x^2 + x - 3)$:

```
: void myeval(todo, x, y, g, H)
> {
>         y = exp(-x^2 + x - 3)
> }
note: argument todo unused
note: argument g unused
note: argument H unused

: S = optimize_init()

: optimize_init_evaluator(S, &myeval())

: optimize_init_params(S, 0)

: x = optimize(S)
Iteration 0:   f(p) = .04978707
Iteration 1:   f(p) = .04978708
Iteration 2:   f(p) = .06381186
Iteration 3:   f(p) = .06392786
Iteration 4:   f(p) = .06392786

: x
.5
```

Notation

We wrote the above in the way that mathematicians think, i.e., optimizing $y = f(x)$. Statisticians, on the other hand, think of optimizing $s = f(b)$. To avoid favoritism, we will write $v = f(p)$ and write the general problem with the following notation:

Maximize or minimize $v = f(p)$,

v: a scalar

p: $1 \times np$

subject to the constraint $Cp' = c$,

C: $nc \times np$ ($nc = 0$ if no constraints)
c: $nc \times 1$

where g, the gradient vector, is $g = f'(p) = \mathrm{d}f/\mathrm{d}p$,

g: $1 \times np$

and H, the Hessian matrix, is $H = f''(p) = \mathrm{d}^2 f/\mathrm{d}p\mathrm{d}p'$

H: $np \times np$

Type d evaluators

You must write an evaluator function to calculate $f()$ before you can use the optimization functions. The example we showed above was of what is called a type d evaluator. Let's stay with that.

The evaluator function we wrote was

```
void myeval(todo, x, y, g, H)
{
        y = exp(-x^2 + x - 3)
}
```

All type d evaluators open the same way,

> *void evaluator(todo, x, y, g, H)*

although what you name the arguments is up to you. We named the arguments the way that mathematicians think, although we could just as well have named them the way that statisticians think:

> *void evaluator(todo, b, s, g, H)*

To avoid favoritism, we will write them as

> *void evaluator(todo, p, v, g, H)*

i.e., we will think in terms of optimizing $v = f(p)$.

Here is the full definition of a type d evaluator:

void evaluator(real scalar todo, real rowvector p, v, g, H)

where v, g, and H are values to be returned:

v:	real scalar
g:	real rowvector
H:	real matrix

evaluator() is to fill in v given the values in p and optionally to fill in g and H, depending on the value of *todo*:

todo	Required action by *evaluator()*
0	calculate $v = f(p)$ and store in v
1	calculate $v = f(p)$ and $g = f'(p)$ and store in v and g
2	calculate $v = f(p)$, $g = f'(p)$, and $H = f''(p)$ and store in v, g, and H

evaluator() may return $v=$. if $f()$ cannot be evaluated at p. Then g and H need not be filled in even if requested.

An evaluator does not have to be able to do all of this. In the first example, `myeval()` could handle only *todo* $= 0$. There are three types of type d evaluators:

d type	Capabilities expected of *evaluator()*
d0	can calculate $v = f(p)$
d1	can calculate $v = f(p)$ and $g = f'(p)$
d2	can calculate $v = f(p)$ and $g = f'(p)$ and $H = f''(p)$

`myeval()` is a type d0 evaluator. Type d0 evaluators are never asked to calculate g or H. Type d0 is the default type but, if we were worried that it was not, we could set the evaluator type before invoking `optimize()` by coding

```
optimize_init_evaluatortype(S, "d0")
```

Here are code outlines of the three types of evaluators:

```
void d0_evaluator(todo, p, v, g, H)
{
        v = ...
}
```

```
void d1_evaluator(todo, p, v, g, H)
{
        v = ...
        if (todo>=1) {
                g = ...
        }
}
```

```
void d2_evaluator(todo, p, v, g, H)
{
        v = ...
        if (todo>=1) {
                g = ...
                if (todo==2) {
                        H = ...
                }
        }
}
```

Example of d0, d1, and d2

We wish to find the p_1 and p_2 corresponding to the maximum of

$$v = \exp(-p_1^2 - p_2^2 - p_1 p_2 + p_1 - p_2 - 3)$$

A d0 solution to the problem would be

```
: void eval0(todo, p, v, g, H)
> {
>         v = exp(-p[1]^2 - p[2]^2 - p[1]*p[2] + p[1] - p[2] - 3)
> }
note: argument todo unused
note: argument g unused
note: argument h unused
: S = optimize_init()
: optimize_init_evaluator(S, &eval0())
```

```
        : optimize_init_params(S, (0,0))
        : p = optimize(S)
        Iteration 0:   f(p) = .04978707   (not concave)
        Iteration 1:   f(p) = .12513024
        Iteration 2:   f(p) = .13495886
        Iteration 3:   f(p) = .13533527
        Iteration 4:   f(p) = .13533528
        : p
                  1    2
            ┌─────────────┐
          1 │  1    -1    │
            └─────────────┘
```

A d1 solution to the problem would be

```
        : void eval1(todo, p, v, g, H)
        > {
        >         v = exp(-p[1]^2 - p[2]^2 - p[1]*p[2] + p[1] - p[2] - 3)
        >         if (todo==1) {
        >                 g[1] = (-2*p[1] - p[2] + 1)*v
        >                 g[2] = (-2*p[2] - p[1] - 1)*v
        >         }
        > }
        note: argument H unused
        : S = optimize_init()
        : optimize_init_evaluator(S, &eval1())
        : optimize_init_evaluatortype(S, "d1")              ← important
        : optimize_init_params(S, (0,0))
        : p = optimize(S)
        Iteration 0:   f(p) = .04978707   (not concave)
        Iteration 1:   f(p) = .12513026
        Iteration 2:   f(p) = .13496887
        Iteration 3:   f(p) = .13533527
        Iteration 4:   f(p) = .13533528
        : p
                  1    2
            ┌─────────────┐
          1 │  1    -1    │
            └─────────────┘
```

The d1 solution is better than the d0 solution because it runs faster and usually is more accurate. Type d1 evaluators require more code, however, and deriving analytic derivatives is not always possible.

A d2 solution to the problem would be

```
        : void eval2(todo, p, v, g, H)
        > {
        >         v = exp(-p[1]^2 - p[2]^2 - p[1]*p[2] + p[1] - p[2] - 3)
        >         if (todo>=1) {
        >                 g[1] = (-2*p[1] - p[2] + 1)*v
        >                 g[2] = (-2*p[2] - p[1] - 1)*v
        >                 if (todo==2) {
        >                         H[1,1] = -2*v + (-2*p[1]-p[2]+1)*g[1]
        >                         H[2,1] = -1*v + (-2*p[2]-p[1]-1)*g[1]
        >                         H[2,2] = -2*v + (-2*p[2]-p[1]-1)*g[2]
        >                         _makesymmetric(H)
        >                 }
        >         }
        > }
        : S = optimize_init()
```

```
        : optimize_init_evaluator(S, &eval2())
        : optimize_init_evaluatortype(S, "d2")           ← important
        : optimize_init_params(S, (0,0))
        : p = optimize(S)
        Iteration 0:   f(p) = .04978707    (not concave)
        Iteration 1:   f(p) = .12513026
        Iteration 2:   f(p) = .13496887
        Iteration 3:   f(p) = .13533527
        Iteration 4:   f(p) = .13533528
        : p
                   1    2
            ┌─────────────┐
          1 │   1    -1   │
            └─────────────┘
```

A d2 solution is best because it runs fastest and usually is the most accurate. Type d2 evaluators require the most code, and deriving analytic derivatives is not always possible.

In the d2 evaluator eval2(), note our use of _makesymmetric(). Type d2 evaluators are required to return H as a symmetric matrix; filling in just the lower or upper triangle is not sufficient. The easiest way to do that is to fill in the lower triangle and then use _makesymmetric() to reflect the lower off-diagonal elements; see [M-5] **makesymmetric( )**.

d1debug and d2debug

In addition to evaluator types "d0", "d1", and "d2" that are specified in optimize_init_evaluatortype(S, *evaluatortype*), there are two more: "d1debug" and "d2debug". They assist in coding d1 and d2 evaluators.

In *Example of d0, d1, and d2* above, we admit that we did not correctly code the functions eval1() and eval2() at the outset, before you saw them. In both cases, that was because we had taken the derivatives incorrectly. The problem was not with our code but with our math. d1debug and d2debug helped us find the problem.

d1debug is an alternative to d1. When you code optimize_init_evaluatortype(S, "d1debug"), the derivatives you calculate are not taken seriously. Instead, optimize() calculates its own numerical derivatives and uses those. Each time optimize() does that, however, it compares your derivatives to the ones it calculated and gives you a report on how they differ. If you have coded correctly, they should not differ by much.

d2debug does the same thing, but for d2 evaluators. When you code optimize_init_evaluatortype(S, "d2debug"), optimize() uses numerical derivatives but, each time, optimize() gives you a report on how much your results for the gradient and for the Hessian differ from the numerical calculations.

For each comparison, optimize() reports just one number: the mreldif() (see [M-5] **reldif( )**) between your results and the numerical ones. When you have done things right, gradient vectors will differ by approximately 1e−12 or less and Hessians will differ by 1e−7 or less.

When differences are large, you will want to see not only the summary comparison but also the full vectors and matrices so that you can compare your results element by element with those calculated numerically. Sometimes the error is in one element and not the others. To do this, set the trace level with optimize_init_tracelevel(S, *tracelevel*) before issuing optimize(). Code optimize_init_tracelevel(S, "gradient") to get a full report on the gradient comparison, or set `optimize_init_tracelevel(S, "hessian")` to get a full report on the gradient comparison and the Hessian comparison.

Type v evaluators

In some statistical applications, you will find v0, v1, and v2 more convenient to code than d0, d1, and d2. The *v* stands for vector.

In statistical applications, one tends to think of a dataset of values arranged in matrix X, the rows of which are observations. A function $h(p, X[i,.])$ can be calculated for each row separately, and it is the sum of those resulting values that forms the function $f(p)$ that is to be maximized or minimized.

The v0, v1, and v2 methods are for such cases.

In a type d0 evaluator, you return scalar $v = f(p)$.

In a type v0 evaluator, you return a column vector v such that $\mathtt{colsum}(v) = f(p)$.

In a type d1 evaluator, you return $v = f(p)$ and you return a row vector $g = f'(p)$.

In a type v1 evaluator, you return v such that $\mathtt{colsum}(v) = f(p)$ and you return matrix g such that $\mathtt{colsum}(g) = f'(p)$.

In a type d2 evaluator, you return $v = f(p)$, $g = f'(p)$, and you return $H = f''(p)$.

In a type v2 evaluator, you return v such that $\mathtt{colsum}(v) = f(p)$, g such that $\mathtt{colsum}(g) = f'(p)$, and you return $H = f''(p)$. This is the same H returned for d2.

The code outline for type v evaluators is the same as those for d evaluators. For instance, the outline for a v2 evaluator is

```
void v2_evaluator(todo, p, v, g, H)
{
        v = ...
        if (todo>=1) {
                g = ...
                if (todo==2) {
                        H = ...
                }
        }
}
```

The above is the same as the outline for d2 evaluators. All that differs is that v and g, which were *real scalar* and *real rowvector* in the d2 case, are now *real colvector* and *real matrix* in the v2 case. The same applies to v1 and v0.

The type v evaluators arise in statistical applications and, in such applications, there are data; i.e., just knowing p is not sufficient to calculate v, g, and H. Actually, that same problem can arise when coding type d evaluators as well.

You can pass extra arguments to evaluators, whether they be d0, d1, or d2 or v0, v1, or v2. The first line of all evaluators, regardless of style, is

void evaluator(todo, p, v, g, H)

If you code

>   optimize_init_argument(S, 1, X)

the first line becomes

>   *void evaluator(todo, p, X, v, g, H)*

If you code

>   optimize_init_argument(S, 1, X)
>   optimize_init_argument(S, 2, Y)

the first line becomes

>   *void evaluator(todo, p, X, Y, v, g, H)*

and so on, up to nine extra arguments. That is, you can specify extra arguments to be passed to your function.

Example of v0, v1, and v2

You have the following data:

```
: x
              1
   1   .35
   2   .29
   3   .3
   4   .3
   5   .65
   6   .56
   7   .37
   8   .16
   9   .26
  10   .19
```

You believe that the data are the result of a beta distribution process with fixed parameters alpha and beta and you wish to obtain the maximum likelihood estimates of alpha and beta (*a* and *b* in what follows). The formula for the density of the beta distribution is

$$\text{density}(x) = \frac{\Gamma(a+b)}{\Gamma(a)\Gamma(b)} x^{a-1} (1-x)^{b-1}$$

The v0 solution to this problem is

```
: void lnbetaden0(todo, p, x, lnf, S, H)
> {
>         a    = p[1]
>         b    = p[2]
>         lnf  = lngamma(a+b) :- lngamma(a) :- lngamma(b) :+
>                (a-1)*log(x) :+ (b-1)*log(1:-x)
> }
note: argument todo unused
note: argument S unused
note: argument H unused
```

[M-5] optimize( ) — Function optimization

```
: S = optimize_init()
: optimize_init_evaluator(S, &lnbetaden0())
: optimize_init_evaluatortype(S, "v0")
: optimize_init_params(S, (1,1))
: optimize_init_argument(S, 1, x)                          ← important
: p = optimize(S)
Iteration 0:   f(p) =           0
Iteration 1:   f(p) =   5.7294728
Iteration 2:   f(p) =   5.7646641
Iteration 3:   f(p) =   5.7647122
Iteration 4:   f(p) =   5.7647122
: p
```

	1	2
1	3.714209592	7.014926315

Note the following:

1. Rather than calling the returned value v, we called it lnf. You can name the arguments as you please.

2. We arranged for an extra argument to be passed by coding optimize_init_argument(S, 1, x). The extra argument is the vector x, which we listed previously for you. In our function, we received the argument as x, but we could have used a different name, just as we used lnf rather than v.

3. We set the evaluator type to "v0".

This being a statistical problem, we should be interested not only in the estimates p but also in their variance. We can get this from the inverse of the negative Hessian, which is the observed information matrix:

```
: optimize_result_V_oim(S)
[symmetric]
```

	1	2
1	2.556301184	
2	4.498194785	9.716647065

The v1 solution to this problem is

```
: void lnbetaden1(todo, p, x, lnf, S, H)
> {
>         a    = p[1]
>         b    = p[2]
>         lnf  = lngamma(a+b) :- lngamma(a) :- lngamma(b) :+
>                (a-1)*log(x) :+ (b-1)*log(1:-x)
>         if (todo >= 1) }
>                 S        = J(rows(x), 2, .)
>                 S[.,1]   = log(x)   :+ digamma(a+b) :- digamma(a)
>                 S[.,2]   = log(1:-x) :+ digamma(a+b) :- digamma(b)
>         }
> }
note: argument H unused
: S = optimize_init()
: optimize_init_evaluator(S, &lnbetaden1())
```

```
: optimize_init_evaluatortype(S, "v1")
: optimize_init_params(S, (1,1))
: optimize_init_argument(S, 1, x)
: p = optimize(S)
Iteration 0:   f(p) =          0
Iteration 1:   f(p) =   5.7297061
Iteration 2:   f(p) =   5.7641349
Iteration 3:   f(p) =   5.7647121
Iteration 4:   f(p) =   5.7647122
: p
```

	1	2
1	3.714209343	7.014925751

```
: optimize_result_V_oim(S)
[symmetric]
```

	1	2
1	2.556299425	
2	4.49819212	9.716643068

Note the following:

1. We called the next-to-last argument of lnbetaden1() S rather than g in accordance with standard statistical jargon. What is being returned is in fact the observation-level scores, which sum to the gradient vector.

2. We called the next-to-last argument S even though that name conflicted with S outside the program, where S is the problem handle. Perhaps we should have renamed the outside S, but there is no confusion on Mata's part.

3. In our program we allocated *S* for ourselves: S = J(rows(x), 2, .). It is worth comparing this with the example of d1 in *Example of d0, d1, and d2*, where we did not need to allocate g. In d1, optimize() preallocates g for us. In v1, optimize() cannot do this because it has no idea how many "observations" we have.

The v2 solution to this problem is

```
: void lnbetaden2(todo, p, x, lnf, S, H)
> {
>         a    = p[1]
>         b    = p[2]
>         lnf  = lngamma(a+b) :- lngamma(a) :- lngamma(b) :+
>                (a-1)*log(x) :+ (b-1)*log(1:-x)
>         if (todo >= 1) {
>                 S       = J(rows(x), 2, .)
>                 S[.,1]  = log(x) :+ digamma(a+b) :- digamma(a)
>                 S[.,2]  = log(1:-x) :+ digamma(a+b) :- digamma(b)
>                 if (todo==2) {
>                         n = rows(x)
>                         H[1,1] = n*(trigamma(a+b) - trigamma(a))
>                         H[2,1] = n*(trigamma(a+b))
>                         H[2,2] = n*(trigamma(a+b) - trigamma(b))
>                         _makesymmetric(H)
>                 }
>         }
> }
: S = optimize_init()
```

```
: optimize_init_evaluator(S, &lnbetaden2())
: optimize_init_evaluatortype(S, "v2")
: optimize_init_params(S, (1,1))
: optimize_init_argument(S, 1, x)
: p = optimize(S)
Iteration 0:   f(p) =          0
Iteration 1:   f(p) =  5.7297061
Iteration 2:   f(p) =  5.7641349
Iteration 3:   f(p) =  5.7647121
Iteration 4:   f(p) =  5.7647122
: p
                   1              2
      1   3.714209343    7.014925751

: optimize_result_V_oim(S)
[symmetric]
                   1              2
      1   2.556299574
      2   4.498192412    9.716643651
```

Functions

optimize_init( )

transmorphic optimize_init()

optimize_init() is used to begin an optimization problem. Store the returned result in a variable name of your choosing; we have used S in this documentation. You pass S as the first argument to the other optimize*() functions.

optimize_init() sets all optimize_init_*() values to their defaults. You may use the query form of the optimize_init_*() to determine an individual default, or you can use optimize_query() to see them all.

The query form of optimize_init_*() can be used before or after optimization performed by optimize().

optimize_init_which( )

void   optimize_init_which(S, {"max" | "min"})

string scalar optimize_init_which(S)

optimize_init_which(S, *which*) specifies whether optimize() is to perform maximization or minimization. Default is maximization if you do not invoke this function.

optimize_init_which(S) returns "max" or "min" according to which is currently set.

optimize_init_evaluator( ) and optimize_init_evaluatortype( )

    void optimize_init_evaluator(*S, pointer(real function) scalar fptr*)

    void optimize_init_evaluatortype(*S, evaluatortype*)

    pointer(real function) scalar optimize_init_evaluator(*S*)

    string scalar   optimize_init_evaluatortype(*S*)

optimize_init_evaluator(*S, fptr*) specifies the function to be called to evaluate $f(p)$. Use of this function is required. If your function is named myfcn(), you code optimize_init_evaluator(*S*, &myfcn()).

optimize_init_evaluatortype(*S, evaluatortype*) specifies the capabilities of the function that has been set using optimize_init_evaluator(). Alternatives for *evaluatortype* are "d0", "d1", "d2", "d1debug", "d2debug", "v0", "v1", "v2", "v1debug", and "v2debug". The default is "d0" if you do not invoke this function.

optimize_init_evaluator(*S*) returns a pointer to the function that has been set.

optimize_init_evaluatortype(*S*) returns the *evaluatortype* currently set.

optimize_init_argument( ) and optimize_init_narguments( )

    void   optimize_init_argument(*S, real scalar k, X*)

    void   optimize_init_narguments(*S, real scalar K*)

    pointer scalar optimize_init_argument(*S, real scalar k*)

    real scalar   optimize_init_narguments(*S*)

optimize_init_argument(*S, k, X*) sets the *k*th extra argument of the evaluator function to be *X*. *X* can be anything, including a view matrix or even a pointer to a function. No copy of *X* is made; it is a pointer to *X* that is stored, so any changes you make to *X* between setting it and *X* being used will be reflected in what is passed to the evaluator function.

optimize_init_narguments(*S, K*) sets the number of extra arguments to be passed to the evaluator function. This function is useless and included only for completeness. The number of extra arguments is automatically set as you use optimize_init_argument().

optimize_init_argument(*S*) returns a pointer to the object that was previously set.

optimize_init_nargs(*S*) returns the number of extra arguments that are passed to the evaluator function.

optimize_init_params( )

    void   optimize_init_params(*S, real rowvector initialvalues*)

    real rowvector optimize_init_params(*S*)

optimize_init_params(*S, initialvalues*) sets the values of *p* to be used at the start of the first iteration. Use of this function is required.

optimize_init_params(*S*) returns the initial values that will be (or were) used.

optimize_init_nmsimplexdeltas( )

void   optimize_init_nmsimplexdeltas(*S, real rowvector delta*)

real rowvector optimize_init_nmsimplexdeltas(*S*)

optimize_init_nmsimplexdeltas(*S, delta*) sets the values of *delta* to be used, along with the initial parameters, to build the simplex required by technique "nm" (Nelder–Mead). Use of this function is required only in the Nelder–Mead case. The values in *delta* must be at least 10 times larger than *ptol*, which is set by optimize_init_conv_ptol(). The initial simplex will be $\{p, p+(d_1, 0), \ldots, 0, p+(0, d_2, 0, \ldots, 0), \ldots, p+(0, 0, \ldots, 0, d_k)\}$.

optimize_init_nmsimplexdeltas(*S*) returns the deltas that will be (or were) used.

optimize_init_technique( )

void   optimize_init_technique(*S, string scalar technique*)

string scalar optimize_init_technique(*S*)

optimize_init_technique(*S, technique*) sets the optimization technique to be used. Current choices are

technique	Description
"nr"	modified Newton–Raphson
"dfp"	Davidon–Fletcher–Powell
"bfgs"	Broyden–Fletcher–Goldfarb–Shanno
"bhhh"	Berndt–Hall–Hall–Hausman
"nm"	Nelder–Mead

"nr" is default if not set.

optimize_init_technique(*S*) returns the technique currently set.

Aside: All techniques require optimize_init_params() be set. Technique "nm" also requires that optimize_init_nmsimplexdeltas() be set. Parameters (and delta) can be set before or after the technique is set.

optimize_init_singularHmethod( )

void   optimize_init_singularHmethod(*S, string scalar method*)

string scalar optimize_init_singularHmethod(*S*)

optimize_init_singularHmethod(*S, method*) specifies what the optimizer should do when, at an iteration step, it finds that *H* is singular. Current choices are

method	Description
"m-marquardt"	modified Marquardt algorithm
"hybrid"	mixture of steepest descent and Newton

"m-marquardt" is default if not set;
"hybrid" is equivalent to ml's difficult option; see [R] **ml**.

optimize_init_technique(*S*) returns the *method* currently set.

optimize_init_conv_maxiter( )

  void optimize_init_conv_maxiter(*S*, *real scalar max*)

  real scalar optimize_init_conv_maxiter(*S*)

optimize_init_conv_maxiter(*S*, *max*) sets the maximum number of iterations to be performed before optimization() is stopped; results are posted to optimize_result_*() just as if convergence were achieved, but optimize_result_converged() is set to 0. The default *max* if not set is c(maxiter), which is probably 16,000; type creturn list in Stata to determine the current default value.

optimize_init_conv_maxiter(*S*) returns the *max* currently set.

optimize_init_conv_ptol( ), optimize_init_conv_vtol( ), optimize_init_conv_nrtol( )

  void optimize_init_conv_ptol(*S*, *real scalar ptol*)

  void optimize_init_conv_vtol(*S*, *real scalar vtol*)

  void optimize_init_conv_nrtol(*S*, *real scalar nrtol*)

  real scalar optimize_init_conv_ptol(*S*)

  real scalar optimize_init_conv_vtol(*S*)

  real scalar optimize_init_conv_nrtol(*S*)

The two-argument form of these functions set the tolerances that control optimize()'s convergence criterion. optimize() performs iterations until the convergence criterion is met or until the number of iterations exceeds optimize_init_conv_maxiter(). When the convergence criterion is met, optimize_result_converged() is set to 1. The default values of *ptol*, *vtol*, and *nrtol* are 1e−6, 1e−7, and 1e−5, respectively.

The single-argument form of these functions return the current values of *ptol*, *vtol*, and *nrtol*.

Optimization criterion: In all cases except optimize_init_technique(*S*)=="nm", i.e., in all cases except Nelder–Mead, i.e., in all cases of derivative-based maximization, the optimization criterion is defined as follows:

Define

  C_ptol: mreldif(p, p_prior) < *ptol*

  C_vtol: reldif(v, v_prior) < *vtol*

  C_nrtol: $g *$ invsym($-H$) $* g' <$ *nrtol*

  $C\_concave$: $-H$ is positive semidefinite

The above definitions apply for maximization. For minimization, think of it as maximization of $-f(p)$. optimize() declares convergence when

  ($C\_ptol\,|\,C\_vtol$) & $C\_concave$ & C_nrtol

For optimize_init_technique(S)=="nm" (Nelder–Mead), the criterion is defined as follows:

Let R be the minimum and maximum values on the simplex and define

C_ptol: mreldif(vertices of R) $<$ *ptol*

C_vtol: reldif(R) $<$ *vtol*

optimize() declares successful convergence when

$C\_ptol \mid C\_vtol$

optimize_init_valueid( )

void   optimize_init_valueid(S, *string scalar id*)

string scalar optimize_init_valueid(S)

By default, optimize() shows an iteration log, a line of which looks like

Iteration 1: f(p) = 5.7641349

See *optimize_init_tracelevel()* directly below.

optimize_init_valueid(S, *id*) sets the string used to identify the value. The default is "f(p)".

optimize_init_valueid(S) returns the *id* currently in use.

optimize_init_tracelevel( )

void   optimize_init_tracelevel(S, *string scalar tracelevel*)

string scalar optimize_init_tracelevel(S)

optimize_init_tracelevel(S, *tracelevel*) sets what is displayed in the iteration log. Allowed values of *tracelevel* are

tracelevel	To be displayed each iteration
"none"	nothing (suppress the log)
"value"	function value
"tolerance"	previous + convergence values
"params"	previous + parameter values
"step"	previous + stepping information
"gradient"	previous + gradient vector
"hessian"	previous + Hessian matrix

"value" is default if not reset.

optimize_init_tracelevel(S) returns the value of *tracelevel* currently set.

optimize_init_constraints( )

> void   optimize_init_constraints(*S*, *real matrix Cc*)
>
> *real matrix* optimize_init_constraints(*S*)

nc linear constraints may be imposed on the *np* parameters in *p* according to $Cp' = c$, C: $nc \times np$ and c: $nc \times 1$. For instance, if there are four parameters and you wish to impose the single constraint $p_1 = p_2$, then $C = (1, -1, 0, 0)$ and $c = (0)$. If you wish to add the constraint $p_4 = 2$, then $C = (1, -1, 0, 0 \backslash 0, 0, 0, 1)$ and $c = (0 \backslash 2)$.

optimize_init_constraints(*S*, *Cc*) allows you to impose such constraints where $Cc = (C, c)$. Use of this function is optional. If no constraints have been set, then Cc is $0 \times np + 1$.

optimize_init_constraints(*S*) returns the current Cc matrix.

optimize_init_verbose( )

> void   optimize_init_verbose(*S*, *real scalar verbose*)
>
> *real scalar* optimize_init_verbose(*S*)

optimize_init_verbose(*S*, *verbose*) sets whether error messages that arise during the execution of optimize() or _optimize() are to be displayed. *verbose*=1 means that they are; 0 means that they are not. The default is 1. Setting *verbose* to 0 is of interest only to users of _optimize(). If you wish to suppress all output, code

> optimize_init_verbose(*S*, 0)
> optimize_init_tracelevel(*S*, "none")

optimize_init_verbose(*S*) returns the current value of *verbose*.

optimize( )

> *real rowvector* optimize(*S*)

optimize(*S*) invokes the optimization process and returns the resulting parameter vector. If something goes wrong, optimize() aborts with error.

Before you can invoke optimize(), you must have defined your evaluator function *evaluator*() and you must have set initial values:

> *S* = optimize_init()
> optimize_init_evaluator(*S*, &*evaluator*())
> optimize_init_params(*S*, (...))

The above assumes that your evaluator function is d0. Often you will also have coded

> optimize_init_evaluatortype(*S*, "..."))

and you may have coded other optimize_init_*() functions as well.

Once optimize() completes, you may use the optimize_result_*() functions. You may also continue to use the optimize_init_*() functions to access initial settings, and you may use them to change settings and restart optimization (i.e., invoke optimize() again) if you wish. If you do that, you will usually want to use the resulting parameter values from the first round of optimization as initial values for the second. If so, do not forget to code

> optimize_init_params(S, optimize_result_params(S))

_optimize()

> *real scalar* _optimize(S)

_optimize(S) performs the same actions as optimize(S) except that, rather than returning the resulting parameter vector, _optimize() returns a real scalar and, rather than aborting if numerical issues arise, _optimize() returns a nonzero value. _optimize() returns 0 if all went well. The returned value is called an error code.

optimize() returns the resulting parameter vector *p*. It can work that way because optimization must have gone well. Had it not, optimize() would have aborted execution.

_optimize() returns an error code. If it is 0, optimization went well and you can obtain the parameter vector by using optimize_result_param(). If optimization did not go well, you can use the error code to diagnose what went wrong and take the appropriate action.

Thus, _optimize(S) is an alternative to optimize(S). Both functions do the same thing. The difference is what happens when there are numerical difficulties.

optimize() and _optimize() work around most numerical difficulties. For instance, the evaluator function you write is allowed to return *v* equal to missing if it cannot calculate the *f*() at the current values of *p*. If that happens during optimization, optimize() and _optimize() will back up to the last value that worked and choose a different direction. optimize(), however, cannot tolerate that happening with the initial values of the parameters because optimize() has no value to back up to. optimize() issues an error message and aborts, meaning that execution is stopped. There can be advantages in that. The calling program need not include complicated code for such instances, figuring that stopping is good enough because a human will know to address the problem.

_optimize(), however, does not stop execution. Rather than aborting, _optimize() returns a nonzero value to the caller, identifying what went wrong.

Programmers implementing advanced systems will want to use _optimize() instead of optimize(). Everybody else should use optimize().

Programmers using _optimize() will also be interested in the functions

> optimize_init_verbose()
> optimize_result_errorcode()
> optimize_result_errortext()
> optimize_result_returncode()

If you perform optimization by using _optimize(), the behavior of all optimize_result_*() functions is altered. The usual behavior is that, if calculation is required and numerical problems arise, the functions abort with error. After _optimize(), however, a properly dimensioned missing result is returned and optimize_result_errorcode() and optimize_result_errortext() are set appropriately.

The error codes returned by _optimize() are listed under the heading *optimize_result_errorcode()* below.

optimize_evaluate( )

> *void* optimize_evaluate(*S*)

optimize_evaluate(*S*) evaluates *f*() at optimize_init_params() and posts results to optimize_result_*() just as if optimization had been performed, meaning that all optimize_result_*() functions are available for use. optimize_result_converged() is set to 1.

The setup for running this function is the same as for running optimize():

> *S* = optimize_init()
> optimize_init_evaluator(*S*, &*evaluator*())
> optimize_init_params(*S*, (...))

Usually, you will have also coded

> optimize_init_evaluatortype(*S*, ...))

The other optimize_init_*() settings do not matter.

_optimize_evaluate( )

> *real scalar* _optimize_evaluate(*S*)

The relationship between _optimize_evaluate() and optimize_evaluate() is the same as that between _optimize() and optimize(); see *_optimize()*.

_optimize_evaluate() returns an error code.

optimize_result_params( )

> *real rowvector* optimize_result_params(*S*)

optimize_result_params(*S*) returns the resulting parameter values. These are the same values that were returned by optimize() itself. There is no computational cost to accessing the results, so rather than coding

> *p* = optimize(*S*)

if you find it more convenient to code

> (void) optimize(*S*)
> ...
> *p* = optimize_result_params(*S*)

then do so.

optimize_result_value() and optimize_result_value0()

> *real scalar* optimize_result_value(*S*)

> *real scalar* optimize_result_value0(*S*)

optimize_result_value(*S*) returns the value of *f*() evaluated at *p* equal to optimize_result_param().

optimize_result_value0(*S*) returns the value of *f*() evaluated at *p* equal to optimize_init_param().

These functions may be called regardless of the evaluator or technique used.

optimize_result_gradient()

> *real rowvector* optimize_result_gradient(*S*)

optimize_result_gradient(*S*) returns the value of the gradient vector evaluated at *p* equal to optimize_result_param(). This function may be called regardless of the evaluator or technique used.

optimize_result_scores()

> *real matrix* optimize_result_scores(*S*)

optimize_result_scores(*S*) returns the value of the scores evaluated at *p* equal to optimize_result_param(). This function may be called only if a type v evaluator is used, but regardless of the technique used.

optimize_result_Hessian()

> *real matrix* optimize_result_Hessian(*S*)

optimize_result_Hessian(*S*) returns the value of the Hessian matrix evaluated at *p* equal to optimize_result_param(). This function may be called regardless of the evaluator or technique used.

optimize_result_V_oim(), optimize_result_V_opg(), optimize_result_V_robust()

> *real matrix* optimize_result_V_oim(*S*)

> *real matrix* optimize_result_V_opg(*S*)

> *real matrix* optimize_result_V_robust(*S*)

These functions return the variance matrix of *p* evaluated at *p* equal to optimize_result_param(). These functions are relevant only for maximization of log-likelihood functions but may be called in any context, including minimization.

optimize_result_V_oim(*S*) returns invsym($-H$), which is the variance matrix obtained from the observed information matrix. For minimization, returned is invsym(H).

optimize_result_V_opg(*S*) returns invsym(*S*′*S*), where *S* is the $N \times np$ matrix of scores. This is known as the variance matrix obtained from the outer product of the gradients. optimize_result_V_opg() is available only when the evaluator function is type v, but regardless of the technique used.

optimize_result_V_robust(*S*) returns $H * \text{invsym}(S'S) * H$, which is the robust estimate of variance, also known as the sandwich estimator of variance. optimize_result_V_robust() is available only when the evaluator function is type v, but regardless of the technique used.

optimize_result_V( ) and optimize_V_Vtype( )

> *real matrix*   optimize_result_V(*S*)
>
> *string scalar*   optimize_result_Vtype(*S*)

optimize_result_V(*S*) returns optimize_result_V_oim(*S*) or optimize_result_V_opg(*S*), depending on which is the natural conjugate for the optimization technique used. If there is no natural conjugate, optimize_result_V_oim(*S*) is returned.

optimize_result_Vtype(*S*) returns "oim" or "opg".

optimize_result_iterations( )

> *real scalar* optimize_result_iterations(*S*)

optimize_result_iterations(*S*) returns the number of iterations used in obtaining results.

optimize_result_converged( )

> *real scalar* optimize_result_converged(*S*)

optimize_result_converged(*S*) returns 1 if results converged and 0 otherwise. See *optimize_init_conv_ptol()* for the definition of convergence.

optimize_result_iterationlog( )

> *real colvector* optimize_result_iterationlog(*S*)

optimize_result_iterationlog(*S*) returns a column vector of the values of *f*() at the start of the final 20 iterations, or, if there were fewer, however many iterations there were. Returned vector is min(optimize_result_iterations(), 20) $\times$ 1.

optimize_result_errorcode( ), ..._errortext( ), and ..._returncode( )

> *real scalar*   optimize_result_errorcode(*S*)
>
> *string scalar*   optimize_result_errortext(*S*)
>
> *real scalar*   optimize_result_returncode(*S*)

These functions are for use after _optimize().

optimize_result_errorcode(*S*) returns the error code of _optimize(), _optimize_evaluate(), or the last optimize_result_*() run after either of the first two functions. The value will be zero if there were no errors. The error codes are listed directly below.

optimize_result_errortext(S) returns a string containing the error message corresponding to the error code. If the error code is zero, the string will be "".

optimize_result_returncode(S) returns the Stata return code corresponding to the error code. The mapping is listed directly below.

In advanced code, these functions might be used as

```
        (void) _optimize(S)
        ...
        if (ec = optimize_result_code(S)) {
                errprintf("{p}\n")
                errprintf("%s\n", optimize_result_errortext(S))
                errprintf("{p_end}\n")
                exit(optimize_result_returncode(S))
                /*NOTREACHED*/
        }
```

The error codes and their corresponding Stata return codes are

Error code	Return code	Error text
1	1400	initial values not feasible
2	412	redundant or inconsistent constraints
3	430	missing values returned by evaluator
4	430	Hessian is not positive semidefinite *or* Hessian is not negative semidefinite
5	430	could not calculate numerical derivatives—discontinuous region with missing values encountered
6	430	could not calculate numerical derivatives—flat or discontinuous region encountered
7	430	could not calculate improvement—discontinuous region encountered
8	430	could not calculate improvement—flat region encountered
9	430	Hessian could not be updated—Hessian is unstable

NOTES: (1) Error 1 can occur only when evaluating $f()$ at initial parameters.

(2) Error 2 can occur only if constraints are specified.

(3) Error 3 can occur only if the technique is "nm".

(4) Error 9 can occur only if technique is "bfgs" or "dfp".

optimize_query()

void optimize_query(S)

optimize_query(S) displays a report on all optimize_init_*() and optimize_result*() values. optimize_query() may be used before or after optimize() and is useful when using optimize() interactively or when debugging a program that calls optimize() or _optimize().

Conformability

All functions have 1×1 inputs and have 1×1 or *void* outputs except the following:

optimize_init_params(*S*, *initialvalues*):
        S: *transmorphic*
        initialvalues: $1 \times np$
        result: *void*

optimize_init_params(*S*):
        S: *transmorphic*
        result: $1 \times np$

optimize_init_argument(*S*, *k*, *X*):
        S: *transmorphic*
        k: 1×1
        X: *anything*
        result: *void*

optimize_init_nmsimplexdeltas(*S*, *delta*):
        S: *transmorphic*
        delta: $1 \times np$
        result: *void*

optimize_init_nmsimplexdeltas(*S*):
        S: *transmorphic*
        result: $1 \times np$

optimize_init_constraints(*S*, *Cc*):
        S: *transmorphic*
        Cc: $nc \times (np + 1)$
        result: *void*

optimize_init_constraints(*S*):
        S: *transmorphic*
        result: $nc \times (np + 1)$

optimize(*S*):
        S: *transmorphic*
        result: $1 \times np$

optimize_result_params(*S*):
        S: *transmorphic*
        result: $1 \times np$

optimize_result_gradient(*S*):
        S: *transmorphic*
        result: $1 \times np$

optimize_result_scores(*S*):
        S: *transmorphic*
        result: $N \times np$

optimize_result_Hessian(*S*):
- *S*: *transmorphic*
- *result*: $np \times np$

optimize_result_V(*S*), optimize_result_V_oim(*S*), optimize_result_V_opg(*S*), optimize_result_V_robust(*S*):
- *S*: *transmorphic*
- *result*: $np \times np$

optimize_result_iterationlog(*S*):
- *S*: *transmorphic*
- *result*: $L \times 1, L \leq 20$

Diagnostics

All functions abort with error when used incorrectly.

optimize() aborts with error if it runs into numerical difficulties. _optimize() does not; it instead returns a nonzero error code.

optimize_evaluate() aborts with error if it runs into numerical difficulties. _optimize_evaluate() does not; it instead returns a nonzero error code.

The optimize_result_*() functions abort with error if they run into numerical difficulties when called after optimize() or optimize_evaluate(). They do not abort when run after _optimize() or _optimize_evaluate(). They instead return a properly dimensioned missing result and set optimize_result_errorcode() and optimize_result_errortext().

References

Berndt, E. K., B. H. Hall, R. E. Hall, and J. A. Hausman. 1974. Estimation and inference in nonlinear structural models. *Annals of Economic and Social Measurement* 3/4: 653–665.

Davidon, W. C. 1959. Variable metric methods for minimization. Atomic Energy Commission Research and Development Report, No. ANL-5990. Argonne National Laboratory: Argonne, IL.

Fletcher, R. 1970. A new approach to variable metric algorithms. *Computer Journal* 13: 317–322.

———. 1980. *Practical Methods of Optimization*. New York: Wiley.

Fletcher, R., and M. J. D. Powell. 1963. A rapidly convergent descent method for minimization. *Computer Journal* 6: 163–168.

Gleick, J. 2003. *Isaac Newton*. New York: Pantheon.

Goldfarb, D. 1970. A family of variable-metric methods derived by variational means. *Mathematics of Computation* 24: 23–26.

Marquardt, D. W. 1963. An algorithm for least squares estimation of nonlinear parameters. *Journal of the Society for Industrial and Applied Mathematics* 11: 431–441.

Nelder, J. A., and R. Mead. 1965. A simplex method for function minimization. *Computer Journal* 7: 308–313.

Newton, I. 1671. *De methodis fluxionum et serierum infinitorum*. Translated by John Colson as *The Method of Fluxions and Infinite Series* (London: Henry Wood Fall, 1736).

Raphson, J. 1690. *Analysis Aequationum Universalis*. London.

Shanno, D. F. 1970. Conditioning of quasi-Newton methods for function minimization. *Mathematics of Computation* 24: 647–656.

Westfall, R. S. 1980. *Never at Rest: A Biography of Isaac Newton.* Cambridge: Cambridge University Press.

Ypma, T. J. 1995. Historical development of the Newton–Raphson method. *SIAM Review* 37: 531–551.

The formula $x_{i+1} = x_i - f(x_i)/f'(x_i)$ and its generalizations for solving $f(x) = 0$ (and its generalizations) are known variously as Newton's method or the Newton–Raphson method. The real history is more complicated than these names imply and has roots in the earlier work of Arabic algebraists and François Viète.

Newton's first formulation dating from about 1669 refers only to solution of polynomial equations and does not use calculus. In his *Philosophiae Naturalis Principia Mathematica*, first published in 1687, the method is used, but not obviously, to solve a nonpolynomial equation. Raphson's work, first published in 1690, also concerns polynomial equations, and proceeds algebraically without using calculus, but lays more stress on iterative calculation and so is closer to present ideas. It was not until 1740 that Thomas Simpson published a more general version explicitly formulated in calculus terms that was applied to both polynomial and nonpolynomial equations and to both single equations and systems of equations. Simpson's work was in turn overlooked in influential later accounts by Lagrange and Fourier, but his contribution also deserves recognition.

Isaac Newton (1643–1727) was an English mathematician, astronomer, physicist, natural philosopher, alchemist, theologian, biblical scholar, historian, politician and civil servant. He was born in Lincolnshire, according to the calendar then in use, in 1642, and studied there and at the University of Cambridge, where he was a fellow of Trinity College and elected Lucasian Professor in 1669. Newton demonstrated the generalized binomial theorem, did major work on power series, and deserves credit with Gottfried Leibniz for the development of calculus: during his lifetime and long afterward, that fact was obscured by a bitter and protracted priority dispute. He described universal gravitation and the laws of motion central to classical mechanics and showed that the motions of objects on Earth and beyond are subject to the same laws. Newton invented the reflecting telescope and developed a theory of color that was based on the fact that a prism splits white light into a visible spectrum. He also studied cooling and the speed of sound and proposed a theory of the origin of stars. Much of his later life was spent in London, including brief spells as member of Parliament and longer periods as master of the Mint and president of the Royal Society. He was knighted in 1705. Although undoubtedly one of the greatest mathematical and scientific geniuses of all time, Newton was also outstandingly contradictory, secretive, and quarrelsome.

Joseph Raphson (1648–1715) was an English or possibly Irish mathematician. No exact dates are known for his birth or death years. He appears to have been largely self-taught and was awarded a degree by the University of Cambridge after the publication of his most notable work, *Analysis Aequationum Universalis* (1690), and his election as a fellow of the Royal Society.

Thomas Simpson (1710–1761) was born in Market Bosworth, Leicestershire, England. He received little formal education and was self-taught in mathematics. Simpson moved to London and worked as a teacher in London coffee houses (as was De Moivre) and then at the Royal Military Academy at Woolwich. He published texts on calculus, astronomy, and probability and is best remembered for his work on interpolation and numerical methods of integration. However, what is now known as "Simpson's rule" was known earlier to Newton. He was also a fellow of the Royal Society.

Also See

[M-4] **mathematical** — Important mathematical functions

[M-4] **statistical** — Statistical functions

Title

[M-5] **panelsetup( )** — Panel-data processing

Syntax

        info = `panelsetup(`*V*, *idcol*`)`

        info = `panelsetup(`*V*, *idcol*, *minobs*`)`

        info = `panelsetup(`*V*, *idcol*, *minobs*, *maxobs*`)`

real rowvector   `panelstats(`*info*`)`

real matrix   `panelsubmatrix(`*V*, *i*, *info*`)`

void   `panelsubview(`*SV*, *V*, *i*, *info*`)`

where,

V:	*real* or *string matrix*, possibly a view
idcol:	*real scalar*
minobs:	*real scalar*
maxobs:	*real scalar*
info:	*real matrix*
i:	*real scalar*
SV:	*matrix* to be created, possibly as view

Description

These functions assist with the processing of panel data. The idea is to make it easy and fast to write loops like

```
for (i=1; i<=number_of_panels; i++) {
    X = matrix corresponding to panel i
    ...
    ... (calculations using X) ...
    ...
}
```

Using these functions, this loop could become

```
st_view(Vid, ., "idvar",      "touse")
st_view(V,   ., ("x1", "x2"), "touse")
info = panelsetup(Vid, 1)
for (i=1; i<=rows(info); i++) {
    X = panelsubmatrix(V, i, info)
    ...
    ... (calculations using X) ...
    ...
}
```

panelsetup(V, $idcol$, ...) sets up panel processing. It returns a matrix (*info*) that is passed to other panel-processing functions.

panelstats(*info*) returns a row vector containing the number of panels, number of observations, minimum number of observations per panel, and maximum number of observations per panel.

panelsubmatrix(V, i, *info*) returns a matrix containing the contents of V for panel i.

panelsubview(SV, V, i, *info*) does nearly the same thing. Rather than returning a matrix, however, it places the matrix in SV. If V is a view, then the matrix placed in SV will be a view.

Remarks

Remarks are presented under the following headings:

>   Definition of panel data
>   Definition of problem
>   Preparation
>   Use of panelsetup( )
>   Using panelstats( )
>   Using panelsubmatrix( )
>   Using panelsubview( )

Definition of panel data

Panel data include multiple observations on subjects, countries, etc.:

subject ID	time ID	x1	x2
1	1	4.2	3.7
1	2	3.2	3.7
1	3	9.2	4.2
2	1	1.7	4.0
2	2	1.9	5.0
3	1	9.5	1.3
⋮	⋮	⋮	⋮

In the above dataset, there are three observations for subject 1, two for subject 2, etc. We labeled the identifier within subject to be time, but that is only suggestive, and in any case, the secondary identifier will play no role in what follows.

If we speak about the first panel, we are discussing the first 3 observations of this dataset. If we speak about the second, that corresponds to observations 4 and 5.

It is common to refer to panel numbers with the letter i. It is common to refer to the number of observations in the ith panel as T_i even when the data within panel have nothing to do with repeated observations over time.

Definition of problem

We want to calculate some statistic on panel data. The calculation amounts to

$$\sum_{i=1}^{K} f(X_i)$$

where the sum is performed across panels, and X_i is the data matrix for panel i. For instance, given the example in the previous section

$$X_1 = \begin{bmatrix} 4.2 & 3.7 \\ 3.2 & 3.7 \\ 9.2 & 4.2 \end{bmatrix}$$

and X_2 is a similarly constructed 2×2 matrix.

Depending on the nature of the calculation, there will be problems for which

1. we want to use all the panels,
2. we want to use only panels for which there are two or more observations, and
3. we want to use the same number of observations in all the panels (balanced panels).

In addition to simple problems of the sort,

$$\sum_{i=1}^{K} f(X_i)$$

you may also need to deal with problems of the form,

$$\sum_{i=1}^{K} f(X_i, Y_i, \ldots)$$

That is, you may need to deal with problems where there are multiple matrices per subject.

We use the sum operator purely for illustration, although it is the most common. Your problem might be

$$F(X_1, Y_1, \ldots, X_2, Y_2, \ldots)$$

Preparation

Before using the functions documented here, create a matrix or matrices containing the data. For illustration, it will be sufficient to create V containing all the data in our example problem:

$$V = \begin{bmatrix} 1 & 1 & 4.2 & 3.7 \\ 1 & 2 & 3.2 & 3.7 \\ 1 & 3 & 9.2 & 4.2 \\ 2 & 1 & 1.7 & 4.0 \\ 2 & 2 & 1.9 & 5.0 \\ 3 & 1 & 9.5 & 1.3 \\ \vdots & \vdots & \vdots & \vdots \end{bmatrix}$$

But you will probably find it more convenient (and we recommend) if you create at least two matrices, one containing the subject identifier and the other containing the x variables (and omit the within-subject "time" identifier altogether):

$$V1 = \begin{bmatrix} 1 \\ 1 \\ 1 \\ 2 \\ 2 \\ 3 \\ \vdots \end{bmatrix} \quad V2 = \begin{bmatrix} 4.2 & 3.7 \\ 3.2 & 3.7 \\ 9.2 & 4.2 \\ 1.7 & 4.0 \\ 1.9 & 5.0 \\ 9.5 & 1.3 \\ \vdots & \vdots \end{bmatrix}$$

In the above, matrix $V1$ contains the subject identifier, and matrix $V2$ contains the data for all the X_i matrices in

$$\sum_{i=1}^{K} f(X_i)$$

If your calculation is

$$\sum_{i=1}^{K} f(X_i, Y_i, \ldots)$$

create additional V matrices, $V3$ corresponding to Y_i, and so on.

To create these matrices, use [M-5] **st_view()**

```
st_view(V1,   ., "idvar",      "touse")
st_view(V2,   ., ("x1", "x2"), "touse")
```

although you could use [M-5] **st_data()** if you preferred. Using st_view() will save memory. You can also construct $V1$, $V2$, ..., however you wish; they are just matrices. Be sure that the matrices align, for example, that row 4 of one matrix corresponds to row 4 of another. We did that above by assuming a *touse* variable had been included (or constructed) in the dataset.

Use of panelsetup()

panelsetup(V, *idcol*, ...) sets up panel processing, returning a $K \times 2$ matrix that contains a row for each panel. The row records the first and last observation numbers (row numbers in V) that correspond to the panel.

For instance, with our example, panelsetup() will return

$$\begin{bmatrix} 1 & 3 \\ 4 & 5 \\ 6 & 9 \\ \vdots & \vdots \end{bmatrix}$$

The first panel is recorded in observations 1 to 3; it contains $3 - 1 + 1 = 3$ observations. The second panel is recorded in observations 4 to 5 and it contains $5 - 4 + 1 = 2$ observations, and so on. We recorded the third panel as being observations 6 to 9, although we did not show you enough of the original data for you to know that 9 was the last observation with ID 3.

panelsetup() has many more capabilities in constructing this result, but it is important to appreciate that returning this observation-number matrix is all that panelsetup() does. This matrix is all that other panel functions need to know. They work with the information produced by panelsetup(), but they will equally well work with any two-column matrix that contains observation numbers. Correspondingly, panelsetup() engages in no secret behavior that ties up memory, puts you in a mode, or anything else. panelsetup() merely produces this matrix.

The number of rows of the matrix panelsetup() returns equals K, the number of panels.

The syntax of panelsetup() is

> *info* = panelsetup(*V*, *idcol*, *minobs*, *maxobs*)

The last two arguments are optional.

The required argument *V* specifies a matrix containing at least the panel identification numbers and required argument *idcol* specifies the column of *V* that contains that ID. Here we will use the matrix *V1*, which contains only the identification number:

> *info* = panelsetup(*V1*, 1)

The two optional arguments are *minobs* and *maxobs*. *minobs* specifies the minimum number of observations within panel that we are willing to tolerate; if a panel has fewer observations, we want to omit it entirely. For instance, were we to specify

> *info* = panelsetup(*V1*, 1, 3)

then the matrix panelsetup() would contain fewer rows. In our example, the returned *info* matrix would contain

$$\begin{bmatrix} 1 & 3 \\ 6 & 9 \\ \vdots & \vdots \end{bmatrix}$$

Observations 4 and 5 are now omitted because they correspond to a two-observation panel, and we said only panels with three or more observations should be included.

We chose three as a demonstration. In fact, it is most common to code

> *info* = panelsetup(*V1*, 1, 2)

because that eliminates the singletons (panels with one observation).

The final optional argument is *maxobs*. For example,

> *info* = panelsetup(*V1*, 1, 2, 5)

means to include only up to five observations per panel. Any observations beyond five are to be trimmed. If we code

$$info = \mathtt{panelsetup}(\mathit{V1},\ 1,\ 3,\ 3)$$

then all the panels contained in *info* would have three observations. If a panel had fewer than three observations, it would be omitted entirely. If a panel had more than three observations, only the first three would be included.

Panel datasets with the same number of observations per panel are said to be balanced. `panelsetup()` also provides panel-balancing capabilities. If you specify *maxobs* as 0, then

1. `panelsetup()` first calculates the `min`(T_i) among the panels with *minobs* observations or more. Call that number *m*.

2. `panelsetup()` then returns `panelsetup`$(\mathit{V1},\ \mathit{idcol},\ m,\ m)$, thus creating balanced panels of size *m* and producing a dataset that has the maximum number of within-panel observations given it has the maximum number of panels.

If we coded

$$info = \mathtt{panelsetup}(\mathit{V1},\ 1,\ 2,\ 0)$$

then `panelsetup()` would create the maximum number of panels with the maximum number of within-panel observations subject to the constraint of no singletons and the panels being balanced.

Using panelstats( )

`panelstats`(*info*) can be used on any two-column matrix that contains observation numbers. `panelstats()` returns a row vector containing

$$\mathtt{panelstats()[1]} = \text{number of panels (same as } \mathtt{rows}(\mathit{info}))$$
$$\mathtt{panelstats()[2]} = \text{number of observations}$$
$$\mathtt{panelstats()[3]} = \min(T_i)$$
$$\mathtt{panelstats()[4]} = \max(T_i)$$

Using panelsubmatrix( )

Having created an *info* matrix using `panelsetup()`, you can obtain the matrix corresponding to the *i*th panel using

$$X = \mathtt{panelsubmatrix}(V,\ i,\ \mathit{info})$$

It is not necessary that `panelsubmatrix()` be used with the same matrix that was used to produce *info*. We created matrix *V1* containing the ID numbers, and we created matrix *V2* containing the *x* variables

```
st_view(V1,    ., "idvar",      "touse")
st_view(V2,    ., ("x1", "x2"), "touse")
```

and we create *info* using *V1*:

$$info = \mathtt{panelsetup}(\mathit{V1},\ 1)$$

We can now create the corresponding X matrix by coding

    X = panelsubmatrix(V2, i, info)

and, had we created a V3 matrix corresponding to Y_i, we could also code

    Y = panelsubmatrix(V3, i, info)

and so on.

Using panelsubview( )

panelsubview() works much like panelsubmatrix(). The difference is that rather than coding

    X = panelsubmatrix(V, i, info)

you code

    panelsubview(X, V, i, info)

The matrix to be defined becomes the first argument of panelsubview(). That is because panelsubview() is designed especially to work with views. panelsubmatrix() will work with views, but panelsubview() does something special. Rather than returning an ordinary matrix (an array, in the jargon), if V is a view, panelsubview() returns a view in its first argument. Views save memory.

Views can save much memory, so it would seem that you would always want to use panelsubview() in place of panelsubmatrix(). What is not always appreciated, however, is that it takes Mata longer to access the data recorded in views, and so there is a tradeoff.

If the panels are likely to be large, you want to use panelsubview(). Conserving memory trumps all other considerations.

In fact, the panels that occur in most datasets are not that large, even when the dataset itself is. If you are going to make many calculations on X, you may wish to use panelsubmatrix().

Both panelsubmatrix() and panelsubview() work with view and nonview matrices. panelsubview() produces a regular matrix when the base matrix V is not a view, just as does panelsubmatrix(). The difference is that panelsubview() will produce a view when V is a view, whereas panelsubmatrix() always produces a nonview matrix.

Conformability

panelsetup(V, idcol, minobs, maxobs):
            V: $r \times c$
        idcol: 1×1
      minobs: 1×1 (optional)
      maxobs: 1×1 (optional)
       result: $K \times 2$, $K =$ number of panels

panelstats(info):
        info: $K \times 2$
      result: 1×4

panelsubmatrix(*V*, *i*, *info*):

        V: $r \times c$
        i: 1×1, $1 \leq i \leq$ rows(*info*)
        info: $K \times 2$
        result: $t \times c$, t = number of obs. in panel

panelsubview(*SV*, *V*, *i*, *info*):

  input:

        SV: irrelevant
        V: $r \times c$
        i: 1×1, $1 \leq i \leq$ rows(*info*)
        info: $K \times 2$
        result: $t \times c$, t = number of obs. in panel

  output:

        SV: $t \times c$, t = number of obs. in panel

Diagnostics

panelsubmatrix(*V*, *i*, *info*) and panelsubview(*SV*, *V*, *i*, *info*) abort with error if $i < 1$ or $i >$ rows(*info*).

panelsetup() can return a 0×2 result.

Also See

[M-4] **utility** — Matrix utility functions

Title

[M-5] **pathjoin( )** — File path manipulation

Syntax

string scalar	pathjoin(*string scalar path1*, *string scalar path2*)
void	pathsplit(*string scalar path, path1, path2*)
string scalar	pathbasename(*string scalar path*)
string scalar	pathsuffix(*string scalar path*)
string scalar	pathrmsuffix(*string scalar path*)
real scalar	pathisurl(*string scalar path*)
real scalar	pathisabs(*string scalar path*)
real scalar	pathasciisuffix(*string scalar path*)
real scalar	pathstatasuffix(*string scalar path*)
string rowvector	pathlist(*string scalar dirlist*)
string rowvector	pathlist()
string rowvector	pathsubsysdir(*string rowvector pathlist*)
string rowvector	pathsearchlist(*string scalar fn*)

Description

pathjoin(*path1, path2*) forms, logically speaking, *path1/path2*, but does so in the appropriate style. For instance, *path1* might be a URL and *path2* a Windows *dirname\filename*, and the two paths will, even so, be joined correctly. All issues of whether *path1* ends with a directory separator, *path2* begins with one, etc., are handled automatically.

pathsplit(*path, path1, path2*) performs the inverse operation, removing the last element of the path (which is typically a filename) and storing it in *path2* and storing the rest in *path1*.

pathbasename(*path*) returns the last element of *path*.

pathsuffix(*path*) returns the file suffix, with leading dot, if there is one, and returns "" otherwise. For instance, pathsuffix("this\that.ado") returns ".ado".

pathrmsuffix(*path*) returns *path* with the suffix removed, if there was one. For instance, pathrmsuffix("this\that.ado") returns "this\that".

pathisurl(*path*) returns 1 if *path* is a URL and 0 otherwise.

pathisabs(*path*) returns 1 if *path* is absolute and 0 if relative. c:\this is an absolute path. this\that is a relative path. URLs are considered to be absolute.

pathasciisuffix(*path*) and pathstatasuffix(*path*) are more for StataCorp use than anything else. pathasciisuffix() returns 1 if the file is known to be ASCII, based on its file suffix. StataCorp uses this function in Stata's net command to decide whether end-of-line characters, which differ across operating systems, should be modified during downloads. pathstatasuffix() is the function used by Stata's net and update commands to decide whether a file belongs in the official directories. pathstatasuffix("example.ado") is true, but pathstatasuffix("example.do") is false because do-files do not go in system directories.

pathlist(*dirlist*) returns a row vector, each element of which contains an element of a semicolon-separated path list *dirlist*. For instance, pathlist("a;b;c") returns ("a", "b", "c").

pathlist() without arguments returns pathlist(c("adopath")), the broken-out elements of the official Stata ado-path.

pathsubsysdir(*pathlist*) returns *pathlist* with any elements that are Stata system directories' shorthands, such as UPDATES, PLUS, PERSONAL, substituted with the actual directory names. For instance, the right way to obtain the official directories over which Stata searches for files is pathsubsysdir(pathlist()).

pathsearchlist(*fn*) returns a row vector. The elements are full paths/filenames specifying all the locations, in order, where Stata would look for *fn* along the official Stata ado-path.

Remarks

Using these functions, you are more likely to produce code that works correctly regardless of operating system.

(*Continued on next page*)

Conformability

pathjoin(*path1*, *path2*):
    path1: 1 × 1
    path2: 1 × 1
    result: 1 × 1

pathsplit(*path*, *path1*, *path2*):
 input:
    path: 1 × 1
 output:
    path1: 1 × 1
    path2: 1 × 1

pathbasename(*path*), pathsuffix(*path*), pathrmsuffix(*path*):
    path: 1 × 1
    result: 1 × 1

pathisurl(*path*), pathisabs(*path*), pathasciisuffix(*path*), pathstatasuffix(*path*):
    path: 1 × 1
    result: 1 × 1

pathlist(*string scalar dirlist*):
    dirlist: 1 × 1 (optional)
    result: 1 × k

pathsubsysdir(*string rowvector pathlist*):
    pathlist: 1 × k
    result: 1 × k

pathsearchlist(*string scalar fn*):
    fn: 1 × 1
    result: 1 × k

Diagnostics

All routines abort with error if the path is too long for the operating system; nothing else causes abort with error.

Also See

[M-4] **io** — I/O functions

Title

[M-5] **pinv( )** — Moore–Penrose pseudoinverse

Syntax

> *numeric matrix*   pinv(*numeric matrix A*)
>
> *numeric matrix*   pinv(*numeric matrix A, rank*)
>
> *numeric matrix*   pinv(*numeric matrix A, rank, real scalar tol*)
>
> *real scalar*   _pinv(*numeric matrix A*)
>
> *real scalar*   _pinv(*numeric matrix A, real scalar tol*)

where the type of *rank* is irrelevant; the rank of *A* is returned there.

To obtain a generalized inverse of a symmetric matrix with a different normalization, see [M-5] **invsym( )**.

Description

pinv(*A*) returns the unique Moore–Penrose pseudoinverse of real or complex, symmetric or nonsymmetric, square or nonsquare matrix *A*.

pinv(*A, rank*) does the same thing, and it returns in *rank* the rank of *A*.

pinv(*A, rank, tol*) does the same thing, and it allows you to specify the tolerance used to determine the rank of *A*, which is also used in the calculation of the pseudoinverse. See [M-5] **svsolve( )** and [M-1] **tolerance** for information on the optional *tol* argument.

_pinv(*A*) and _pinv(*A, tol*) do the same thing as pinv( ), except that *A* is replaced with its inverse and the rank is returned.

Remarks

The Moore–Penrose pseudoinverse is also known as the Moore–Penrose inverse and as the generalized inverse. Whatever you call it, the pseudoinverse $A*$ of A satisfies four conditions,

$$\begin{aligned} A(A*)A &= A \\ (A*)A(A*) &= A* \\ (A(A*))' &= A(A*) \\ ((A*)A)' &= (A*)A \end{aligned}$$

where the transpose operator $'$ is understood to mean the conjugate transpose when A is complex. Also, if A is of full rank, then

[M-5] pinv( ) — Moore–Penrose pseudoinverse

$$A^* = A^{-1}$$

pinv(A) is logically equivalent to svsolve(A, I(rows(A))); see [M-5] **svsolve( )** for details and for use of the optional *tol* argument.

Conformability

pinv(*A*, *rank*, *tol*):

   input:

        A: $r \times c$

        tol: 1×1 (optional)

   output:

        rank: 1×1 (optional)

        result: $c \times r$

_pinv(*A*, *tol*):

   input:

        A: $r \times c$

        tol: 1×1 (optional)

   output:

        A: $c \times r$

        result: 1×1 (containing rank)

Diagnostics

The inverse returned by these functions is real if *A* is real and is complex if *A* is complex.

pinv(*A*, *rank*, *tol*) and _pinv(*A*, *tol*) return missing results if *A* contains missing values.

pinv( ) and _pinv( ) also return missing values if the algorithm for computing the SVD, [M-5] **svd( )**, fails to converge. This is a near zero-probability event. Here *rank* also is returned as missing.

See [M-5] **svsolve( )** and [M-1] **tolerance** for information on the optional *tol* argument.

Reference

James, I. 2002. *Remarkable Mathematicians: From Euler to von Neumann*. Cambridge: Cambridge University Press.

Moore, E. H. 1920. On the reciprocal of the general algebraic matrix. *Bulletin of the American Mathematical Society* 26: 394–395.

Penrose, R. 1955. A generalized inverse for matrices. *Proceedings, Cambridge Philosophical Society* 51: 406–413.

Also See

[M-5] **invsym( )** — Symmetric real matrix inversion

[M-5] **cholinv( )** — Symmetric, positive-definite matrix inversion

[M-5] **luinv( )** — Square matrix inversion

[M-5] **qrinv( )** — Generalized inverse of matrix via QR decomposition

[M-5] **svd( )** — Singular value decomposition

[M-5] **fullsvd( )** — Full singular value decomposition

[M-4] **matrix** — Matrix functions

[M-4] **solvers** — Functions to solve AX=B and to obtain A inverse

---

Eliakim Hastings Moore (1862–1932) was born in Marietta, Ohio. He studied mathematics and astronomy at Yale and was awarded a Ph.D. for a thesis on n-dimensional geometry. After a year studying in Germany and teaching posts at Northwestern and Yale, he settled at the University of Chicago in 1892. Moore worked on algebra, including fields and groups, the foundations of geometry and the foundations of analysis, algebraic geometry, number theory, and integral equations. He was an inspiring teacher and a great organizer in American mathematics, playing an important part in the early years of the American Mathematical Society.

Roger Penrose (1931–) was born in Colchester in England. His father was a statistically minded medical geneticist and his mother was a doctor. Penrose studied mathematics at University College London and Cambridge and published an article on generalized matrix inverses in 1955. He taught and researched at several universities in Great Britain and the United States before being appointed Rouse Ball Professor of Mathematics at Oxford in 1973. Penrose is perhaps best known for papers ranging from cosmology and general relativity (including work with Stephen Hawking) to pure mathematics (including results on tilings of the plane) and for semipopular and wide-ranging books making controversial connections between physics, computers, mind, and consciousness. He was knighted in 1994.

Title

[M-5] polyeval( ) — Manipulate and evaluate polynomials

Syntax

numeric vector	`polyeval(`*numeric rowvector c, numeric vector x*`)`
numeric rowvector	`polysolve(`*numeric vector y, numeric vector x*`)`
numeric rowvector	`polytrim(`*numeric vector c*`)`
numeric rowvector	`polyderiv(`*numeric rowvector c, real scalar i*`)`
numeric rowvector	`polyinteg(`*numeric rowvector c, real scalar i*`)`
numeric rowvector	`polyadd(`*numeric rowvector c_1, numeric rowvector c_2*`)`
numeric rowvector	`polymult(`*numeric rowvector c_1, numeric rowvector c_2*`)`
void	`polydiv(`*numeric rowvector c_1, numeric rowvector c_2, c_q, c_r*`)`
complex rowvector	`polyroots(numeric rowvector c)`

In the above, row vector c contains the coefficients for a $\text{cols}(c)-1$ degree polynomial. For instance,

$$c = (4, 2, 1)$$

records the polynomial

$$4 + 2x + x^2$$

Description

`polyeval(`c, x`)` evaluates polynomial c at each value recorded in x, returning the results in a p-conformable-with-x vector. For instance, `polyeval((4,2,1), (3\5))` returns $(4+2*3+3^2 \setminus 4+2*5+5^2) = (19\setminus39)$.

`polysolve(`y, x`)` returns the minimal-degree polynomial c fitting $y = $ `polyeval(`c, x`)`. Solution is via Lagrange's interpolation formula.

`polytrim(`c`)` returns polynomial c with trailing zeros removed. For instance, `polytrim((1,2,3,0))` returns `(1,2,3)`. `polytrim((0,0,0,0))` returns `(0)`. Thus if $n = $ `cols(polytrim(`c`))`, then c records an $(n-1)$th degree polynomial.

`polyderiv(`c, i`)` returns the polynomial that is the ith derivative of polynomial c. For instance, `polyderiv((4,2,1), 1)` returns `(2,2)` (the derivative of $4+2x+x^2$ is $2+2x$). The value of the first derivative of polynomial c at x is `polyeval(polyderiv(`c`,1),`x`)`.

`polyinteg(`c, i`)` returns the polynomial that is the ith integral of polynomial c. For instance, `polyinteg((4,2,1), 1)` returns `(0,4,1,.3333)` (the integral of $4+2x+x^2$ is $0+4x+x^2+.3333x^3$). The value of the integral of polynomial c at x is `polyeval(polyinteg(`c`,1),` x`)`.

polyadd(c_1, c_2) returns the polynomial that is the sum of the polynomials c_1 and c_2. For instance, polyadd((2,1), (3,5,1)) is (5,6,1) (the sum of $2+x$ and $3+5x+x^2$ is $5+6x+x^2$).

polymult(c_1, c_2) returns the polynomial that is the product of the polynomials c_1 and c_2. For instance, polymult((2,1), (3,5,1)) is (6,13,7,1) (the product of $2+x$ and $3+5x+x^2$ is $6+13x+7x^2+x^3$).

polydiv(c_1, c_2, c_q, c_r) calculates polynomial c_1/c_2, storing the quotient polynomial in c_q and the remainder polynomial in c_r. For instance, polydiv((3,5,1), (2,1), c_q, c_r) returns c_q=(3,1) and c_r=(-3); i.e.,

$$\frac{3+5x+x^2}{2+x} = 3+x \text{ with a remainder of } -3$$

or

$$3+5x+x^2 = (3+x)(2+x) - 3$$

polyroots(c) find the roots of polynomial c and returns them in complex row vector (complex even if c is real). For instance, polyroots((3,5,1)) returns (-4.303+0i, -.697+0i) (the roots of $3+5x+x^2$ are -4.303 and $-.697$).

Remarks

Given the real or complex coefficients c that define an $n-1$ degree polynomial in x, polyroots(c) returns the $n-1$ roots for which

$$0 = c_1 + c_2 x^1 + c_3 x^2 + \cdots + c_n x^{n-1}$$

polyroots(c) obtains the roots by calculating the eigenvalues of the companion matrix. The $(n-1) \times (n-1)$ companion matrix for the polynomial defined by c is

$$C = \begin{bmatrix} -c_{n-1}s & -c_{n-2}s & \cdots & -c_2 s & -c_1 s \\ 1 & 0 & \cdots & 0 & 0 \\ 0 & 1 & \cdots & 0 & 0 \\ \vdots & \vdots & \ddots & \vdots & \vdots \\ 0 & 0 & \cdots & 1 & 0 \\ 0 & 0 & \cdots & 0 & 1 \end{bmatrix}$$

where $s = 1/c_n$ if c is real and

$$s = C\left(\frac{\text{Re}(c_n)}{\text{Re}(c_n)^2 + \text{Im}(c_n)^2}, \frac{-\text{Im}(c_n)}{\text{Re}(c_n)^2 + \text{Im}(c_n)^2}\right)$$

otherwise.

As in all nonsymmetric eigenvalue problems, the returned roots are complex and sorted from largest to smallest, see [M-5] **eigensystem( )**.

Conformability

polyeval(c, x):
   c: $1 \times n$, $n > 0$
   x: $r \times 1$ or $1 \times c$
  result: $r \times 1$ or $1 \times c$

polysolve(y, x):
   y: $n \times 1$ or $1 \times n$, $n \geq 1$
   x: $n \times 1$ or $1 \times n$
  result: $1 \times k$, $1 \leq k \leq n$

polytrim(c):
   c: $1 \times n$
  result: $1 \times k$, $1 \leq k \leq n$

polyderiv(c, i):
   c: $1 \times n$, $n > 0$
   i: 1×1, i may be negative
  result: $1 \times \max(1, n - i)$

polyinteg(c, i):
   c: $1 \times n$, $n > 0$
   i: 1×1, i may be negative
  result: $1 \times \max(1, n + i)$

polyadd(c_1, c_2):
   c_1: $1 \times n_1$, $n_1 > 0$
   c_2: $1 \times n_2$, $n_2 > 0$
  result: $1 \times \max(n_1, n_2)$

polymult(c_1, c_2):
   c_1: $1 \times n_1$, $n_1 > 0$
   c_2: $1 \times n_2$, $n_2 > 0$
  result: $1 \times n_1 + n_2 - 1$

polydiv(c_1, c_2, c_q, c_r):
  input:
   c_1: $1 \times n_1$, $n_1 > 0$
   c_2: $1 \times n_2$, $n_2 > 0$
  output:
   c_q: $1 \times k_1$, $1 \leq k_1 \leq \max(n_1 - n_2 + 1, 1)$
   c_r: $1 \times k_2$, $1 \leq k_2 \leq \max(n_1 - n_2, 1)$

polyroots(c):
   c: $1 \times n_i$, $n > 0$
  result: $1 \times k - 1$, $k = \text{cols}(\text{polytrim}(c))$

Diagnostics

All functions abort with an error if a polynomial coefficient row vector is void, but they do not necessarily give indicative error messages as to the problem. Polynomial coefficient row vectors may contain missing values.

polyderiv(c, i) returns c when $i = 0$. It returns polyinteg(c, $-i$) when $i < 0$. It returns (0) when i is missing (think of missing as positive infinity).

polyinteg(c, i) returns c when $i = 0$. It returns polyderiv(c, $-i$) when $i < 0$. It aborts with error if i is missing (think of missing as positive infinity).

polyroots(c) returns a vector of missing values if any element of c equals missing.

Also See

[M-4] **mathematical** — Important mathematical functions

Title

[M-5] **printf( )** — Format output

Syntax

>   *void*   printf(*string scalar fmt, r1, r2, ..., rN*)

>   *string scalar* sprintf(*string scalar fmt, r1, r2, ..., rN*)

where *fmt* may contain a mix of text and *%fmts*, such as

>   printf("The result is %9.2f, adjusted\n", result)
>   printf("%s = %9.0g\n", name, value)

There must be a one-to-one correspondence between the *%fmts* in *fmt* and the number of results to be displayed.

Along with the usual *%fmts* that Stata provides (see [D] **format**), also provided are

format	meaning
%f	%11.0g, compressed
%g	%11.0g, compressed
%e	%11.8e, compressed
%s	%#s, # = whatever necessary

Compressed means that, after the indicated format is applied, all leading and trailing blanks are removed.

C programmers, be warned: %d is Stata's (old) calendar date format (equivalent to modern Stata's %td format) and not an integer format; use %f for formatting integers.

The following character sequences are given a special meaning when contained within *fmt*:

character sequence	meaning
%%	one %
\n	newline
\r	carriage return
\t	tab
\\	one \

Description

>   printf() displays output at the terminal.

sprintf() returns a string that can then be displayed at the terminal, written to a file, or used in any other way a string might be used.

Remarks

Remarks are presented under the following headings:

> *printf( )*
> *sprintf( )*

printf( )

printf() displays output at the terminal. A program might contain the line

        printf("the result is %f\n", result)

and display the output

        the result is 5.213

or it might contain the lines

        printf("{txt}{space 13}{c |}   Coef.   Std. Err.\n")
        printf("{hline 13}{c +}{hline 24}\n")
        printf("{txt}%12s {c |} {res}%10.0g   %10.0g\n",
            varname[i], coef[i], se[i])

and so display the output

```
             |      Coef.    Std. Err.
        -----+------------------------
         mpg |  -.0059541    .0005921
```

Do not forget to include \n at the end of lines. When \n is not included, the line continues. For instance, the code

        printf("{txt}{space 13}{c |}   Coef.   Std. Err.\n")
        printf("{hline 13}{c +}{hline 24}\n")
        printf("{txt}%12s {c |} {res}", varname[i])
        printf("%10.0g", coef[i])
        printf(" ")
        printf("%10.0g", se[i])
        printf("\n")

produces the same output as shown above.

Although users are unaware of it, Stata buffers output. This makes Stata faster. A side effect of the buffering, however, is that output may not appear when you want it to appear. Consider the code fragment

        for (n=1; !converged(b, b0); n++) {
            printf("iteration %f: diff = %12.0g\n", n, b-b0)
            b0 = b
            ... *new calculation of b* ...
        }

One of the purposes of the iteration output is to keep the user informed that the code is indeed working, yet as the above code is written, the user probably will not see the iteration messages as they occur. Instead, nothing will appear for a while, and then, unexpectedly, many iteration messages will appear as Stata, buffers full, decides to send to the terminal the waiting output.

To force output to be displayed, use [M-5] **displayflush( )**:

```
for (n=1; !converged(b, b0); n++) {
        printf("iteration %f:   diff = %12.0g\n", n, b-b0)
        displayflush()
        b0 = b
        ... new calculation of b ...
}
```

It is only in situations like the above that use of `displayflush()` is necessary. In other cases, it is better to let Stata decide when output buffers should be flushed. (Ado-file programmers: you have never had to worry about this because, at the ado-level, all output is flushed as it is created. Mata, however, is designed to be fast and so `printf()` does not force output to be flushed until it is efficient to do so.)

sprintf( )

The difference between `sprintf()` and `printf()` is that, whereas `printf()` sends the resulting string to the terminal, `sprintf()` returns it. Since Mata displays the results of expressions that are not assigned to variables, `sprintf()` used by itself also displays output:

```
: sprintf("the result is %f\n", result)
  the result is 5.2130a
```

The outcome is a little different from that produced by `printf()` because the output-the-unassigned-expression routine indents results by 2 and displays all the characters in the string (the 0a at the end is the \n newline character). Also, the output-the-unassigned-expression routine does not honor SMCL, electing instead to display the codes:

```
: sprintf("{txt}the result is {res}%f", result)
  {txt}the result is {res}5.213
```

The purpose of `sprintf()` is to create strings that will then be used with `printf()`, with [M-5] **display( )**, with `fput()` (see [M-5] **fopen( )**), or with some other function.

Pretend that we are creating a dynamically formatted table. One of the columns in the table contains integers, and we want to create a %*fmt* that is exactly the width required. That is, if the integers to appear in the table are 2, 9, and 20, we want to create a %2.0f format for the column. We assume the integers are in the column vector dof in what follows:

```
max = 0
for (i=1; i<=rows(dof); i++) {
        len = strlen(sprintf("%f", dof[i])
        if (len>max) max = len
}
fmt = sprintf("%%%f.0f", max)
```

We used sprintf() twice in the above. We first used sprintf() to produce the string representation of the integer dof[i], and we used the %f format so that the length would be whatever was necessary, and no more. We obtained in max the maximum length. If dof contained 2, 9, and 20, by the end of our loop, max will contain 2. We finally used sprintf() to create the %#.0f format that we wanted: %2.0f.

The format string %%%f.0f in the final sprintf() is a little difficult to read. The first two percent signs amount to one real percent sign, so in the output we now have % and we are left with %f.0f. The %f is a format—it is how we are to format max—and so in the output we now have %2, and we are left with .0f. .0f is just a string, so the final output is %2.0f.

Conformability

printf(*fmt*, r_1, r_2, ..., r_N)
    fmt: 1×1
    r_1: 1×1
    r_2: 1×1
    ...
    r_N: 1×1
    result: void

sprintf(*fmt*, r_1, r_2, ..., r_N)
    fmt: 1×1
    r_1: 1×1
    r_2: 1×1
    ...
    r_N: 1×1
    result: 1×1

Diagnostics

printf() and sprintf() abort with error if a %*fmt* is misspecified, if a numeric %*fmt* corresponds to a string result or a string %*fmt* to a numeric result, or there are too few or too many %*fmts* in *fmt* relative to the number of *results* specified.

Also See

[M-5] **displayas( )** — Set display level

[M-5] **displayflush( )** — Flush terminal-output buffer

[M-4] **io** — I/O functions

Title

[M-5] **qrd( )** — QR decomposition

Syntax

void	qrd(*numeric matrix A, Q, R*)
void	hqrd(*numeric matrix A, H, tau, R_1*)
void	_hqrd(*numeric matrix A, tau, R_1*)
numeric matrix	hqrdmultq(*numeric matrix H, numeric matrix tau, numeric matrix X, real scalar transpose*)
numeric matrix	hqrdmultq1t(*numeric matrix H, numeric matrix tau, numeric matrix X*)
numeric matrix	hqrdq(*numeric matrix H, numeric matrix tau*)
numeric matrix	hqrdq1(*numeric matrix H, numeric matrix tau*)
numeric matrix	hqrdr(*numeric matrix H*)
numeric matrix	hqrdr1(*numeric matrix H*)
void	qrdp(*numeric matrix A, Q, R, real rowvector p*)
void	hqrdp(*numeric matrix A, H, tau, R_1, real rowvector p*)
void	_hqrdp(*numeric matrix A, tau, R_1, real rowvector p*)
void	_hqrdp_la(*numeric matrix A, tau, real rowvector p*)

Description

qrd(*A, Q, R*) calculates the QR decomposition of A: $m \times n$, $m \geq n$, returning results in Q and R.

hqrd(*A, H, tau, R_1*) calculates the QR decomposition of A: $m \times n$, $m \geq n$, but rather than returning Q and R, returns the Householder vectors in H and the scale factors *tau*—from which Q can be formed—and returns an upper-triangular matrix in R_1 that is a submatrix of R; see *Remarks* below for its definition. Doing this saves calculation and memory, and other routines allow you to manipulate these matrices:

1. hqrdmultq(*H, tau, X, transpose*) returns QX or $Q'X$ on the basis of the Q implied by H and *tau*. QX is returned if *transpose* $= 0$, and $Q'X$ is returned otherwise.

2. hqrdmultq1t(*H, tau, X*) returns $Q'_1 X$ on the basis of the $Q1$ implied by H and *tau*.

3. hqrdq(*H, tau*) returns the Q matrix implied by H and *tau*. This function is rarely used.

4. hqrdq1(*H, tau*) returns the Q_1 matrix implied by H and *tau*. This function is rarely used.

5. hqrdr(H) returns the full R matrix. This function is rarely used. (It may surprise you that hqrdr() is a function of H and not R_1. R_1 also happens to be stored in H, and there is other useful information there, as well.)

6. hqrdr1(H) returns the R_1 matrix. This function is rarely used.

_hqrd(A, tau, R_1) does the same thing as hqrd(A, H, tau, R_1), except that it overwrites H into A and so conserves even more memory.

qrdp(A, Q, R, p) is similar to qrd(A, Q, R): it returns the QR decomposition of A in Q and R. The difference is that this routine allows for pivoting. New argument p specifies whether a column is available for pivoting and, on output, p is overwritten with a permutation vector that records the pivoting actually performed. On input, p can be specified as . (missing)—meaning all columns are available for pivoting—or p can be specified as an $n \times 1$ column vector containing 0s and 1s, with 1 meaning the column is fixed and so may not be pivoted.

hqrdp(A, H, tau, R_1, p) is a generalization of hqrd(A, H, tau, R_1) just as qrdp() is a generalization of qrd().

_hqrdp(A, tau, R_1, p) does the same thing as hqrdp(A, H, tau, R_1, p), except that _hqrdp() overwrites H into A.

_hqrdp_la() is the interface into the [M-1] **LAPACK** routine that performs the QR calculation; it is used by all the above routines. Direct use of _hqrdp_la() is not recommended.

Remarks

Remarks are presented under the following headings:

> QR decomposition
> Avoiding calculation of Q
> Pivoting
> Least-squares solutions with dropped columns

QR decomposition

The decomposition of square or nonsquare matrix A can be written

$$A = QR \tag{1}$$

where Q is an orthogonal matrix ($Q'Q = I$), and R is upper triangular. qrd(A, Q, R) will make this calculation:

```
: A
         1    2
    1    7    4
    2    9    6
    3    9    6
    4    7    2
    5    3    1

: Q = R = .
: qrd(A, Q, R)
: Ahat = Q*R
: mreldif(Ahat, A)
  3.55271e-16
```

Avoiding calculation of Q

In fact, you probably do not want to use qrd(). Calculating the necessary ingredients for Q is not too difficult, but going from those necessary ingredients to form Q is devilish. The necessary ingredients are usually all you need, which are the Householder vectors and their scale factors, known as H and *tau*. For instance, one can write down a mathematical function $f(H, tau, X)$ that will calculate QX or $Q'X$ for some matrix X.

Also, QR decomposition is often carried out on violently nonsquare matrices $A: m \times n, m \gg n$. We can write

$$\underset{m \times n}{A} = \begin{bmatrix} \underset{m \times n}{Q_1} & \underset{m \times m-n}{Q_2} \end{bmatrix} \begin{bmatrix} \underset{n \times n}{R_1} \\ \underset{m-n \times n}{R_2} \end{bmatrix} = \underset{m \times n}{Q_1 R_1} + \underset{m \times n}{Q_2 R_2}$$

R_2 is zero, and thus

$$\underset{m \times n}{A} = \begin{bmatrix} \underset{m \times n}{Q_1} & \underset{m \times m-n}{Q_2} \end{bmatrix} \begin{bmatrix} \underset{n \times n}{R_1} \\ \underset{m-n \times n}{0} \end{bmatrix} = \underset{m \times n}{Q_1 R_1}$$

Thus it is enough to know Q_1 and R_1. Rather than defining QR decomposition as

$$A = QR, \qquad Q : m \times m, \quad R : m \times n \tag{1}$$

it is better to define it as

$$A = Q_1 R_1 \qquad Q_1 : m \times n \quad R_1 : n \times n \tag{1'}$$

To appreciate the savings, consider the reasonable case where $m = 4,000$ and $n = 3$:

$$A = QR, \qquad Q : 4,000 \times 4,000, \quad R : 4,000 \times 3$$

versus,

$$A = Q_1 R_1 \qquad Q_1 : 4,000 \times 3 \qquad R_1 : 3 \times 3$$

Memory consumption is reduced from 125,094 kilobytes to 94 kilobytes, a 99.92% saving!

Combining the arguments, we need not save Q because Q_1 is sufficient, we need not calculate Q_1 because H and *tau* are sufficient, and we need not store R because R_1 is sufficient.

That is what hqrd(A, H, tau, R_1) does. Having used hqrd(), if you need to multiply the full Q by some matrix X, you can use hqrdmultq(). Having used hqrd(), if you need the full Q, you can use hqrdq() to obtain it, but by that point you will be making the devilish calculation you sought to avoid and so you might as well have used qrd() to begin with. If you want Q_1, you can use hqrdq1(). Finally, having used hqrd(), if you need R or R_1, you can use hqrdr() and hqrdr1():

```
: A
        1   2
    ┌─       ─┐
  1 │  7   4  │
  2 │  9   6  │
  3 │  9   6  │
  4 │  7   2  │
  5 │  3   1  │
    └─       ─┘

: H = tau = R1 = .
: hqrd(A, H, tau, R1)
: Ahat = hqrdq1(H, tau) * R1           // i.e., Q1*R1
: mreldif(Ahat, A)
  3.55271e-16
```

Pivoting

The QR decomposition with column pivoting solves

$$AP = QR \tag{2}$$

or, if you prefer,

$$AP = Q_1 R_1 \tag{2'}$$

where P is a permutation matrix; see [M-1] **permutation**. We can rewrite this as

$$A = QRP' \tag{3}$$

and

$$A = Q_1 R_1 P' \tag{3'}$$

Column pivoting can improve the numerical accuracy. The functions qrdp(A, Q, R, p) and hqrdp(A, H, tau, R_1, p) perform pivoting and return the permutation matrix P in permutation vector form:

```
: A
        1   2
    ┌─       ─┐
  1 │  7   4  │
  2 │  9   6  │
  3 │  9   6  │
  4 │  7   2  │
  5 │  3   1  │
    └─       ─┘

: Q = R = p = .
: qrdp(A, Q, R, p)
: Ahat = (Q*R)[., invorder(p)]          // i.e., QRP'
: mreldif(Ahat, A)
  3.55271e-16
: H = tau = R1 = p = .
: hqrdp(A, H, tau, R1, p)
: Ahat = (hqrdq1(H, tau)*R1)[., invorder(p)]    // i.e., Q1*R1*P'
: mreldif(Ahat, A)
  3.55271e-16
```

Before calling qrdp() or hqrdp(), we set p equal to missing, specifying that all columns could be pivoted. We could just as well have set p equal to (0, 0), which would have stated that both columns were eligible for pivoting.

When pivoting is disallowed, and when A is not of full column rank, the order in which columns appear affects the kind of generalized solution produced; later columns are, in effect, dropped. When pivoting is allowed, the columns are reordered based on numerical accuracy considerations. In the rank-deficient case, you no longer know ahead of time which columns will be dropped, because you do not know in what order the columns will appear. Generally, you do not care, but there are occasions when you do.

In such cases, you can specify which columns are eligible for pivoting and which are not—you specify p as a vector and if p_i==1, the ith column may not be pivoted. The p_i==1 columns are (conceptually) moved to appear first in the matrix, and the remaining columns are ordered optimally after that. The permutation vector that is returned in p accounts for all of this.

Least-squares solutions with dropped columns

Least-square solutions are one popular use of QR decomposition. We wish to solve for x

$$Ax = b \quad (A : m \times n, \quad m \geq n) \quad (4)$$

The problem is that there is no solution to (4) when $m > n$ because we have more equations than unknowns. Then we want to find x such that $(Ax - b)'(Ax - b)$ is minimized.

If A is of full column rank then it is well known that the least-squares solution for x is given by solveupper(R_1, $Q_1'b$) where solveupper() is an upper-triangular solver; see [M-5] **solvelower( )**.

If A is of less than full column rank and we do not care which columns are dropped, then we can use the same solution: solveupper(R_1, $Q_1'b$).

Adding pivoting to the above hardly complicates the issue; the solution becomes solveupper(R_1, $Q_1'b$)[invorder(p)].

For both cases, the full details are

```
: A
            1    2    3
       1    3    9    1
       2    3    8    1
       3    3    7    1
       4    3    6    1

: b
            1
       1    7
       2    3
       3   12
       4    0

: H = tau = R1 = p = .
: hqrdp(A, H, tau, R1, p)
: q1b = hqrdmultq1t(H, tau, b)                    // i.e., Q1'b
: xhat = solveupper(R1, q1b)[invorder(p)]
```

                : xhat

```
                       1
       ┌─────────────────────┐
    1  │    -1.166666667     │
    2  │             1.2     │
    3  │               0     │
       └─────────────────────┘
```

The *A* matrix in the above example has less than full column rank; the first column contains a variable with no variation and the third column contains the data for the intercept. The solution above is correct, but we might prefer a solution that included the intercept. To do that, we need to specify that the third column cannot be pivoted:

```
        : p = (0, 0, 1)
        : H = tau = R1 = .
        : hqrdp(A, H, tau, R1, p)
        : q1b = hqrdmultq1t(H, tau, b)
        : xhat = solveupper(R1, q1b)[invorder(p)]
        : xhat
                       1
       ┌─────────┐
    1  │    0    │
    2  │  1.2    │
    3  │ -3.5    │
       └─────────┘
```

Conformability

qrd(*A*, *Q*, *R*):
    input:
        A: $m \times n$, $m \geq n$
    output:
        Q: $m \times m$
        R: $m \times n$

hqrd(*A*, *H*, *tau*, R_1):
    input:
        A: $m \times n$, $m \geq n$
    output:
        H: $m \times n$
        tau: $1 \times n$
        R_1: $n \times n$

_hqrd(*A*, *tau*, R_1):
    input:
        A: $m \times n$, $m \geq n$
    output:
        A: $m \times n$ (contains *H*)
        tau: $1 \times n$
        R_1: $n \times n$

hqrdmultq(*H*, *tau*, *X*, *transpose*):

    H: $m \times n$
    tau: $1 \times n$
    X: $m \times c$
    result: $m \times c$

hqrdmultq1t(*H*, *tau*, *X*):

    H: $m \times n$
    tau: $1 \times n$
    X: $m \times c$
    result: $n \times c$

hqrdq(*H*, *tau*):

    H: $m \times n$
    tau: $1 \times n$
    result: $m \times m$

hqrdq1(*H*, *tau*):

    H: $m \times n$
    tau: $1 \times n$
    result: $m \times n$

hqrdr(*H*):

    H: $m \times n$
    result: $m \times n$

hqrdr1(*H*):

    H: $m \times n$
    result: $n \times n$

qrdp(*A*, *Q*, *R*, *p*):

  input:
    A: $m \times n$, $m \geq n$
    p: 1×1 or $1 \times n$
  output:
    Q: $m \times m$
    R: $m \times n$
    p: $1 \times n$

hqrdp(*A*, *H*, *tau*, R_1, *p*):

  input:
    A: $m \times n$, $m \geq n$
    p: 1×1 or $1 \times n$
  output:
    H: $m \times n$
    tau: $1 \times n$
    R_1: $n \times n$
    p: $1 \times n$

_hqrdp(A, tau, R_1, p):

   input:
        A: $m \times n$, $m \geq n$
        p: 1×1 or $1 \times n$

   output:
        A: $m \times n$ (contains H)
        tau: $1 \times n$
        R_1: $n \times n$
        p: $1 \times n$

_hqrdp_la(A, tau, p):

   input:
        A: $m \times n$, $m \geq n$
        p: 1×1 or $1 \times n$

   output:
        A: $m \times n$ (contains H)
        tau: $1 \times n$
        p: $1 \times n$

Diagnostics

qrd(A, ...), hqrd(A, ...), _hqrd(A, ...), qrdp(A, ...), hqrdp(A, ...), and _hqrdp(A, ...) return missing results if A contains missing values. That is, Q will contain all missing values. R will contain missing values on and above the diagonal. p will contain the integers 1, 2, ...

_hqrd(A, ...) and _hqrdp(A, ...) abort with error if A is a view.

hqrdmultq(H, tau, X, *transpose*) and hqrdmultq1t(H, tau, X) return missing results if X contains missing values.

> Alston Scott Householder (1904–1993) was born in Rockford, Illinois, and grew up in Alabama. He studied philosophy at Northwestern and Cornell, and then mathematics, earning a doctorate in the calculus of variations from the University of Chicago. Householder worked on mathematical biology for several years at Chicago, but in 1946 he joined the Mathematics Division of Oak Ridge National Laboratory. There he moved into numerical analysis, specializing in linear equations and eigensystems and helping to unify the field through reviews and symposia. His last post was at the University of Tennessee.

Also See

[M-5] **qrsolve( )** — Solve AX=B for X using QR decomposition

[M-5] **qrinv( )** — Generalized inverse of matrix via QR decomposition

[M-4] **matrix** — Matrix functions

Title

[M-5] **qrinv( )** — Generalized inverse of matrix via QR decomposition

Syntax

> *numeric matrix*   qrinv(*numeric matrix A*)
>
> *numeric matrix*   qrinv(*numeric matrix A, rank*)
>
> *numeric matrix*   qrinv(*numeric matrix A, rank, real scalar tol*)
>
> *real scalar*   _qrinv(*numeric matrix A*)
>
> *real scalar*   _qrinv(*numeric matrix A, real scalar tol*)

where the type of *rank* is irrelevant; the rank of *A* is returned there.

Description

qrinv(*A*, ...) returns the inverse or generalized inverse of real or complex matrix A: $m \times n$, $m \geq n$. If optional argument *rank* is specified, the rank of A is returned there.

_qrinv(*A*, ...) does the same thing except that, rather than returning the result, it overwrites the original matrix A with the result. _qrinv( ) returns the rank of A.

In both cases, optional argument *tol* specifies the tolerance for determining singularity; see *Remarks* below.

Remarks

qrinv( ) and _qrinv( ) are most often used on square and possibly rank-deficient matrices but may be used on nonsquare matrices that have more rows than columns. Also see [M-5] **pinv( )** for an alternative. See [M-5] **luinv( )** for a more efficient way to obtain the inverse of full-rank, square matrices, and see [M-5] **invsym( )** for inversion of real, symmetric matrices.

When A is of full rank, the inverse calculated by qrinv( ) is essentially the same as that computed by the faster luinv( ). When A is singular, qrinv( ) and _qrinv( ) compute a generalized inverse, A^*, which satisfies

$$A(A^*)A = A$$
$$(A^*)A(A^*) = A^*$$

This generalized inverse is also calculated for nonsquare matrices that have more rows than columns and, then returned is a least-squares solution. If A is $m \times n$, $m \geq n$, and if the rank of A is equal to n, then $(A^*)A = I$, ignoring roundoff error.

qrinv(*A*) is implemented as qrsolve(*A*, I(rows(*A*))); see [M-5] **qrsolve( )** for details and for use of the optional *tol* argument.

Conformability

qrinv(*A*, *rank*, *tol*):

  input:
      A: $m \times n$, $m \geq n$
      tol: 1×1 (optional)
  output:
      rank: 1×1 (optional)
      result: $n \times m$

_qrinv(*A*, *tol*):

  input:
      A: $m \times n$, $m \geq n$
      tol: 1×1 (optional)
  output:
      A: $n \times m$
      result: 1×1 (containing rank)

Diagnostics

The inverse returned by these functions is real if *A* is real and is complex if *A* is complex.

qrinv(*A*, ...) and _qrinv(*A*, ...) return a result containing missing values if *A* contains missing values.

_qrinv(*A*, ...) aborts with error if *A* is a view.

See [M-5] **qrsolve()** and [M-1] **tolerance** for information on the optional *tol* argument.

Also See

[M-5] **invsym()** — Symmetric real matrix inversion

[M-5] **cholinv()** — Symmetric, positive-definite matrix inversion

[M-5] **luinv()** — Square matrix inversion

[M-5] **pinv()** — Moore–Penrose pseudoinverse

[M-5] **qrsolve()** — Solve AX=B for X using QR decomposition

[M-5] **solve_tol()** — Tolerance used by solvers and inverters

[M-4] **matrix** — Matrix functions

[M-4] **solvers** — Functions to solve AX=B and to obtain A inverse

Title

[M-5] **qrsolve()** — Solve AX=B for X using QR decomposition

Syntax

  numeric matrix qrsolve(*A*, *B*)

  numeric matrix qrsolve(*A*, *B*, *rank*)

  numeric matrix qrsolve(*A*, *B*, *rank*, *tol*)

  real scalar _qrsolve(*A*, *B*)

  real scalar _qrsolve(*A*, *B*, *tol*)

where

  A: *numeric matrix*
  B: *numeric matrix*
  rank: irrelevant; *real scalar* returned
  tol: *real scalar*

Description

qrsolve(*A*, *B*, ...) uses QR decomposition to solve $AX = B$ and returns X. When A is singular or nonsquare, qrsolve() computes a least-squares generalized solution. When *rank* is specified, in it is placed the rank of A.

_qrsolve(*A*, *B*, ...), does the same thing, except that it destroys the contents of A and it overwrites B with the solution. Returned is the rank of A.

In both cases, *tol* specifies the tolerance for determining whether A is of full rank. *tol* is interpreted in the standard way—as a multiplier for the default if *tol* > 0 is specified and as an absolute quantity to use in place of the default if *tol* ≤ 0 is specified; see [M-1] **tolerance**.

Remarks

qrsolve(*A*, *B*, ...) is suitable for use with square and possibly rank-deficient matrix A, or when A has more rows than columns. When A is square and full rank, qrsolve() returns the same solution as lusolve() (see [M-5] **lusolve()**), up to roundoff error. When A is singular, qrsolve() returns a generalized (least-squares) solution.

Remarks are presented under the following headings:

  Derivation
  Relationship to inversion
  Tolerance

Derivation

We wish to solve for X

$$AX = B \qquad (1)$$

Perform QR decomposition on A so that we have $A = QRP'$. Then (1) can be rewritten

$$QRP'X = B$$

Premultiplying by Q' and remembering that $Q'Q = QQ' = I$, we have

$$RP'X = Q'B \qquad (2)$$

Define

$$Z = P'X \qquad (3)$$

Then (2) can be rewritten

$$RZ = Q'B \qquad (4)$$

It is easy to solve (4) for Z because R is upper triangular. Having Z, we can obtain X via (3), because $Z = P'X$, premultiplied by P (and if we remember that $PP' = I$), yields

$$X = PZ$$

For more information on QR decomposition, see [M-5] **qrd()**.

Relationship to inversion

For a general discussion, see *Relationship to inversion* in [M-5] **lusolve()**.

For an inverse based on QR decomposition, see [M-5] **qrinv()**. qrinv(*A*) amounts to qrsolve(*A*, I(rows(*A*))), although it is not actually implemented that way.

Tolerance

The default tolerance used is

$$eta = \texttt{1e-13} * \texttt{trace(abs}(R))/\texttt{rows}(R)$$

where R is the upper-triangular matrix of the QR decomposition; see *Derivation* above. When A is less than full rank, by, say, d degrees of freedom, then R is also rank deficient by d degrees of freedom and the bottom d rows of R are essentially zero. If the ith diagonal element of R is less than or equal to *eta*, then the ith row of Z is set to zero. Thus if the matrix is singular, qrsolve() provides a generalized solution.

If you specify *tol* > 0, the value you specify is used to multiply *eta*. You may instead specify *tol* ≤ 0, and then the negative of the value you specify is used in place of *eta*; see [M-1] **tolerance**.

Conformability

qrsolve(*A*, *B*, *rank*, *tol*):

  input:
      A: $m \times n$, $m \geq n$
      B: $m \times k$
      tol: 1×1 (optional)

  output:
      rank: 1×1 (optional)
      result: $n \times k$

_qrsolve(*A*, *B*, *tol*):

  input:
      A: $m \times n$, $m \geq n$
      B: $m \times k$
      tol: 1×1 (optional)

  output:
      A: 0×0
      B: $n \times k$
      result: 1×1

Diagnostics

qrsolve(*A*, *B*, ...) and _qrsolve(*A*, *B*, ...) return a result containing missing if *A* or *B* contain missing values.

_qrsolve(*A*, *B*, ...) aborts with error if *A* or *B* are views.

Also See

[M-5] **qrinv( )** — Generalized inverse of matrix via QR decomposition

[M-5] **qrd( )** — QR decomposition

[M-5] **solvelower( )** — Solve AX=B for X, A triangular

[M-5] **cholsolve( )** — Solve AX=B for X using Cholesky decomposition

[M-5] **lusolve( )** — Solve AX=B for X using LU decomposition

[M-5] **svsolve( )** — Solve AX=B for X using singular value decomposition

[M-5] **solve_tol( )** — Tolerance used by solvers and inverters

[M-4] **matrix** — Matrix functions

[M-4] **solvers** — Functions to solve AX=B and to obtain A inverse

Title

[M-5] **quadcross( )** — Quad-precision cross products

Syntax

> *real matrix* quadcross(*X*, *Z*)
>
> *real matrix* quadcross(*X*, *w*, *Z*)
>
> *real matrix* quadcross(*X*, *xc*, *Z*, *zc*)
>
> *real matrix* quadcross(*X*, *xc*, *w*, *Z*, *zc*)
>
> *real matrix* quadcrossdev(*X*, *x*, *Z*, *z*)
>
> *real matrix* quadcrossdev(*X*, *x*, *w*, *Z*, *z*)
>
> *real matrix* quadcrossdev(*X*, *xc*, *x*, *Z*, *zc*, *z*)
>
> *real matrix* quadcrossdev(*X*, *xc*, *x*, *w*, *Z*, *zc*, *z*)

where

> *X*: real matrix *X*
> *xc*: real scalar *xc*
> *x*: real rowvector *x*
> *w*: real vector *w*
> *Z*: real matrix *Z*
> *zc*: real scalar *zc*
> *z*: real rowvector *z*

Description

quadcross( ) makes calculations of the form

> $X'X$
>
> $X'Z$
>
> $X'\mathrm{diag}(w)X$
>
> $X'\mathrm{diag}(w)Z$

This function mirrors cross( ) (see [M-5] **cross( )**), the difference being that sums are formed in quad precision rather than in double precision, so quadcross( ) is more accurate.

quadcrossdev() makes calculations of the form

$(X: -x)'(X: -x)$

$(X: -x)'(Z: -z)$

$(X: -x)'\mathtt{diag}(w)(X: -x)$

$(X: -x)'\mathtt{diag}(w)(Z: -z)$

This function mirrors crossdev() (see [M-5] **crossdev( )**), the difference being that sums are formed in quad precision rather than in double precision, so quadcrossdev() is more accurate.

Remarks

The returned result is double precision, but the sum calculations made in creating that double-precision result were made in quad precision.

Conformability

quadcross() has the same conformability requirements as cross(); see [M-5] **cross( )**.

quadcrossdev() has the same conformability requirements as crossdev(); see [M-5] **crossdev( )**.

Diagnostics

See *Diagnostics* under [M-5] **cross( )** and [M-5] **crossdev( )**.

Also See

[M-5] **cross( )** — Cross products

[M-5] **crossdev( )** — Deviation cross products

[M-4] **statistical** — Statistical functions

[M-4] **utility** — Matrix utility functions

Title

[M-5] **range( )** — Vector over specified range

Syntax

>   *numeric colvector*   range(*a, b, numeric scalar delta*)
>
>   *numeric colvector*   rangen(*a, b, real scalar n*)

where *a* and *b* are numeric scalars.

Description

range(*a, b, delta*) returns a column vector going from *a* to *b* in steps of abs(*delta*) ($b \geq a$) or $-$abs(*delta*) ($b < a$).

rangen(*a, b, n*) returns a round(*n*) $\times$ 1 column vector going from *a* to *b* in round(*n*)-1 steps. *a* may be less than, equal to, or greater than *b*.

Remarks

range(0, 1, .25) returns (0 \ .25 \ .5 \ .75 \ 1). The sign of the third argument does not matter; range(0, 1, -.25) returns the same thing. range(1, 0, .25) and range(1, 0, -.25) return (1 \ .75 \ .5 \ .25 \ 0).

rangen(0, .5, 6) returns (0 \ .1 \ .2 \ .3 \ .4 \ .5). rangen(.5, 0, 6) returns (.5 \ .4 \ .3 \ .2 \ .1 \ 0).

range( ) and rangen( ) may be used with complex arguments. range(1, 1i, .4) returns (1 \ .75+.25i \ .5+.5i \ .25+.75i \ 1i). rangen(1, 1i, 5) returns the same thing. For range( ), only the distance of *delta* from zero matters, so range(1, 1i, .4i) would produce the same result, as would range(1, 1i, .25+.312i).

Conformability

range(*a, b, delta*):
        a: 1 $\times$ 1
        b: 1 $\times$ 1
   delta: 1 $\times$ 1
  result: 1 $\times$ 1, if $a = b$
                   max(1+abs(*b-a*)/abs(*delta*), 2) $\times$ 1, otherwise

rangen(*a, b, n*):
        a: 1 $\times$ 1
        b: 1 $\times$ 1
        n: *n* $\times$ 1

Diagnostics

range(a, b, *delta*) aborts with error if a, b, or *delta* contains missing, if abs($b-a$)/abs(*delta*) results in overflow, or if 1+abs($b-a$)/abs(*delta*) results in a vector that is too big given the amount of memory available.

range(a, b, *delta*) returns a 1×1 result when $a = b$. In all other cases, the result is 2×1 or longer.

rangen(a, b, n) aborts with error if round(n) is less than 0 or missing.

Also See

[M-4] **standard** — Functions to create standard matrices

Title

[M-5] **rank( )** — Rank of matrix

Syntax

>*real scalar*   rank(*numeric matrix A*)
>
>*real scalar*   rank(*numeric matrix A, real scalar tol*)

Description

rank(*A*) and rank(*A, tol*) return the rank of *A*: $m \times n$.

Remarks

The row rank of a matrix A: $m \times n$ is the number of rows of A that are linearly independent. The column rank is the number of columns that are linearly independent. The terms row rank and column rank, however, are used merely for emphasis. The ranks are equal, and the result is simply called the rank of A.

rank( ) calculates the rank by counting the number of nonzero singular values of the SVD of A, where nonzero is interpreted relative to a tolerance. rank( ) uses the same tolerance as pinv( ) (see [M-5] **pinv( )**) and as svsolve( ) (see [M-5] **svsolve( )**), and optional argument *tol* is specified in the same way as with those functions.

Thus if you were going to use rank( ) before calculating an inverse using pinv( ), it would be better to skip rank( ) altogether and proceed to the pinv( ) step, because pinv( ) will return the rank, calculated as a by-product of calculating the inverse. Using rank( ) ahead of time, the SVD would be calculated twice.

rank( ) in general duplicates calculations; and, worse, if you are not planning on using pinv( ) or svsolve( ) but rather are planning on using some other function, the rank returned by rank( ) may disagree with the implied rank of whatever numerical method you subsequently use because each numerical method has its own precision and tolerances.

All that said, rank( ) is useful in interactive and pedagogical situations.

Conformability

rank(*A, tol*):
>| | | |
>|--------|--------------|------------|
>| *A*: | $m \times n$ | |
>| *tol*: | 1×1 | (optional) |
>| *result*: | 1×1 | |

Diagnostics

rank(*A*) returns missing if *A* contains missing values.

Also See

[M-5] **svd( )** — Singular value decomposition

[M-5] **fullsvd( )** — Full singular value decomposition

[M-5] **pinv( )** — Moore–Penrose pseudoinverse

[M-4] **matrix** — Matrix functions

Title

[M-5] Re( ) — Extract real or imaginary part

Syntax

>   *real matrix*   Re(*numeric matrix Z*)

>   *real matrix*   Im(*numeric matrix Z*)

Description

Re(Z) returns a real matrix containing the real part of Z. Z may be real or complex.

Im(Z) returns a real matrix containing the imaginary part of Z. Z may be a real or complex. If Z is real, Im(Z) returns a matrix of zeros.

Conformability

Re(Z), Im(Z):
>   Z: $r \times c$
>   *result*: $r \times c$

Diagnostics

Re(Z), if Z is real, literally returns Z and not a copy of Z. This makes execution of Re() applied to real arguments instant.

Also See

[M-5] **C( )** — Make complex

[M-4] **scalar** — Scalar mathematical functions

[M-4] **utility** — Matrix utility functions

Title

[M-5] **reldif( )** — Relative/absolute difference

Syntax

> *real matrix*   `reldif(`*numeric matrix X, numeric matrix Y*`)`
>
> *real scalar*   `mreldif(`*numeric matrix X, numeric matrix Y*`)`
>
> *real scalar*   `mreldifsym(`*numeric matrix X*`)`
>
> *real scalar*   `mreldifre(`*numeric matrix X*`)`

Description

`reldif(`X, Y`)` returns the relative difference defined by

$$r = \frac{|X - Y|}{|Y| + 1}$$

calculated element by element.

`mreldif(`X, Y`)` returns the maximum relative difference and is equivalent to `max(reldif(`X, Y`))`.

`mreldifsym(`X`)` is equivalent to `mreldif(`X', X`)` and so is a measure of how far the matrix is from being symmetric (Hermitian).

`mreldifre(`X`)` is equivalent to `mreldif(Re(`X`), `X`)` and so is a measure of how far the matrix is from being real.

Conformability

```
reldif(X, Y):
        X:       r × c
        y:       r × c
   result:       r × c

mreldif(X, Y):
        X:       r × c
        y:       r × c
   result:       1 × 1

mreldifsym(X):
        X:       n × n
   result:       1 × 1

mreldifre(X):
        X:       r × c
   result:       1 × 1
```

Diagnostics

The relative difference function treats equal missing values as having a difference of 0 and different missing values as having a difference of missing (.):

reldif(., .) == reldif(.a, .a) == ··· == reldif(.z, .z) == 0

reldif(., .a) == reldif(., .z) == ··· == reldif(.y, .z) == .

Also See

[M-4] **utility** — Matrix utility functions

Title

[M-5] rows( ) — Number of rows and number of columns

Syntax

    real scalar   rows(*transmorphic matrix P*)

    real scalar   cols(*transmorphic matrix P*)

    real scalar   length(*transmorphic matrix P*)

Description

rows(*P*) returns the number of rows of *P*.

cols(*P*) returns the number of columns of *P*.

length(*P*) returns rows(*P*)*cols(*P*).

Remarks

length(*P*) is typically used with vectors, as in

```
for (i=1; i<=length(x); i++) {
        ... x[i] ...
}
```

Conformability

rows(*P*), cols(*P*), length(*P*):

      P: $r \times c$
  result: 1×1

Diagnostics

rows(*P*), cols(*P*), and length(*P*) return a result that is greater than or equal to zero.

Also See

[M-4] **utility** — Matrix utility functions

Title

[M-5] **rowshape( )** — Reshape matrix

Syntax

>*transmorphic matrix*   `rowshape`(*transmorphic matrix T, real scalar r*)
>
>*transmorphic matrix*   `colshape`(*transmorphic matrix T, real scalar c*)

Description

`rowshape`(*T, r*) returns *T* transformed into a matrix with `trunc`(*r*) rows.

`colshape`(*T, c*) returns *T* having `trunc`(*c*) columns.

In both cases, elements are assigned sequentially with the column index varying more rapidly. See [M-5] **vec( )** for a function that varies the row index more rapidly.

Remarks

Remarks are presented under the following headings:

>*Example of rowshape( )*
>*Example of colshape( )*

Example of rowshape( )

```
: A
          1    2    3    4
      ┌─────────────────────┐
    1 │ 11   12   13   14   │
    2 │ 21   22   23   24   │
    3 │ 31   32   33   34   │
    4 │ 41   42   43   44   │
      └─────────────────────┘

: rowshape(A,2)
          1    2    3    4    5    6    7    8
      ┌────────────────────────────────────────────┐
    1 │ 11   12   13   14   21   22   23   24      │
    2 │ 31   32   33   34   41   42   43   44      │
      └────────────────────────────────────────────┘
```

Example of colshape( )

```
: colshape(A, 2)
          1    2
      ┌───────────┐
    1 │ 11   12   │
    2 │ 13   14   │
    3 │ 21   22   │
    4 │ 23   24   │
    5 │ 31   32   │
    6 │ 33   34   │
    7 │ 41   42   │
    8 │ 43   44   │
      └───────────┘
```

Conformability

rowshape(T, r):
> T: $r_0 \times c_0$
> r: 1×1
>   result: $r \times r_0 c_0/r$

colshape(T, c):
> T: $r_0 \times c_0$
> c: 1×1
>   result: $r_0 c_0/c \times c$

Diagnostics

Let r_0 and c_0 be the number of rows and columns of T.

rowshape() aborts with error if $r_0 \times c_0$ is not evenly divisible by trunc(r).

colshape() aborts with error if $r_0 \times c_0$ is not evenly divisible by trunc(c).

Also See

[M-4] **manipulation** — Matrix manipulation

Title

[M-5] runningsum( ) — Running sum of vector

Syntax

>   *numeric vector*   runningsum(*numeric vector x* [*, missing*])
>
>   *numeric vector*   quadrunningsum(*numeric vector x* [*, missing*])
>
>   *void*   _runningsum(*y, numeric vector x* [*, missing*])
>
>   *void*   _quadrunningsum(*y, numeric vector x* [*, missing*])

where optional argument *missing* is a *real scalar* that determines how missing values in *x* are treated:

1. Specifying *missing* as 0 is equivalent to not specifying the argument; missing values in *x* are treated as contributing 0 to the sum.

2. Specifying *missing* as 1 specifies that missing values in *x* are to be treated as missing values and turn the sum to missing.

Description

runningsum(*x*) returns a vector of the same dimension as *x* containing the running sum of *x*. Missing values are treated as contributing zero to the sum.

runningsum(*x, missing*) does the same but lets you specify how missing values are treated. runningsum(*x*, 0) is the same as runningsum(*x*). runningsum(*x*, 1) specifies that missing values are to turn the sum to missing where they occur.

quadrunningsum(*x*) and quadrunningsum(*x, missing*) do the same but perform the accumulation in quad precision.

_runningsum(*y, x* [*, missing*]) and _quadrunningsum(*y, x* [*, missing*]) work the same way, except that rather than returning the running-sum vector, they store the result in *y*. This method is slightly more efficient when *y* is a view.

Remarks

The running sum of (1, 2, 3) is (1, 3, 6).

All functions return the same type as the argument, real if argument is real, complex if complex.

Conformability

runningsum(*x*, *missing*), quadrunningsum(*x*, *missing*):

    x: $r \times 1$ or $1 \times c$
    missing: 1×1 (optional)
    result: $r \times 1$ or $1 \times c$

_runningsum(*y*, *x*, *missing*), _quadrunningsum(*y*, *x*, *missing*):

   input:
      x: $r \times 1$ or $1 \times c$
      y: $r \times 1$ or $1 \times c$ (contents irrelevant)
      missing: 1×1 (optional)
   output:
      y: $r \times 1$ or $1 \times c$

Diagnostics

If *missing* = 0, missing values are treated as contributing zero to the sum; they do not turn the sum to missing. Otherwise, missing values turn the sum to missing.

_runningsum(*y*, *x*, *missing*) and _quadrunningsum(*y*, *x*, *missing*) abort with error if *y* is not p-conformable with *x* and of the same eltype. The contents of *y* are irrelevant.

Also See

[M-5] **sum( )** — Sums

[M-4] **mathematical** — Important mathematical functions

[M-4] **utility** — Matrix utility functions

Title

[M-5] **select()** — Select rows or columns

Syntax

> *transmorphic matrix* select(*transmorphic matrix X, real vector v*)
>
> *void* st_select(*A, transmorphic matrix X, real vector v*)

Description

select(*X, v*) returns *X*

1. omitting the rows for which $v[i]==0$ (*v* a column vector) or
2. omitting the columns for which $v[j]==0$ (*v* a row vector).

st_select(*A, X, v*) does the same thing, except that the result is placed in *A* and, if *X* is a view, *A* will be a view.

Remarks

Remarks are presented under the following headings:

> *Examples*
> *Using st_select()*

Examples

1. To select rows 1, 2, and 4 of $5 \times c$ matrix X,

    submat = select(X, (1\1\0\1\0))

    See [M-2] **subscripts** for another solution, submat = X[(1\2\4), .].

2. To select columns 1, 2, and 4 of $r \times 5$ matrix X,

    submat = select(X, (1,1,0,1,0))

    See [M-2] **subscripts** for another solution, submat = X[., (1,2,4)].

3. To select rows of X for which the first element is positive,

    submat = select(X, X[.,1]:>0)

4. To select columns of X for which the first element is positive,

    submat = select(X, X[1,.]:>0)

5. To select rows of X for which there are no missing values,

    submat = select(X, rowmissing(A):==0)

6. To select rows and columns of square matrix X for which the diagonal elements are positive,

$$\begin{aligned}\text{pos} &= \text{diagonal(X):>0}\\ \text{submat} &= \text{select(X, pos)}\\ \text{submat} &= \text{select(submat, pos')}\end{aligned}$$

or, equivalently,

$$\begin{aligned}\text{pos} &= \text{diagonal(X):>0}\\ \text{submat} &= \text{select(select(X, pos), pos')}\end{aligned}$$

Using st_select()

Coding

$$\text{st\_select(submat, X, v)} \qquad (1)$$

produces the same result as coding

$$\text{submat} = \text{st\_select(X, v)} \qquad (2)$$

The difference is in how the result is stored. If X is a view (it need not be), then (1) will produce submat as a view or, if you will, a subview, whereas in (2), submat will always be a regular (nonview) matrix.

When X is a view, (1) executes more quickly than (2) and produces a result that consumes less memory.

See [M-5] **st_view()** for a description of views.

Conformability

select(X, v):

$$\begin{array}{lll} X: & r_1 \times c_1 & \\ v: & r_1 \times 1 & \text{or} \quad 1 \times c_1 \\ \textit{result}: & r_2 \times c_1 & \text{or} \quad r_1 \times c_2, \quad r_2 \leq r_1, c_2 \leq c_1 \end{array}$$

st_select(A, X, v):

input:

$$\begin{array}{lll} X: & r_1 \times c_1 & \\ v: & r_1 \times 1 & \text{or} \quad 1 \times c_1 \end{array}$$

output:

$$A: \quad r_2 \times c_1 \quad \text{or} \quad r_1 \times c_2, \quad r_2 \leq r_1, c_2 \leq c_1$$

Diagnostics

None.

Also See

[M-5] **st_subview( )** — Make view from view

[M-2] **op_colon** — Colon operators

[M-2] **subscripts** — Use of subscripts

[M-4] **utility** — Matrix utility functions

Title

[M-5] **setbreakintr( )** — Break-key processing

Syntax

> *real scalar*   setbreakintr(*real scalar val*)
>
> *real scalar*   querybreakintr()
>
> *real scalar*   breakkey()
>
> *void*   breakkeyreset()

Description

setbreakintr(*val*) turns the break-key interrupt off (*val*==0) or on (*val*!=0) and returns the value of the previous break-key mode, 1, it was on, or 0, it was off.

querybreakintr() returns 1 if the break-key interrupt is on and 0 otherwise.

breakkey() (for use in setbreakintr(0) mode) returns 1 if the break key has been pressed since it was last reset.

breakkeyreset() (for use in setbreakintr(0) mode) resets the break key.

Remarks

Remarks are presented under the following headings:

> *Default break-key processing*
> *Suspending the break-key interrupt*
> *Break-key polling*

Default break-key processing

By default, if the user presses *Break*, Mata stops execution and returns control to the console, setting the return code to 1.

To obtain this behavior, there is nothing you need do. You do not need to use these functions.

Suspending the break-key interrupt

The default behavior is known as interrupt-on-break and is also known as setbreakintr(1) mode.

The alternative is break-key suspension, also known as setbreakintr(0) mode.

For instance, you have several steps that must be performed in their entirety or not at all. The way to do this is

```
val = setbreakintr(0)
...
... (critical code) ...
...
(void) setbreakintr(val)
```

The first line stores in *val* the current break-key processing mode and then sets the mode to break-key suspension. The critical code then runs. If the user presses *Break* during the execution of the critical code, that will be ignored. Finally, the code restores the previous break-key processing mode.

Break-key polling

In coding large, interactive systems, you may wish to adopt the break-key polling style of coding rather than interrupt-on-break. In this alternative style of coding, you turn off interrupt-on-break:

```
val = setbreakintr(0)
```

and, from then on in your code, wherever you are willing to interrupt your code, you ask (poll whether) the break key has been pressed:

```
...
if (breakkey()) {
        ...
}
...
```

In this style of coding, you must decide where and when you are going to reset the break key, because once the break key has been pressed, `breakkey()` will continue to return 1 every time it is called. To reset the break key, code,

```
breakkeyreset()
```

You can also adopt a mixed style of coding, using interrupt-on-break in some places and polling in others. Function `querybreakintr()` can then be used to determine the current mode.

Conformability

```
setbreakintr(val):
        val:      1 × 1
        result:   1 × 1
querybreakintr(), breakkey():
        result:   1 × 1
breakkeyreset():
        result:   void
```

Diagnostics

setbreakintr(1) aborts with break if the break key has been pressed since the last setbreakintr(0) or breakkeyreset(). Code breakkeyreset() before setbreakintr(1) if you do not want this behavior.

After coding setbreakintr(1), remember to restore setbreakintr(0) mode. It is not, however, necessary, to restore the original mode if exit() or _error() is about to be executed.

breakkey(), once the break key has been pressed, continues to return 1 until breakkeyreset() is executed.

There is absolutely no reason to use breakkey() in setbreakintr(0) mode, because the only value it could return is 0.

Also See

[M-5] **error( )** — Issue error message

[M-4] **programming** — Programming functions

Title

[M-5] **sign( )** — Sign and complex quadrant functions

Syntax

>*real matrix*   sign(*real matrix R*)
>
>*real matrix*   quadrant(*complex matrix Z*)

Description

sign(*R*) returns the elementwise sign of *R*. sign( ) is defined

argument range	sign(*arg*)
arg ≥ .	.
arg < 0	−1
arg = 0	0
arg > 0	1

quadrant(*Z*) returns a real matrix recording the quadrant of each complex entry in *Z*. quadrant( ) is defined

argument range		quadrant(*arg*)
Re(*arg*)	Im(*arg*)	
Re ≥ .		.
Re = 0	Im = 0	.
Re > 0	Im ≥ 0	1
Re ≤ 0	Im > 0	2
Re < 0	Im ≤ 0	3
Re ≥ 0	Im < 0	4

quadrant(1+0i)==1, quadrant(-1+0i)==3
quadrant(0+1i)==2, quadrant(0-1i)==4

Conformability

sign(*R*):
>   *R*:   *r* × *c*
>   *result*: *r* × *c*

quadrant(*Z*):
>   *Z*:   *r* × *c*
>   *result*: *r* × *c*

Diagnostics

sign(*R*) returns missing when *R* is missing.

quadrant(*Z*) returns missing when *Z* is missing.

Also See

[M-5] **dsign( )** — FORTRAN-like DSIGN( ) function

[M-4] **scalar** — Scalar mathematical functions

Title

[M-5] **sin( )** — Trigonometric and hyperbolic functions

Syntax

numeric matrix	sin(*numeric matrix Z*)
numeric matrix	cos(*numeric matrix Z*)
numeric matrix	tan(*numeric matrix Z*)
numeric matrix	asin(*numeric matrix Z*)
numeric matrix	acos(*numeric matrix Z*)
numeric matrix	atan(*numeric matrix Z*)
real matrix	atan2(*real matrix X, real matrix Y*)
real matrix	asinr(*real matrix R*)
real matrix	acosr(*real matrix R*)
real matrix	atanr(*real matrix R*)
real matrix	arg(*complex matrix Z*)
numeric matrix	sinh(*numeric matrix Z*)
numeric matrix	cosh(*numeric matrix Z*)
numeric matrix	tanh(*numeric matrix Z*)
numeric matrix	asinh(*numeric matrix Z*)
numeric matrix	acosh(*numeric matrix Z*)
numeric matrix	atanh(*numeric matrix Z*)
real scalar	pi()

Description

sin(*Z*), cos(*Z*), and tan(*Z*) return the appropriate trigonometric function. Angles are measured in radians. All return real if argument is real, complex if argument is complex.

asin(*Z*) returns arcsine in the range $-\pi/2$ to $\pi/2$. Returned value is real if argument is real, complex if argument is complex.

acos(*Z*) returns arccosine in the range 0 to π. Returned value is real if argument is real, complex if argument is complex.

atan(*Z*) returns arctangent in the range $-\pi/2$ to $\pi/2$. Returned value is real if argument is real, complex if argument is complex.

Warning: asin(), acos(), and atan() base their result on whether the input is real or complex. Thus acos(2) == ., whereas acos(2+0i) == -1.317i.

atan2(*X*, *Y*) returns the radian value in the range $(-\pi, \pi]$ of the angle of the vector determined by (*X,Y*), the result being in the range $[0, \pi]$ for quadrants 1 and 2 and $[0, -\pi)$ in quadrants 4 and 3. *X* and *Y* are real; also see related function arg() below; atan2(*X*, *Y*) is equivalent to arg(C(*X*, *Y*)).

asinr(*R*), acosr(*R*), and atanr(*R*) are the real-only variants of asin(), acos(), and atan(). asin(), acos(), and atan() examine the received argument and, if it is real, call the appropriate real function. Thus use of asinr(), acosr(), and atanr() in place of asin(), acos(), and atan() when arguments are known to be real will speed execution, but not by much. We recommend use of asin(), acos(), and atan() in all cases.

arg(*Z*) returns the arctangent of Im(*Z*)/Re(*Z*) in the correct quadrant, the result being in the range $(-\pi, \pi]$; $[0, \pi]$ in quadrants 1 and 2 and $[0, -\pi)$ in quadrants 4 and 3. Also see related function atan2() above; arg(*Z*) is equivalent to atan2(Re(*Z*), Im(*Z*)).

sinh(*Z*), cosh(*Z*), and tanh(*Z*) return the hyperbolic sine, cosine, and tangent, respectively. Returned value is real if argument is real, complex if complex.

asinh(*Z*), acosh(*Z*), and atanh(*Z*) return the inverse hyperbolic sine, cosine, and tangent, respectively. Returned value is real if argument is real, complex if complex. See warning above for asin(), etc. This warning applies here.

pi() returns the value of π.

Conformability

atan2(*X*, *Y*):
    X: $r_1 \times c_1$
    Y: $r_2 \times c_2$,   *X* and *Y* r-conformable
  result: $\max(r_1, r_2) \times \max(c_1, c_2)$

pi() returns a 1×1 scalar.

All other functions return a matrix of the same dimension as input, containing element-by-element calculated results.

Diagnostics

tan(*Z*) returns missing when cos(*Z*)==0.

All functions return missing for real arguments when the result would be complex but return the appropriate complex result when given a complex argument.

Also See

[M-4] **scalar** — Scalar mathematical functions

Title

[M-5] **sizeof( )** — Number of bytes consumed by object

Syntax

>   *real scalar*   sizeof(*transmorphic matrix A*)

Description

sizeof(*A*) returns the number of bytes consumed by *A*.

Remarks

sizeof(*A*) returns the same number as shown by mata describe; see [M-3] **mata describe**.

A 500 × 5 real matrix consumes 20,000 bytes:

```
: sizeof(mymatrix)
  20000
```

A 500 × 5 view matrix, however, consumes only 24 bytes:

```
: sizeof(myview)
  24
```

To obtain the number of bytes consumed by a function, pass a dereferenced function pointer:

```
: sizeof(*&myfcn())
  320
```

Conformability

sizeof(*A*):
>   *A*:   r × c
>   *result*: 1 × 1

Diagnostics

The number returned by sizeof(*A*) does not include any overhead, which usually amounts to 64 bytes, but can be less (as small as zero in the case of recently used scalars).

If *A* is a pointer matrix, the number returned reflects the amount of memory required to store *A* itself and does not include the memory consumed by its siblings.

Also See

>   [M-4] **programming** — Programming functions

Title

> [M-5] **solve_tol( )** — Tolerance used by solvers and inverters

Syntax

> *real scalar* solve_tol(*numeric matrix Z, real scalar usertol*)

Description

solve_tol(*Z, usertol*) returns the tolerance used by many Mata solvers to solve $AX = B$ and by many Mata inverters to obtain A^{-1}. *usertol* is the tolerance specified by the user or is missing value if the user did not specify a tolerance.

Remarks

The tolerance used by many Mata solvers to solve $AX = B$ and by many Mata inverters to obtain A^{-1} is

$$eta = s * \frac{\text{trace(abs}(Z))}{n} \qquad (1)$$

where $s = 1e{-}13$ or a value specified by the user, n is the min(rows(Z), cols(Z)), and Z is a matrix related to A, usually by some form of decomposition, but could be A itself (for instance, if A were triangular). See, for instance, [M-5] **solvelower( )** and [M-5] **cholsolve( )**.

When *usertol* > 0 and *usertol* $<$. is specified, solvetol() returns *eta* calculated with $s =$ *usertol*.

When *usertol* ≤ 0 is specified, solvetol() returns $-$*usertol*.

When *usertol* $\geq$. is specified, solvetol() returns a default result, calculated as

1. If external real scalar _solvetolerance does not exist, as is usually the case, the value of *eta* is returned using $s = 1e{-}13$.

2. If external real scalar _solvetolerance does exist,

    a. If _solvetolerance > 0, the value of *eta* is returned using $s =$ solvetolerance.

    b. If _solvetolerance ≤ 0, $-$_solvetolerance is returned.

Conformability

solve_tol(*Z, usertol*):
           Z: $r \times c$
    usertol: 1×1
     result: 1×1

Diagnostics

solve_tol(Z, *usertol*) skips over missing values in Z in calculating (1); n is defined as the number of nonmissing elements on the diagonal.

Also See

[M-4] **utility** — Matrix utility functions

Title

[M-5] **solvelower( )** — Solve AX=B for X, A triangular

Syntax

    numeric matrix   `solvelower(`*A*, *B* [, *rank* [, *tol* [, *d*]]]`)`

    numeric matrix   `solveupper(`*A*, *B* [, *rank* [, *tol* [, *d*]]]`)`

    real scalar   `_solvelower(`*A*, *B* [, *tol* [, *d*]]`)`

    real scalar   `_solveupper(`*A*, *B* [, *tol* [, *d*]]`)`

where

    A: *numeric matrix*
    B: *numeric matrix*
  rank: irrelevant; *real scalar* returned
    tol: *real scalar*
    d: *numeric scalar*

Description

These functions are used in the implementation of the other solve functions; see [M-5] **lusolve( )**, [M-5] **qrsolve( )**, and [M-5] **svsolve( )**.

`solvelower(`*A*, *B*, ...`)` and `_solvelower(`*A*, *B*, ...`)` solve lower-triangular systems.

`solveupper(`*A*, *B*, ...`)` and `_solveupper(`*A*, *B*, ...`)` solve upper-triangular systems.

Functions without a leading underscore—`solvelower()` and `solveupper()`—return the solution; *A* and *B* are unchanged.

Functions with a leading underscore—`_solvelower()` and `_solveupper()`—return the solution in *B*.

All four functions produce a generalized solution if *A* is singular. The functions without an underscore place the rank of *A* in *rank*, if the argument is specified. The underscore functions return the rank.

Determination of singularity is made via *tol*. *tol* is interpreted in the standard way—as a multiplier for the default if *tol* > 0 is specified and as an absolute quantity to use in place of the default if *tol* ≤ 0 is specified.

All four functions allow *d* to be optionally specified. Specifying *d* = . is equivalent to not specifying *d*.

If *d* ≠ . is specified, that value is used as if it appeared on the diagonal of *A*. The four functions do not in fact require that *A* be triangular; they merely look at the lower or upper triangle and pretend that the opposite triangle contains zeros. This feature is useful when a decomposition utility has stored both the lower and upper triangles in one matrix, because one need not take apart the combined matrix. In such cases, it sometimes happens that the diagonal of the matrix corresponds to one matrix but not the other, and that for the other matrix, one merely knows that the diagonal elements are, say, 1. Then you can specify *d* = 1.

Remarks

The triangular-solve functions documented here exploit the triangular structure in A and solve for X by recursive substitution.

When A is of full rank, these functions provide the same solution as the other solve functions, such as [M-5] **lusolve( )**, [M-5] **qrsolve( )**, and [M-5] **svsolve( )**. The `solvelower()` and `solveupper()` functions, however, will produce the answer more quickly because of the large computational savings.

When A is singular, however, you may wish to consider whether you want to use these triangular-solve functions. The triangular-solve functions documented here reach a generalized solution by setting $B_{ij} = 0$, for all j, when A_{ij} is zero or too small (as determined by *tol*). The method produces a generalized inverse, but there are many generalized inverses, and this one may not have the other properties you want.

Remarks are presented under the following headings:

    Derivation
    Tolerance

Derivation

We wish to solve

$$AX = B \tag{1}$$

when A is triangular. Let us consider the lower-triangular case first. `solvelower()` is up to handling full matrices for B and X, but let us assume X: $n \times 1$ and B: $m \times 1$:

$$\begin{bmatrix} a_{11} & 0 & 0 \ldots & 0 \\ a_{21} & 0 & 0 \ldots & 0 \\ \vdots & \vdots & \ddots & \vdots \\ a_{m1} & a_{m2} & a_{m3} & a_{mn} \end{bmatrix} \begin{bmatrix} x_1 \\ x_2 \\ \vdots \\ x_n \end{bmatrix} = \begin{bmatrix} b_1 \\ b_2 \\ \vdots \\ b_m \end{bmatrix}$$

The first equation to be solved is

$$a_{11} x_1 = b_1$$

and the solution is simply

$$x_1 = \frac{b_1}{a_{11}} \tag{2}$$

The second equation to be solved is

$$a_{21} x_1 + a_{22} x_2 = b_2$$

and because we have already solved for x_1, the solution is simply

$$x_2 = \frac{b_2 - a_{21} x_1}{a_{22}} \tag{3}$$

We proceed similarly for the remaining rows of A. If there are additional columns in B and X, we can then proceed to handling each remaining column just as we handled the first column above.

In the upper-triangular case, the formulas are similar except that you start with the last row of A.

Tolerance

In (2) and (3), we divide by the diagonal elements of A. If element a_{ii} is less than *eta* in absolute value, the corresponding x_i is set to zero. *eta* is given by

$$eta = \text{1e-13} * \texttt{trace(abs}(A))/\texttt{rows}(A)$$

If you specify *tol* > 0, the value you specify is used to multiply *eta*. You may instead specify *tol* ≤ 0, and then the negative of the value you specify is used in place of *eta*; see [M-1] **tolerance**.

Conformability

solvelower(*A*, *B*, *rank*, *tol*, *d*), solveupper(*A*, *B*, *rank*, *tol*, *d*):

   input:

	A:	$n \times n$	
	B:	$n \times k$	
	tol:	1×1	(optional)
	d:	1×1	(optional)

   output:

	rank:	1×1	(optional)
	result:	$n \times k$	

_solvelower(*A*, *B*, *tol*, *d*), _solveupper(*A*, *B*, *tol*, *d*):

   input:

	A:	$n \times n$	
	B:	$n \times k$	
	tol:	1×1	(optional)
	d:	1×1	(optional)

   output:

	B:	$n \times k$	
	result:	1×1	(contains rank)

Diagnostics

solvelower(*A*, *B*, ...), _solvelower(*A*, *B*, ...), solveupper(*A*, *B*, ...), and _solveupper(*A*, *B*, ...) do not verify that the upper (lower) triangle of *A* contains zeros; they just use the lower (upper) triangle of *A*.

_solvelower(*A*, *B*, ...) and _solveupper(*A*, *B*, ...) do not abort with error if *B* is a view but can produce results subject to considerable roundoff error.

Also See

[M-5] **cholsolve( )** — Solve AX=B for X using Cholesky decomposition

[M-5] **lusolve( )** — Solve AX=B for X using LU decomposition

[M-5] **qrsolve( )** — Solve AX=B for X using QR decomposition

[M-5] **svsolve( )** — Solve AX=B for X using singular value decomposition

[M-5] **solve_tol( )** — Tolerance used by solvers and inverters

[M-4] **matrix** — Matrix functions

Title

[M-5] **sort( )** — Reorder rows of matrix

Syntax

transmorphic matrix	`sort`(*transmorphic matrix X, real rowvector idx*)
void	`_sort`(*transmorphic matrix X, real rowvector idx*)
transmorphic matrix	`jumble`(*transmorphic matrix X*)
void	`_jumble`(*transmorphic matrix X*)
real colvector	`order`(*transmorphic matrix X, real rowvector idx*)
real colvector	`unorder`(*real scalar n*)
void	`_collate`(*transmorphic matrix X, real colvector p*)

where

1. *X* may not be a pointer matrix.
2. *p* must be a permutation column vector, a $1 \times c$ vector containing the integers 1, 2, ..., *c* in some order.

Description

`sort`(*X*, *idx*) returns *X* with rows in ascending or descending order of the columns specified by *idx*. For instance, `sort`(*X*, 1) sorts *X* on its first column; `sort`(*X*, (1,2)) sorts *X* on its first and second columns (meaning rows with equal values in their first column are ordered on their second column). In general, the *i*th sort key is column `abs`(*idx*[*i*]). Order is ascending if *idx*[*i*] > 0 and descending otherwise. Ascending and descending are defined in terms of [M-5] **abs( )** (length of elements) for complex.

`_sort`(*X*, *idx*) does the same as `sort`(*X*, *idx*), except that *X* is sorted in place.

`jumble`(*X*) returns *X* with rows in random order. For instance, to shuffle a deck of cards numbered 1 to 52, one could code `jumble(1::52)`. See `uniformseed()` in [M-5] **uniform( )** for information on setting the random-number seed.

`_jumble`(*X*) does the same as `jumble`(*X*), except that *X* is jumbled in place.

`order`(*X*, *idx*) returns the permutation vector—see [M-1] **permutation**—that would put *X* in ascending (descending) order of the columns specified by *idx*. A row-permutation vector is a $1 \times c$ column vector containing the integers 1, 2, ..., *c* in some order. Vectors (1\2\3), (1\3\2), (2\1\3), (2\3\1), (3\1\2), and (3\2\1) are examples. Row-permutation vectors are used to specify the order in which the rows of a matrix *X* are to appear. If *p* is a row-permutation vector, *X*[*p*, .] returns *X* with its rows in the order of *p*; *p* = (3\2\1) would reverse the rows of *X*. `order`(*X*, *idx*) returns the row-permutation vector that would sort *X* and, as a matter of fact, `sort`(*X*, *idx*) is implemented as *X*[`order`(*X*, *idx*), .].

unorder(n) returns a $1 \times n$ permutation vector for placing the rows in random order. Random numbers are calculated by uniform(); see uniformseed() in [M-5] **uniform( )** for information on setting the random-number seed. jumble() is implemented in terms of unorder(): jumble(X) is equivalent to X[unorder(rows(X)), .].

_collate(X, p) is equivalent to $X = X[p, .]$; it changes the order of the rows of X. _collate() is used by _sort() and _jumble() and has the advantage over subscripting in that no extra memory is required when the result is to be assigned back to itself. Consider

$$X = X[p, .]$$

There will be an instant after $X[p, .]$ has been calculated but before the result has been assigned back to X when two copies of X exist. _collate(X, p) avoids that. _collate() is not a substitute for subscripting in all cases; _collate() requires p be a permutation vector.

Remarks

If X is complex, the ordering is defined in terms of [M-5] **abs( )** of its elements.

Also see invorder() and revorder() in [M-5] **invorder( )**. Let p be the permutation vector returned by order():

$$p = \text{order}(X, \ldots)$$

Then $X[p, .]$ are the sorted rows of X. revorder() can be used to reverse sort order: X[revorder(p), .] are the rows of X in the reverse of the order of $X[p, .]$. invorder() provides the inverse transform: If $Y = X[p, .]$, then $X = Y$[invorder(p), .].

Conformability

sort(X, idx), jumble(X):
        X: $r_1 \times c_1$
        idx: $1 \times c_2$, $c_2 \leq c_1$
        result: $r_1 \times c_1$

_sort(X, idx), _jumble(X):
        X: $r_1 \times c_1$
        idx: $1 \times c_2$, $c_2 \leq c_1$
        result: *void*; X row order modified

order(X, idx):
        X: $r_1 \times c_1$
        idx: $1 \times c_2$, $c_2 \leq c_1$
        result: $r_1 \times 1$

unorder(n):
        n: 1×1
        result: $n \times 1$

_collate(X, p):
        X: $r \times c$
        p: $r \times 1$
        result: *void*; X row order modified

Diagnostics

sort(*X*, *idx*) aborts with error if any element of abs(*idx*) is less than 1 or greater than rows(*X*) or if cols(*idx*) > 500.

_sort(*X*, *idx*) aborts with error if any element of abs(*idx*) is less than 1 or greater than rows(*X*), or if cols(*idx*) > 500, or if *X* is a view.

_jumble(*X*) aborts with error if *X* is a view.

order(*X*, *idx*) aborts with error if any element of abs(*idx*) is less than 1 or greater than rows(*X*) or if cols(*idx*) > 500.

unorder(*n*) aborts with error if $n < 1$.

_collate(*X*, *p*) aborts with error if *p* is not a permutation vector or if *X* is a view.

Also See

[M-5] **invorder( )** — Permutation vector manipulation

[M-5] **uniqrows( )** — Obtain sorted, unique values

[M-4] **manipulation** — Matrix manipulation

Title

[M-5] **spline3()** — Cubic spline interpolation

Syntax

> *real matrix*   spline3(*real vector x, real vector y*)
>
> *real vector*   spline3eval(*real matrix spline_info, real vector x*)

Description

spline3(*x, y*) returns the coefficients of a cubic natural spline S(*x*). The elements of *x* must be strictly monotone increasing.

spline3eval(*spline_info, x*) uses the information returned by spline3() to evaluate and return the spline at the abscissas *x*. Elements of the returned result are set to missing if outside the range of the spline. *x* is assumed to be monotonically increasing.

Remarks

spline3() and spline3eval() is a translation into Mata of Herriot and Reinsch (CUBNATSPLINE)

For *xx* in $[x_i, x_{i+1})$:

$$S(xx) = \{(d_i t + c_i)t + b_i\}t + y_i$$

with $t = xx - x_i$.

spline3() returns (b, c, d, x, y) or, if x and y are row vectors, (b, c, d, x', y').

Conformability

spline3(*x, y*):
    x: $n \times 1$ or $1 \times n$
    y: $n \times 1$ or $1 \times n$
    result: $n \times 5$

spline3eval(*spline_info, x*):
    info: $n \times 5$
    x: $m \times 1$ or $1 \times m$
    result: $m \times 1$ or $1 \times m$

Diagnostics

spline3(*x, y*) requires that *x* be in ascending order.

spline3eval(*spline_info, x*) that requires *x* be in ascending order.

Reference

Algorithm CUBNATSPLINE, 472-P 5-0, Collected Algorithms from CACM, *Procedures for Natural Spline Interpolation* [E1] John G. Herriot, Computer Science Department, Stanford University and Christian H. Reinsch, Mathematisches Institut der Technischen Universität, 8 München 2, Germany.

Also See

[M-4] **mathematical** — Important mathematical functions

Title

[M-5] **sqrt( )** — Square root

Syntax

> *numeric matrix*   sqrt(*numeric matrix Z*)

Description

sqrt(*Z*) returns the elementwise square root of *Z*.

Conformability

sqrt(*Z*)
>   *Z*: $r \times c$
>   *result*: $r \times c$

Diagnostics

sqrt(*Z*) returns missing when *Z* is real and $Z < 0$; i.e., sqrt(-4) = . but sqrt(-4+0i) = 2i.

Also See

[M-5] **cholesky( )** — Cholesky square-root decomposition

[M-4] **scalar** — Scalar mathematical functions

Title

[M-5] **st_addobs( )** — Add observations to current Stata dataset

Syntax

> void   st_addobs(*real scalar n*)
>
> void   st_addobs(*real scalar n*, *real scalar nofill*)
>
> *real scalar*   _st_addobs(*real scalar n*)
>
> *real scalar*   _st_addobs(*real scalar n*, *real scalar nofill*)

Description

st_addobs(*n*) adds *n* observations to the current Stata dataset.

st_addobs(*n*, *nofill*) does the same thing but saves computer time by not filling in the additional observations with the appropriate missing-value code if *nofill* $\neq$ 0. st_addobs(*n*, 0) is equivalent to st_addobs(*n*). Use of st_addobs() with *nofill* $\neq$ 0 is not recommended. If you specify *nofill* $\neq$ 0, it is your responsibility to ensure that the added observations ultimately are filled in or removed before control is returned to Stata.

_st_addobs(*n*) and _st_addobs(*n*, *nofill*) perform the same action as st_addobs(*n*) and st_addobs(*n*, *nofill*), except that they return 0 if successful and the appropriate Stata return code otherwise (otherwise usually being caused by insufficient memory). Where _st_addobs() would return nonzero, st_addobs() aborts with error.

Remarks

There need not be any variables defined to add observations. If you are attempting to create a dataset from nothing, you can add the observations first and then add the variables, or you can add the variables and then add the observations. Use st_addvar() (see [M-5] **st_addvar( )**) to add variables.

Conformability

st_addobs(*n*, *nofill*):
         n: 1×1
     nofill: 1×1 (optional)
     result: *void*

_st_addobs(*n*, *nofill*):
         n: 1×1
     nofill: 1×1 (optional)
     result: 1×1

Diagnostics

st_addobs(n[, *nofill*]) and _st_addobs(n[, *nofill*]) abort with error if $n < 0$. They do nothing if $n = 0$.

st_addobs() aborts with error if there is insufficient memory to add the requested number of observations.

_st_addobs() aborts with error if $n < 0$ but otherwise returns the appropriate Stata return code if the observations cannot be added. If they are added, 0 is returned.

st_addobs() and _st_addobs() do not set st_updata() (see [M-5] **st_updata( )**); you must set it if you want it set.

Also See

[M-4] **stata** — Stata interface functions

Title

[M-5] **st_addvar( )** — Add variable to current Stata dataset

Syntax

> *real rowvector*   st_addvar(*type*, *name*)
>
> *real rowvector*   st_addvar(*type*, *name*, *nofill*)
>
> *real rowvector*   _st_addvar(*type*, *name*)
>
> *real rowvector*   _st_addvar(*type*, *name*, *nofill*)

where

> *type*: *string scalar* or *rowvector* containing "byte", "int", "long", "float", "double", or "str#"
> or
> *real scalar* or *rowvector* containing # (interpreted as str#)
>
> *name*: *string rowvector* containing new variable names
>
> *nofill*: *real scalar* containing 0 or non-0

Description

st_addvar(*type*, *name*) adds new variable *name*(s) of type *type* to the Stata dataset. Returned are the variable indices of the new variables. st_addvar() aborts with error (and adds no variables) if any of the variables already exist or cannot be added for other reasons.

st_addvar(*type*, *name*, *nofill*) does the same thing. *nofill* $\neq$ 0 specifies that the variables' values are not to be filled in with missing values. st_addvar(*type*, *name*, 0) is the same as st_addvar(*type*, *name*). Use of *nofill* $\neq$ 0 is not, in general, recommended. See *Remarks* below.

_st_addvar() does the same thing as st_addvar() except that, rather than aborting with error if the new variable cannot be added, returned is a 1 × 1 scalar containing the negative of the appropriate Stata return code.

Remarks

Remarks are presented under the following headings:

> Creating a new variable
> Creating new variables
> Creating new string variables
> Creating a new temporary variable
> Creating temporary variables
> Handling errors
> Using nofill

Creating a new variable

To create new variable myvar as a double, code

```
idx = st_addvar("double", "myvar")
```

or

```
(void) st_addvar("double", "myvar")
```

You use the first form if you will subsequently need the variable's index number, or you use the second form otherwise.

Creating new variables

You can add more than one variable. For instance,

```
idx = st_addvar("double", ("myvar1","myvar2"))
```

adds two new variables, both of type double.

```
idx = st_addvar(("double","float") ("myvar1","myvar2"))
```

also adds two new variables, but this time, myvar1 is double and myvar2 is float.

Creating new string variables

Creating string variables is no different from any other type:

```
idx = st_addvar(("str10","str5"), ("myvar1","myvar2"))
```

creates myvar1 as a str10 and myvar2 as a str5.

There is, however, another way to specify the types.

```
idx = st_addvar((10,5), ("myvar1","myvar2"))
```

also creates myvar1 as a str10 and myvar2 as a str5.

```
idx = st_addvar(10, ("myvar1","myvar2"))
```

creates both variables as str10s.

Creating a new temporary variable

Function st_tempname() (see [M-5] **st_tempname()**) returns temporary variable names. To create a temporary variable as a double, code

```
idx = st_addvar("double", st_tempname())
```

or code

```
(void) st_addvar("double", name=st_tempname())
```

You use the first form if you will subsequently need the variable's index, or you use the second form if you will subsequently need the variable's name. You will certainly need one or the other. If you will need both, code

```
idx = st_addvar("double", name=st_tempname())
```

Creating temporary variables

`st_tempname()` can return a vector of temporary variable names.

        idx = st_addvar("double", st_tempname(5))

creates five temporary variables, each of type `double`.

Handling errors

There are three common reasons why `st_addvar()` might fail: the variable name is invalid or a variable under that name already exists or there is insufficient memory to add another variable. If there is a problem adding a variable, `st_addvar()` will abort with error. If you wish to avoid the traceback log and just have Stata issue an error, use `_st_addvar()` and code

        if ((idx = _st_addvar("double", "myvar"))<0) exit(error(-idx))

If you are adding multiple variables, look at the first element of what `_st_addvar()` returns:

        if ((idx = _st_addvar(types, names))[1]<0) exit(error(-idx))

Using nofill

The three-argument versions of `st_addvar()` and `_st_addvar()` allow you to avoid filling in the values of the newly created variable. Filling in those values with missing really is a waste of time if the next thing you are going to do is fill in the values with something else. On the other hand, it is important that all the observations be filled in on the new variable before control is returned to Stata, and this includes returning to Stata because of subsequent error or the user pressing *Break*. Thus use of *nofill* $\neq 0$ is not, in general, recommended. Filling in values really does not take that long.

If you are determined to save the computer time, however, see [M-5] **setbreakintr()**. To do things right, you need to set the break key off, create your variable, fill it in, and turn break-key processing back on.

There is, however, a case in which use of *nofill* $\neq 0$ is acceptable and such effort is not required: when you are creating a temporary variable. Temporary variables vanish in any case, and it does not matter whether they are filled in before they vanish.

Temporary variables in fact vanish not when Mata ends but when the ado-file calling Mata ends, if there is an ado-file. We will assume there is an ado-file because that is the only case in which you would be creating a temporary variable anyway. Because they do not disappear until later, there is the possibility of there being an issue if the variable is not filled in. If we assume, however, that your Mata program is correctly written and does fill in the variable ultimately, then the chances of a problem are minimal. If the user presses *Break* or there is some other problem in your program that causes Mata to abort, the ado-file will be aborted, too, and the variable will vanish.

Let us add that Stata will not crash if a variable is not filled in, even if it regains control. The danger is that the user will look at the variable or, worse, use it and be baffled by what he or she sees, which might concern not only odd values but also NaNs and worse.

Conformability

st_addvar(*type*, *name*, *nofill*):
    type: 1×1 or $1 \times k$
    name: $1 \times k$
    nofill: 1×1 (optional)
    result: $1 \times k$

_st_addvar(*type*, *name*, *nofill*):
    type: 1×1 or $1 \times k$
    name: $1 \times k$
    nofill: 1×1 (optional)
    result: $1 \times k$ or, if error, 1×1

Diagnostics

st_addvar(*type*, *name*, *nofill*) aborts with error if

1. *type* is not equal to a valid Stata variable type and it is not a number that would form a valid str# variable type;

2. *name* is not a valid variable name;

3. a variable named *name* already exists;

4. there is insufficient memory to add another variable.

_st_addvar(*type*, *name*, *nofill*) aborts with error for reason 1 above, but otherwise, it returns the negative value of the appropriate Stata return code.

Both functions, when creating multiple variables, create either all the variables or none of them. Whether creating one variable or many, if variables are created, st_updata() (see [M-5] **st_updata()**) is set unless all variables are temporary; see [M-5] **st_tempname()**.

Reference

Gould, W. 2006. Mata Matters: Creating new variables—sounds boring, isn't. *Stata Journal* 6: 112–123.

Also See

[M-5] **st_store()** — Modify values stored in current Stata dataset

[M-5] **st_tempname()** — Temporary Stata names

[M-4] **stata** — Stata interface functions

Title

[M-5] **st_data( )** — Load copy of current Stata dataset

Syntax

real scalar	_st_data(*real scalar i, real scalar j*)	
real matrix	st_data(*real matrix i, rowvector j*)	(1,2)
real matrix	st_data(*real matrix i, rowvector j, scalar selectvar*)	(1,2,3)
string scalar	_st_sdata(*real scalar i, real scalar j*)	
string matrix	st_sdata(*real matrix i, rowvector j*)	(1,2)
string matrix	st_sdata(*real matrix i, rowvector j, scalar selectvar*)	(1,2,3)

where

1. *i* may be specified as a 1×1 scalar, as a 1×1 scalar containing missing, as a column vector of observation numbers, as a row vector specifying an observation range, or as a $k \times 2$ matrix specifying both.

   a. st_data(1, 2) returns the first observation on the second variable.

   b. st_data(., 2) returns all observations on the second variable.

   c. st_data((1\2\5), 2) returns observations 1, 2, and 5 on the second variable.

   d. st_data((1,5), 2) returns observations 1 through 5 on the second variable.

   e. st_data((1,5\7,9), 2) returns observations 1 through 5 and observations 7 through 9 on the second variable.

   When a range is specified, any element of the range (i_1, i_2) may be specified to contribute zero observations if $i_2 = i_1 - 1$.

2. *j* may be specified as a real row vector or as a string row vector.

   a. st_data(2, .) returns the values of all variables, second observation, of the Stata dataset.

   b. st_data(2, 1) returns the value of the first variable, second observation.

   c. st_data(2, (3,1,9)) returns the values of the third, first, and ninth variables of the second observation.

   d. st_data(2, ("mpg", "weight")) returns the values of variables mpg and weight, second observation.

   e. st_data(2, ("gnp", "l.gnp")) returns the values of gnp and the lag of gnp, second observation.

3. *selectvar* may be specified as real or as a string. Observations for which *selectvar* $\neq 0$ will be selected. If *selectvar* is real, it is interpreted as a variable number. If string, *selectvar* should contain the name of a Stata variable.

Specifying *selectvar* as "" or as missing (.) has the same result as not specifying *selectvar* at all.

Specifying *selectvar* as 0 means that observations with missing values of the variables specified by *j* are to be excluded.

Description

_st_data(*i*, *j*) returns the numeric value of the *i*th observation of the *j*th Stata variable. Observations are numbered 1 through st_nobs(). Variables are numbered 1 through st_nvar().

st_data(*i*, *j*) is similar to _st_data(*i*, *j*) except

1. *i* may be specified as a vector or matrix to obtain multiple observations simultaneously,

2. *j* may be specified using names or indices (indices are faster), and

3. *j* may be specified as a vector to obtain multiple variables simultaneously.

The net effect is that st_data() can return a scalar (the value of one variable in one observation), a row vector (the value of many variables in an observation), a column vector (the value of a variable in many observations), or a matrix (the value of many variables in many observations).

st_data(*i*, *j*, *selectvar*) works like st_data(*i*, *j*) except that only observations for which *selectvar* $\neq 0$ are returned.

_st_sdata() and st_sdata() are the string variants of _st_data() and st_data(). _st_data() and st_data() are for use with numeric variables; they return missing (.) when used with string variables. _st_sdata() and st_sdata() are for use with string variables; they return empty string ("") when used with numeric variables.

Remarks

Remarks are presented under the following headings:

> Description of _st_data() and _st_sdata()
> Description of st_data() and st_sdata()
> Details of observation subscripting using st_data() and st_sdata()

Description of _st_data() and _st_sdata()

_st_data() returns one variable's value in one observation. You refer to variables and observations by their numbers. The first variable in the Stata dataset is 1; the first observation is 1.

_st_data(1, 1)	value of 1st obs., 1st variable
_st_data(1, 2)	value of 1st obs., 2nd variable
_st_data(2, 1)	value of 2nd obs., 1st variable

_st_sdata() works the same way. _st_data() is for use with numeric variables, and _st_sdata() is for use with string variables.

_st_data() and _st_sdata() are the fastest way to obtain the value of a variable in one observation.

[M-5] st_data( ) — Load copy of current Stata dataset

Description of st_data( ) and st_sdata( )

st_data( ) can be used just like _st_data( ), and used that way, it produces the same result.

Variables, however, can be referred to by their names or their numbers:

st_data(1, 1)	value of 1st obs., 1st variable
st_data(1, 2)	value of 1st obs., 2nd variable
st_data(2, 1)	value of 2nd obs., 1st variable
st_data(1, "mpg")	value of 1st obs, variable mpg
st_data(2, "mpg")	value of 2nd obs, variable mpg

Also, you may specify more than one variable:

st_data(2, (1,2,3))	value of 2nd obs., variables 1, 2, and 3
st_data(2, ("mpg","weight","displ"))	value of 2nd obs., variables mpg, weight, and displ

Used this way, st_data( ) returns a row vector.

Similarly, you may obtain multiple observations:

st_data((1\2\3), 10)	values of obs. 1, 2, and 3, variable 10
st_data((1,5), 10)	values of obs. 1 through 5, variable 10
st_data((1,5)\(7,9), 10)	values of obs. 1 through 5 and 7 through 9, variable 10

st_sdata( ) works the same way as st_data( ).

Details of observation subscripting using st_data( ) and st_sdata( )

1. i may be specified as a scalar: the specified, single observation is returned. i must be between 1 and st_nobs( ).

2. i may be specified as a scalar containing missing value: all observations are returned.

3. i may be specified as a column vector: the specified observations are returned. Each element of i must be between 1 and st_nobs( ) or may be missing. Missing is interpreted as st_nobs( ).

4. i may be specified as a row vector with two columns: the specified range of observations is returned; (c_1, c_2) returns the $c_2 - c_1 + 1$ observations c_1 through c_2.

    $c_2 - c_1 + 1$ must evaluate to a number greater than or equal to 0. In general, c_1 and c_2 must be between 1 and st_nobs( ), but if $c_2 - c_1 + 1 = 0$, then c_1 may be between 1 and st_nobs( ) + 1 and c_2 may be between 0 and st_nobs( ). Regardless, $c_1 ==$. or $c_2 ==$. is interpreted as st_nobs( ).

5. i may be specified as a $k \times 2$ matrix: $((1,5)\backslash(7,7)\backslash(20,30))$ specifies observations 1 through 5, 7, and 20 through 30.

Conformability

_st_data(i, j), _st_sdata(i, j):
        i: 1×1
        j: 1×1
   result: 1×1

st_data(i, j), st_sdata(i, j):
        i: $n \times 1$ or $n_2 \times 2$
        j: $1 \times k$
   result: $n \times k$

st_data(i, j, selectvar), st_sdata(i, j, selectvar):
        i: $n \times 1$ or $n_2 \times 2$
        j: $1 \times k$
  selectvar: 1×1
   result: $(n - e) \times k$, where e is number of observations excluded by selectvar

Diagnostics

_st_data(i, j) returns missing (.) if i or j is out of range; it does not abort with error.

_st_sdata(i, j) returns "" if i or j is out of range; it does not abort with error.

st_data(i, j) and st_sdata(i, j) abort with error if any element of i or j is out of range. j may be specified as a vector of variable names or a vector of variable indices. If names are specified, abbreviations are allowed. If you do not want this, use st_varindex() (see [M-5] **st_varindex()**) to translate variable names into variable indices.

Also See

[M-5] **st_view()** — Make matrix that is a view onto current Stata dataset

[M-5] **st_store()** — Modify values stored in current Stata dataset

[M-4] **stata** — Stata interface functions

Title

[M-5] **st_dir( )** — Obtain list of Stata objects

Syntax

string colvector   st_dir(*cat, subcat, pattern*)

string colvector   st_dir(*cat, subcat, pattern, adorn*)

where

cat: *string scalar* containing "local", "global", "r()", "e()", "s()", or "char"

subcat: *string scalar* containing "macro", "numscalar", "strscalar", "matrix", or, if *cat*=="char", "_dta" or a name.

pattern: *string scalar* containing a pattern as defined in [M-5] **strmatch( )**

adorn: *string scalar* containing 0 or non-0

The valid *cat–subcat* combinations and their meanings are

cat	subcat	meaning
"local"	"macro"	Stata's local macros
"global"	"macro"	Stata's global macros
"global"	"numscalar"	Stata's numeric scalars
"global"	"strscalar"	Stata's string scalars
"global"	"matrix"	Stata's matrices
"r()"	"macro"	macros in r()
"r()"	"numscalar"	numeric scalars in r()
"r()"	"matrix"	matrices in r()
"e()"	"macro"	macros in e()
"e()"	"numscalar"	numeric scalars in e()
"e()"	"matrix"	matrices in e()
"s()"	"macro"	macros in s()
"char"	"_dta"	characteristics in _dta[]
"char"	"*name*"	characteristics in variable *name*[]

st_dir( ) returns an empty list if an invalid *cat–subcat* combination is specified.

[M-5] st_dir( ) — Obtain list of Stata objects

Description

st_dir(*cat*, *subcat*, *pattern*) and st_dir(*cat*, *subcat*, *pattern*, *adorn*) return a column vector containing the names matching *pattern* of the Stata objects described by *cat–subcat*.

Argument *adorn* is optional; not specifying it is equivalent to specifying *adorn* = 0. By default, simple names are returned. If *adorn* $\neq$ 0 is specified, the name is adorned in the standard Stata way used to describe the object. Say that one is listing the macros in e() and one of the elements is e(cmd). By default, the returned vector will contain an element equal to "cmd". With *adorn* $\neq$ 0, the element will be "e(cmd)".

For many objects, the adorned and unadorned forms of the names are the same.

Conformability

st_dir(*cat*, *subcat*, *pattern*, *adorn*):
        cat: 1 $\times$ 1
    subcat: 1 $\times$ 1
    pattern: 1 $\times$ 1
    adorn: 1 $\times$ 1   (optional)
    result: k $\times$ 1

Diagnostics

st_dir(*cat*, *subcat*, *pattern*) and st_dir(*cat*, *subcat*, *pattern*, *adorn*) abort with error if *cat* or *subcat* is invalid. If the combination is invalid, however, J(0,1,"") is returned. *subcat*==*name* is considered invalid unless *cat*=="char".

st_dir() aborts with error if any of its arguments are views.

Also See

[M-4] **stata** — Stata interface functions

Title

[M-5] **st_dropvar( )** — Drop variables or observations

Syntax

>   *void*   st_dropvar(*transmorphic rowvector vars*)
>
>   *void*   st_dropobsin(*real matrix range*)
>
>   *void*   st_dropobsif(*real colvector select*)
>
>   *void*   st_keepvar(*transmorphic rowvector vars*)
>
>   *void*   st_keepobsin(*real matrix range*)
>
>   *void*   st_keepobsif(*real colvector select*)

Description

st_dropvar(*vars*) drops the variables specified. *vars* is a row vector that may contain either variable names or variable indices. st_dropvar(.) drops all variables and observations.

st_dropobsin() and st_dropobsif() have to do with dropping observations.

st_dropobsin(*range*) specifies the observations to be dropped:

> st_dropobsin(5) drops observation 5.
>
> st_dropobsin((5,9)) drops observations 5 through 9.
>
> st_dropobsin((5\8\12)) drops observations 5 and 8 and 12.
>
> st_dropobsin((5,7\8,11\13,13)) drops observations 5 through 7, 8 through 11, and 13.
>
> st_dropobsin(.) drops all observations (but not the variables).
>
> st_dropobsin(J(0,1,.)) drops no observations (or variables).

st_dropobsif(*select*) specifies a st_nobs() $\times$ 1 vector. Observations i for which $select_i \neq 0$ are dropped.

st_keepvar(), st_keepobsin(), and st_keepobsif() do the same thing, except that the variables and observations to be kept are specified.

Remarks

To drop all variables and observations, code any of the following:

        st_dropvar(.)
        st_keepvar(J(1,0,.))
        st_keepvar(J(1,0,""))

All do the same thing. Dropping all the variables clears the dataset.

Dropping all the observations, however, leaves the variables in place.

Conformability

st_dropvar(*vars*), st_keepvar(*vars*):
    vars: $1 \times k$
    result: void

st_dropobsin(*range*), st_keepobsin(*range*):
    range: $k \times 1$ or $k \times 2$
    result: void

st_dropobsif(*select*), st_keepobsif(*select*):
    select: st_nobs() $\times 1$
    result: void

Diagnostics

st_dropvar(*vars*) and st_keepvar(*vars*) abort with error if any element of *vars* is missing unless *vars* is 1×1, in which case they drop or keep all the variables.

st_dropvar(*vars*) and st_keepvar(*vars*) abort with error if any element of *vars* is not a valid variable index or name, or if *vars* is a view. If *vars* is specified as names, abbreviations are not allowed.

st_dropvar() and st_keepvar() set st_updata() (see [M-5] **st_updata( )**) unless all variables dropped are temporary; see [M-5] **st_tempname( )**.

st_dropobsin(*range*) and st_keepobsin(*range*) abort with error if any element of *range* is missing unless *range* is 1×1, in which case they drop or keep all the observations.

st_dropobsin(*range*) and st_keepobsin(*range*) abort with error if any element of *range* is not a valid observation number (is not between 1 and st_nobs() [see [M-5] **st_nvar( )**] inclusive) or if *range* is a view.

st_dropobsif(*select*) and st_keepobsif(*select*) abort with error if *select* is a view.

st_dropobsin(), st_dropobsif(), st_keepobsin(), and st_keepobsif() set st_updata() if any observations are removed from the data.

Be aware that, after dropping any variables or observations, any previously constructed views (see [M-5] **st_view( )**) are probably invalid because views are internally stored in terms of variable and observation numbers. Subsequent use of an invalid view may lead to unexpected results or an abort with error.

Also See

[M-4] **stata** — Stata interface functions

Title

> [M-5] **st_global( )** — Obtain strings from and put strings into global macros

Syntax

>> *string scalar* st_global(*string scalar name*)
>>
>> *void* st_global(*string scalar name*, *string scalar contents*)
>
> where
>
> 1. *name* is to contain
>
>   a. global macro such as "myname"
>
>   b. r() macro such as "r(names)"
>
>   c. e() macro such as "e(cmd)"
>
>   d. s() macro such as "s(vars)"
>
>   e. c() macro such as "c(current_date)"
>
>   f. dataset characteristic such as "_dta[date]"
>
>   g. variable characteristic such as "mpg[note]"
>
> 2. st_global(*name*, "") deletes, and it deletes even if the *name* is not a macro. For instance, perhaps r(N) is a numeric scalar; st_global(r(N), "") will delete it. Perhaps e(X) is matrix; st_global(e(X), "") will delete it.

Description

st_global(*name*) returns the contents of the specified Stata global.

st_global(*name*, *contents*) sets or resets the contents of the specified Stata global. If the Stata global did not previously exist, a new global is created. If the global did exist, the new contents replace the old.

Remarks

Mata provides a suite of functions for obtaining and setting the contents of global macros, local macros, saved results, etc. It can sometimes be confusing to know which you should use. The table on the following page will help.

[M-5] st_global() — Obtain strings from and put strings into global macros

Stata component/action	function call
Local macro	
obtain contents	*contents* = st_local("*name*")
create/set/replace	st_local("*name*", *contents*)
delete	st_local("*name*", "")
Global macro	
obtain contents	*contents* = st_global("*name*")
create/set/replace	st_global("*name*", *contents*)
delete	st_global("*name*", "")
Global numeric scalar	
obtain contents	*value* = st_numscalar("*name*")
create/set/replace	st_numscalar("*name*", *value*)
delete	st_numscalar("*name*", J(0,0,.))
Global string scalar	
obtain contents	*contents* = st_strscalar("*name*")
create/set/replace	st_strscalar("*name*", *contents*)
delete	st_strscalar("*name*", J(0,0,""))
Global matrix	
obtain contents	*matrix* = st_matrix("*name*") *rowlabel* = st_matrixrowstripe("*name*") *collabel* = st_matrixcolstripe("*name*")
create/set/replace	st_matrix("*name*", *matrix*) st_matrixrowstripe("*name*", *rowlabel*) st_matrixcolstripe("*name*", *collabel*)
replace	st_replacematrix("*name*", *matrix*)
delete	st_matrix("name", J(0,0,.))
Characteristic	
obtain contents	*contents* = st_global("*name*[*name*]")
create/set/replace	st_global("*name*[*name*]", *contents*)
delete	st_global("*name*[*name*]", "")

[M-5] st_global( ) — Obtain strings from and put strings into global macros

Stata component/action	function call
r( ) results	
macro	
obtain contents	*contents* = st_global("r(*name*)")
create/set/replace	st_global("r(*name*)", *contents*)
numeric scalar	
obtain contents	*value* = st_numscalar("r(*name*)")
create/set/replace	st_numscalar("r(*name*)", *value*)
matrix	
obtain contents	*matrix* = st_matrix("r(*name*)")
	rowlabel = st_matrixrowstripe("r(*name*)")
	collabel = st_matrixcolstripe("r(*name*)")
create/set/replace	st_matrix("r(name)", *matrix*)
	st_matrixrowstripe("r(*name*)", *rowlabel*)
	st_matrixcolstripe("r(*name*)", *collabel*)
replace	st_replacematrix("r(*name*)", *matrix*)
IN ALL CASES	
delete	st_global("r(*name*)", "")
to delete all of r( )	st_rclear( )
e( ) results	same as r( ) results, but code e(*name*) and st_eclear( )
s( ) results	
macro	
obtain contents	*contents* = st_global("s(*name*)")
create/set/replace	st_global("s(*name*)", *contents*)
delete	st_global("s(*name*)", "")
to delete all of s( )	st_sclear( )
c( ) results	
macro	
obtain contents	*contents* = st_global("c(*name*)")
numeric scalar	
obtain contents	*value* = st_numscalar("c(*name*)")

See [M-5] **st_local( )**, [M-5] **st_numscalar( )**, [M-5] **st_matrix( )**, and [M-5] **st_rclear( )**.

Conformability

st_global(*name*):
       name: 1 × 1
       result: 1 × 1

st_global(*name*, *contents*):
       name: 1 × 1
       contents: 1 × 1
       result: *void*

Diagnostics

st_global(*name*) returns "" if the name contained in *name* is not defined. st_global(*name*) aborts with error if the name is malformed, such as st_global("invalid name").

st_global(*name*, *contents*) aborts with error if the name contained in *name* is malformed. The maximum length of strings in Mata is significantly longer than in Stata. st_global() truncates what is stored at the appropriate maximum length if that is necessary.

Also See

[M-5] **st_rclear( )** — Clear r( ), e( ), or s( )

[M-4] **stata** — Stata interface functions

Title

[M-5] st_isfmt( ) — Whether valid %fmt

Syntax

>*real scalar*   st_isfmt(*string scalar s*)
>
>*real scalar*   st_isnumfmt(*string scalar s*)
>
>*real scalar*   st_isstrfmt(*string scalar s*)

Description

st_isfmt(*s*) returns 1 if *s* contains a valid Stata %*fmt* and 0 otherwise.

st_isnumfmt(*s*) returns 1 if *s* contains a valid Stata numeric %*fmt* and 0 otherwise.

st_isstrfmt(*s*) returns 1 if *s* contains a valid Stata string %*fmt* and 0 otherwise.

Conformability

st_isfmt(*s*), st_isnumfmt(*s*), st_isstrfmt(*s*):
>   *s*: 1×1
>   *result*: 1×1

Diagnostics

st_isfmt(*s*), st_isnumfmt(*s*), and st_isstrfmt(*s*) abort with error if *s* is a view.

Also See

[M-4] **stata** — Stata interface functions

Title

[M-5] st_isname( ) — Whether valid Stata name

Syntax

> *real scalar* st_isname(*string scalar s*)
>
> *real scalar* st_islmname(*string scalar s*)

Description

st_isname(*s*) returns 1 if *s* contains a valid Stata name and 0 otherwise.

st_islmname(*s*) returns 1 if *s* contains a valid Stata local-macro name and 0 otherwise.

Conformability

st_isname(*s*), st_islmname(*s*):
    s: 1×1
  result: 1×1

Diagnostics

st_isname(*s*) aborts with error if *s* is a view (but st_islmname() does not).

Also See

[M-4] **stata** — Stata interface functions

Title

> [M-5] **st_local( )** — Obtain strings from and put strings into Stata macros

Syntax

>   *string scalar*   st_local(*string scalar name*)
>
>   *void*   st_local(*string scalar name*, *string scalar contents*)

Note: st_local(*name*, "") deletes.

Description

st_local(*name*) returns the contents of the specified local macro.

st_local(*name*, *contents*) sets or resets the contents of the specified local macro. If the macro did not previously exist, a new macro is created. If it did previously exist, the new contents replace the old.

Remarks

See [M-5] **st_global( )** and [M-5] **st_rclear( )**.

Conformability

>   st_local(*name*):
>   *name*: 1×1
>   *result*: 1×1
>   st_local(*name*, *contents*):
>   *name*: 1×1
>   *contents*: 1×1
>   *result*:   *void*

Diagnostics

st_local(*name*) returns "" if the name contained in *name* is not defined. st_local(*name*) aborts with error if the name is malformed.

st_local(*name*, *contents*) aborts with error if the name contained in *name* is malformed.

Also See

[M-4] **stata** — Stata interface functions

Title

> **[M-5] st_macroexpand( )** — Expand Stata macros in string

Syntax

>   *string scalar*   st_macroexpand(*string scalar s*)
>
>   *real scalar*   _st_macroexpand(*S*, *string scalar s*)

Note: the type of *S* does not matter; it is replaced and becomes a string scalar.

Description

st_macroexpand(*s*) returns *s* with any quoted or dollar sign–prefixed macros expanded.

_st_macroexpand(*S*, *s*) places in *S* the contents of *s* with any quoted or dollar sign–prefixed macros expanded and returns a Stata return code (it returns 0 if all went well).

Remarks

Be careful coding string literals containing quoted or prefixed macros because macros are also expanded at compile time. For instance, consider

>   s = st_macroexpand("regress `varlist'")

`varlist' will be substituted with its value at compile time. What you probably want is

>   s = st_macroexpand("regress " + "`" + "varlist" + "'")

Conformability

st_macroexpand(*s*):
>   *s*: 1×1
>   *result*: 1×1

_st_macroexpand(*S*, *s*):
>   *input*:
>   *s*: 1×1
>   *output*:
>   *S*: 1×1
>   *result*: 1×1

Diagnostics

st_macroexpand(*s*) aborts with error if *s* is too long (exceedingly unlikely) or if macro expansion fails (also unlikely).

_st_macroexpand(*S*, *s*) aborts with error if *s* is too long.

Also See

[M-4] **stata** — Stata interface functions

Title

[M-5] **st_matrix( )** — Obtain and put Stata matrices

Syntax

real matrix	`st_matrix(`*string scalar name*`)`
string matrix	`st_matrixrowstripe(`*string scalar name*`)`
string matrix	`st_matrixcolstripe(`*string scalar name*`)`
void	`st_matrix(`*string scalar name*`, `*real matrix X*`)`
void	`st_matrixrowstripe(`*string scalar name*`, `*string matrix s*`)`
void	`st_matrixcolstripe(`*string scalar name*`, `*string matrix s*`)`
void	`st_replacematrix(`*string scalar name*`, `*real matrix X*`)`

where

1. All functions allow *name* to be
   a. global matrix name such as `"mymatrix"`,
   b. `r()` matrix such as `"r(Z)"`, or
   c. `e()` matrix such as `"e(V)"`.

2. `st_matrix(`*name*`)` returns `J(0,0,.)` if the matrix does not exist.

3. The contents of `r()` and `e()` may be replaced except for the estimation results `e(b)` and `e(V)`.

4. `st_matrix(`*name*`, X)` deletes the specified Stata matrix if $X==J(0,0,.)$.

Description

`st_matrix(`*name*`)` returns the contents of Stata's matrix *name*, or it returns `J(0,0,.)` if the matrix does not exist.

`st_matrixrowstripe(`*name*`)` returns the row stripe associated with the matrix *name*, or it returns `J(0,2,"")` if the matrix does not exist.

`st_matrixcolstripe(`*name*`)` returns the column stripe associated with the matrix *name*, or it returns `J(0,2,"")` if the matrix does not exist.

`st_matrix(`*name*`, X)` sets or resets the contents of the Stata matrix *name* to be X. If the matrix did not previously exist, a new matrix is created. If the matrix did exist, the new contents replace the old. Either way, the row and column stripes are also reset to contain `"r1"`, `"r2"`, ..., and `"c1"`, `"c2"`, ...

st_matrix(*name*, *X*) deletes the Stata matrix *name* when *X* is 0 × 0: st_matrix(*name*, J(0,0,.)) deletes Stata matrix *name* or does nothing if *name* does not exist.

st_matrixrowstripe(*name*, *s*) and st_matrixcolstripe(*name*, *s*) change the contents to be *s* of the row and column stripe associated with the already existing Stata matrix *name*. In either case, *s* must be $n \times 2$, where n = the number of rows (columns) of the underlying matrix.

st_matrixrowstripe(*name*, *s*) and st_matrixcolstripe(*name*, *s*) reset the row and column stripe to be "r1", "r2", ..., and "c1", "c2", ..., when *s* is 0 × 2 (i.e., J(0,2,"")).

st_replacematrix(*name*, *X*) resets the contents of the Stata matrix *name* to be *X*. The existing Stata matrix must have the same number of rows and columns as *X*. The row stripes and column stripes remain unchanged.

Remarks

Remarks are presented under the following headings:

> Processing Stata's row and column stripes
> Stata's matsize is irrelevant

Also see [M-5] **st_global( )** and [M-5] **st_rclear( )**.

Processing Stata's row and column stripes

Both row stripes and column stripes are presented in the same way: each row of *s* represents the *eq:op.name* associated with a row or column of the underlying matrix. The first column records *eq*, and the second column records *op.name*. For instance, given the following Stata matrix

	turn	L.turn	eq2:turn	eq2:L.turn
mpg	1	2	3	4
L.mpg	5	6	7	8
eq2:mpg	9	10	11	12
eq2:L.mpg	13	14	15	16

st_matrixrowstripe(*name*) returns the 4 × 2 string matrix

""	"mpg"
""	"L.mpg"
"eq2"	"mpg"
"eq2"	"L.mpg"

and st_matrixcolstripe(*name*) returns

""	"turn"
""	"L.turn"
"eq2"	"turn"
"eq2"	"L.turn"

Stata's matsize is irrelevant

Matrices in Stata are limited to matsize (see [R] **matsize**), a number between 40 and 11,000. Mata matrices have no such limits.

When getting a matrix, the matsize limit plays no role.

When putting a matrix, the matsize limit is ignored; meaning that, to use the matrix in Stata, the user may have to reset matsize or, if the matrix is too large, the user may not be able to use the matrix at all.

Conformability

st_matrix(*name*):
    name: 1×1
    result: $m \times n$ (0×0 if not found)

st_matrixrowstripe(*name*):
    name: 1×1
    result: $m \times 2$ (0×2 if not found)

st_matrixcolstripe(*name*):
    name: 1×1
    result: $n \times 2$ (0×2 if not found)

st_matrix(*name*, *X*):
    name: 1×1
    X: $r \times c$ (0×0 means delete)
    result: *void*

st_matrixrowstripe(*name*, *s*):
    name: 1×1
    s: $r \times 2$ (0×2 means default "r1", "r2", ...)
    result: *void*

st_matrixcolstripe(*name*, *s*):
    name: 1×1
    s: $c \times 2$ (0×2 means default "c1", "c2", ...)
    result: *void*

st_replacematrix(*name*, *X*):
    name: 1×1
    X: $m \times n$ (0×0 means delete)
    result: *void*

Diagnostics

st_matrix(*name*), st_matrixrowstripe(*name*), and st_matrixcolstripe(*name*) abort with error if *name* is malformed. Also,

1. st_matrix(*name*) returns J(0,0,.) if Stata matrix *name* does not exist.

2. st_matrixrowstripe(*name*) and st_matrixcolstripe(*name*) return J(0,2,"") if Stata matrix *name* does not exist. There is no possibility that matrix *name* might exist and not have row and column stripes.

st_matrix(*name*, X), st_matrixrowstripe(*name*, s), and st_matrixcolstripe(*name*, s) abort with error if *name* is malformed. Also,

1. st_matrixrowstripe(*name*, s) aborts with error if rows(s) is not equal to the number of rows of Stata matrix *name* and rows(s)!=0, or if cols(s)!=2.

2. st_matrixcolstripe(*name*, s) aborts with error if cols(s) is not equal to the number of columns of Stata matrix *name* and cols(s)!=0, or if cols(s)!=2.

st_replacematrix(*name*, X) aborts with error if Stata matrix *name* does not have the same number of rows and columns as X. st_replacematrix() also aborts with error if Stata matrix *name* does not exist and X!=J(0,0,.); st_replacematrix() does nothing if the matrix does not exist and X==J(0,0,.). st_replacematrix() aborts with error if *name* is malformed.

Also See

[M-5] **st_rclear()** — Clear r( ), e( ), or s( )

[M-4] **stata** — Stata interface functions

Title

[M-5] **st_numscalar( )** — Obtain values from and put values into Stata scalars

Syntax

    real   st_numscalar(*string scalar name*)

    void   st_numscalar(*string scalar name*, *real value*)

    string   st_strscalar(*string scalar name*)

    void   st_strscalar(*string scalar name*, *string value*)

where

1. All functions allow *name* to be

   a. global scalar such as "myname",

   b. r() scalar such as "r(mean)",

   c. e() scalar such as "e(N)", or

   d. c() scalar such as "c(namelen)".

2. st_numscalar(*name*) and st_strscalar(*name*) return a 1 × 1 result containing the value of *name* or they return a 0 × 0 if *name* does not exist.

3. st_numscalar(*name*, *value*) and st_strscalar(*name*, *value*) allow *value* to be 1 × 1 containing the new value with which *name* is to be created or replaced, or 0 × 0, specifying that *name* is to be deleted.

Description

st_numscalar(*name*) returns the value of the specified Stata numeric scalar, or it returns J(0,0,.) if the scalar does not exist.

st_numscalar(*name*, *value*) sets or resets the value of the specified numeric scalar, assuming *value* != J(0,0,.). st_numscalar(*name*, *value*) deletes the specified scalar if *value* == J(0,0,.). st_numscalar("x", J(0,0,.)) erases the scalar x, or it does nothing if scalar x did not exist.

st_strscalar(*name*) returns the value of the specified Stata string scalar, or it returns J(0,0,"") if the scalar does not exist.

st_strscalar(*name*, *value*) sets or resets the value of the specified scalar, assuming *value* != J(0,0,""). st_strscalar(*name*, *value*) deletes the specified scalar if *value* == J(0,0,""). st_strscalar("x", J(0,0,"")) erases the scalar x, or it does nothing if scalar x did not exist.

Concerning deletion of a scalar, it does not matter whether you code st_numscalar(*name*, J(0,0,.)) or st_strscalar(*name*, J(0,0,"")); both yield the same result.

Remarks

See [M-5] **st_global( )** and [M-5] **st_rclear( )**.

Conformability

st_numscalar(*name*), st_strscalar(*name*):
    name: 1×1
    result: 1×1 or 0×0

st_numscalar(*name*, *value*), st_strscalar(*name*, *value*):
    name: 1×1
    value: 1×1 or 0×0
    result: *void*

Diagnostics

st_numscalar(*name*) and st_strscalar(*name*) return J(0,0,.) or J(0,0,"") if Stata scalar *name* does not exist. They abort with error, however, if the name is malformed.

st_numscalar(*name*, *value*) and st_strscalar(*name*, *value*) abort with error if the name is malformed.

Also See

[M-5] **st_rclear( )** — Clear r( ), e( ), or s( )

[M-4] **stata** — Stata interface functions

Title

[M-5] **st_nvar( )** — Numbers of variables and observations

Syntax

>*real scalar*   st_nvar()

>*real scalar*   st_nobs()

Description

st_nvar() returns the number of variables defined in the dataset currently loaded in Stata.

st_nobs() returns the number of observations defined in the dataset currently loaded in Stata.

Conformability

st_nvar(), st_nobs():
>*result*: 1×1

Diagnostics

None.

Also See

[M-4] **stata** — Stata interface functions

Title

[M-5] **st_rclear( )** — Clear r( ), e( ), or s( )

Syntax

>   *void* st_rclear()
>
>   *void* st_eclear()
>
>   *void* st_sclear()

Description

st_rclear() clears Stata's r() saved results.

st_eclear() clears Stata's e() saved results.

st_sclear() clears Stata's s() saved results.

Remarks

Returning results in r(), e(), or s() is one way of communicating results calculated in Mata back to Stata; see [M-1] **ado**. See [R] **saved results** for a description of e(), r(), and s().

Use st_rclear(), st_eclear(), or st_sclear() to clear results, and then use st_global() to define macros, st_numscalar() to define scalars, and st_matrix() to define Stata matrices in r(), e(), or s(). For example,

```
st_rclear()
st_global("r(name)", "tab")        see [M-5] st_global( )
st_numscalar("r(N)", n1+n2)        see [M-5] st_numscalar( )
st_matrix("r(table)", X+Y)         see [M-5] st_matrix( )
```

It is not necessary to clear before saving, but it is considered good style unless it is your intention to add to previously saved results.

If a saved result already exists, st_global(), st_numscalar(), and st_matrix() may be used to redefine it and even to redefine it to a different type. For instance, continuing with our example, later in the same code might appear

```
if (...) {
    st_matrix("r(name)", X)
}
```

Saved result r(name) was previously defined as a macro containing "tab", and, even so, can now be redefined to become a matrix.

If you want to eliminate a particular saved result, use st_global() to change its contents to "":

```
st_global("r(name)", "")
```

Do this regardless of the type of the saved result. Here we use st_global() to clear saved result r(name), which might be a macro and might be a matrix.

Conformability

st_rclear(), st_eclear(), and st_sclear() take no arguments and return void.

Diagnostics

st_rclear(), st_eclear(), and st_sclear() cannot fail.

Also See

[M-5] **st_global()** — Obtain strings from and put strings into global macros

[M-5] **st_numscalar()** — Obtain values from and put values into Stata scalars

[M-5] **st_matrix()** — Obtain and put Stata matrices

[M-4] **stata** — Stata interface functions

Title

[M-5] st_store( ) — Modify values stored in current Stata dataset

Syntax

    void   _st_store(*real scalar i, real scalar j, real scalar x*)

    void   st_store(*real matrix i, rowvector j, real matrix X*)   (1,2)

    void   st_store(*real matrix i, rowvector j, scalar selectvar, real matrix X*) (1,2,3)

    void   _st_sstore(*real scalar i, real scalar j, string scalar s*)

    void   st_sstore(*real matrix i, rowvector j, string matrix X*)   (1,2)

    void   st_sstore(*real matrix i, rowvector j, scalar selectvar, string matrix X*) (1,2,3)

where

1. *i* may be specified in the same way as with st_data( ).
2. *j* may be specified in the same way as with st_data( ), except that time-series operators may not be specified.
3. *selectvar* may be specified in the same way as with st_data( ).

See [M-5] **st_data( )**.

Description

These functions mirror _st_data( ), st_data( ), and st_sdata( ). Rather than returning the contents from the Stata dataset, these commands change those contents to be as given by the last argument.

Remarks

See [M-5] **st_data( )**.

Conformability

_st_store(i, j, x), _st_sstore(i, j, x):
    i: 1×1
    j: 1×1
    x: 1×1
    result: *void*

st_store(i, j, X), st_sstore(i, j, X):
    i: $n \times 1$ or $n_2 \times 2$
    j: $1 \times k$
    X: $n \times k$
    result: *void*

st_store(*i*, *j*, *selectvar*, *X*), st_sstore(*i*, *j*, *selectvar*, *X*):

i:	$n \times 1$ or $n_2 \times 2$
j:	$1 \times k$
selectvar:	1×1
X:	$(n - e) \times k$, where *e* is number of observations excluded by *selectvar*
result:	void

Diagnostics

_st_store(*i*, *j*, *x*) and _st_sstore(*i*, *j*, *s*) do nothing if *i* or *j* is out of range; they do not abort with error.

st_store(*i*, *j*, *X*) and st_sstore(*i*, *j*, *s*) abort with error if any element of *i* or *j* is out of range. *j* may be specified as a vector of variable names or as a vector of variable indices. If names are specified, abbreviations are allowed. If you do not want this, use st_varindex() (see [M-5] **st_varindex( )**) to translate variable names into variable indices.

st_store() and st_sstore() abort with error if *X* is not p-conformable with the matrix that st_data() (st_sdata()) would return.

Also See

[M-5] **st_data( )** — Load copy of current Stata dataset

[M-5] **st_addvar( )** — Add variable to current Stata dataset

[M-4] **stata** — Stata interface functions

Title

[M-5] **st_subview( )** — Make view from view

Syntax

>*void* st_subview(*X*, *transmorphic matrix V*, *real matrix i*, *real matrix j*)

where

1. The type of *X* does not matter; it is replaced.

2. *V* is typically a view, but that is not required. *V*, however, must be real or string.

Description

st_subview(*X*, *V*, *i*, *j*) creates new view matrix *X* from existing view matrix *V*. *V* is to have been created from a previous call to st_view() (see [M-5] **st_view( )**) or st_subview().

Although st_subview() is intended for use with view matrices, it may also be used when *V* is a regular matrix. Thus code may be written in such a way that it will work without regard to whether a matrix is or is not a view.

i may be specified as a 1 × 1 scalar, a 1 × 1 scalar containing missing, as a column vector of row numbers, as a row vector specifying a row-number range, or as a $k \times 2$ matrix specifying both:

   a. st_subview(*X*,*V*, 1,2) makes *X* equal to the first row of the second column of *V*.

   b. st_subview(*X*,*V*, .,2) makes *X* equal to all rows of the second column of *V*.

   c. st_subview(*X*,*V*, (1\2\5),2) makes *X* equal to rows 1, 2, and 5 of the second column of *V*.

   d. st_subview(*X*,*V*, (1,5),2) makes *X* equal to rows 1 through 5 of the second column of *V*.

   e. st_subview(*X*,*V*, (1,5\7,9),2) makes *X* equal to rows 1 through 5 and 7 through 9 of the second column of *V*.

   f. When a range is specified, any element of the range (i_1,i_2) may be set to contribute zero observations if $i_2 = i_1 - 1$. For example, (1,0) is not an error and neither is (1,0\5,7).

j may be specified in the same way as *i*, except transposed, to specify the selected columns:

   a. st_subview(*X*,*V*, 2,.) makes *X* equal to all columns of the second row of *V*.

   b. st_subview(*X*,*V*, 2,(1,2,5)) makes *X* equal to columns 1, 2, and 5 of the second row of *V*.

   c. st_subview(*X*,*V*, 2,(1\5)) makes *X* equal to columns 1 through 5 of the second row of *V*.

   d. st_subview(*X*,*V*, 2,((1\5),(7\9))) makes *X* equal to columns 1 through 5 and 7 through 9 of the second row of *V*.

e. When a range is specified, any element of the range ($j_1 \backslash j_2$) may be set to contribute zero columns if $j_2 = j_1 - 1$. For example, (1\0) is not an error and neither is ((1\0),(5\7)).

Obviously, notations for i and j can be specified simultaneously:

a. st_subview(X,V, .,.) makes X a duplicate of V.

b. st_subview(X,V, .,(1\5)) makes X equal to columns 1 through 5 of all rows of X.

c. st_subview(X,V, (10,25),(1\5)) makes X equal to columns 1 through 5 of rows 10 through 25 of X.

Also, st_subview() may be used to create views with duplicate variables or observations from V.

Remarks

Say that you need to make a calculation on matrices X and Y, which might be views. Perhaps the calculation is invsym($X'X$)*$X'Y$. Regardless, you start as follows:

```
st_view(X, ., ("v2","v3","v4"), 0)
st_view(Y, ., ("v1","v7")      , 0)
```

You are already in trouble. You smartly coded fourth argument as 0, meaning exclude the missing values, but you do not know that the same observations were excluded in the manufacturing of X as in the manufacturing of Y.

If you had previously created a touse variable in your dataset marking the observations to be used in the calculation, one solution would be

```
st_view(X, ., ("v2","v3","v4"), "touse")
st_view(Y, ., ("v1","v7")      , "touse")
```

That solution is recommended, but let's assume you did not do that. The other solution is

```
st_view(M, ., ("v2","v3","v4", "v1","v7"), 0)
st_subview(X, M, ., (1,2,3))
st_subview(Y, M, ., (4,5))
```

The first call to st_view() will eliminate observations with missing values on any of the variables, and the second two st_subview() calls will create the matrices you wanted, obtaining them from the correctly formed M. Basically, the two st_subview() calls amount to the same thing as

```
X = M[., (1,2,3)]
Y = M[., (4,5)]
```

but you do not want to code that because then matrices X and Y would contain copies of the data, and you are worried that the dataset might be large.

For a second example, let's pretend that you are processing a panel dataset and making calculations from matrix X within panel. Your code looks something like

```
st_view(id, ., "panelid", 0)
for (i=1; i<=rows(id); i=j+1) {
    j = endobs(id, i)
    st_view(X, (i,j), ("v1", "v2", ...), 0)
    ...
}
```

where you have previously written function endobs() to be

```
scalar endobs(vector id, scalar i)
{
    scalar  j
    for (j=i+1; j<=rows(id); j++) {
        if (id[j]!=id[i]) return(j-1)
    }
    return(rows(id))
}
```

In any case, there could be a problem. Missing values of variable panelid might not align with missing values of variables v1, v2, etc. The result could be that observation and row numbers are not in accordance or that there appears to be a group that, in fact, has all missing data. The right way to handle the problem is

```
st_view(M, ., ("panelid", "v1", "v2", ...), 0)
st_subview(id, M, ., 1)
for (i=1; i<=rows(id); i=j+1) {
    j = endobs(id, i)
    st_subview(X, M, (i,j), (2\cols(M)))
    ...
}
```

Conformability

st_subview(X, V, i, j):

input:

    V: $r \times c$
    i: $1 \times 1, n \times 1,$ or $n_2 \times 2$
    j: $1 \times 1, 1 \times k,$ or $2 \times k_2$

output:

    X: $n \times k$

Diagnostics

st_subview(X, V, i, j) aborts with error if i or j are out of range. i and j refer to row and column numbers of V, not observation and variable numbers of the underlying Stata dataset.

Also See

[M-5] **st_view()** — Make matrix that is a view onto current Stata dataset

[M-5] **select()** — Select rows or columns

[M-4] **stata** — Stata interface functions

Title

[M-5] **st_tempname( )** — Temporary Stata names

Syntax

>   *string scalar*   st_tempname()
>
>   *string rowvector*   st_tempname(*real scalar n*)
>
>   *string scalar*   st_tempfilename()
>
>   *string rowvector*   st_tempfilename(*real scalar n*)

Description

st_tempname() returns a Stata temporary name, the same as would be returned by Stata's tempvar and tempname commands; see [P] **macro**.

st_tempname(*n*) returns *n* temporary Stata names, $n \geq 0$.

st_tempfilename() returns a Stata temporary filename, the same as would be returned by Stata's tempfile command; see [P] **macro**.

st_tempfilename(*n*) returns *n* temporary filenames, $n \geq 0$.

Remarks

Remarks are presented under the following headings:

>   Creating temporary objects
>   When temporary objects will be eliminated

Creating temporary objects

st_tempname()s can be used to name Stata's variables, matrices, and scalars. Although in Stata a distinction is drawn between tempvars and tempnames, there is no real distinction, and so st_tempname() handles both in Mata. For instance, one can create a temporary variable by coding

```
idx = st_addvar("double", st_tempname())
```

See [M-5] **st_addvar( )**.

One creates a temporary file by coding

```
fh = fopen(st_tempfilename(), "w")
```

See [M-5] **fopen( )**.

When temporary objects will be eliminated

Temporary objects do not vanish when the Mata function ends, nor when Mata itself ends. They are removed when the ado-file (or do-file) calling Mata terminates.

Forget Mata for a minute. Stata eliminates temporary variables and files when the program that created them ends. That same rule applies to Mata: Stata eliminates them, not Mata, and that means that the ado-file or do-file that called Mata will eliminate them when that ado-file or do-file ends. Temporary variables and files are not eliminated by Mata when the Mata function ends. Thus Mata functions can create temporary objects for use by their ado-file callers, should that prove useful.

Conformability

st_tempname(), st_tempfilename():
    result: 1×1

st_tempname(*n*), st_tempfilename(*n*):
    n: 1×1
    result: $1 \times n$

Diagnostics

st_tempname(*n*) and st_tempfilename(*n*) abort with error if $n < 0$ and return J(1,0,"") if $n = 0$.

Also See

[M-5] **st_addvar()** — Add variable to current Stata dataset

[M-4] **stata** — Stata interface functions

Title

[M-5] **st_tsrevar( )** — Create time-series op.varname variables

Syntax

> *real rowvector* st_tsrevar(*string rowvector s*)

Description

st_tsrevar(*s*) is the equivalent of Stata's [TS] **tsrevar** programming command: it generates temporary variables containing the evaluation of any *op.varname* combinations appearing in *s*.

Remarks

In *s* you have a row vector of Stata variable names, some of which might contain time-series *op.varname* combinations (such as l.gnp). Say that *s* contains

> s = ("gnp", "r", "l.gnp")

You want to create a view of the data in *V*, but you cannot code

> st_view(V, ., s)

because st_view() (see [M-5] **st_view( )**) does not allow time series–operated variables. The solution is to code

> st_view(V, ., st_tsrevar(s))

st_tsrevar() will create temporary variables for each *op.varname* combination (l.gnp in our example), and then return the Stata variable indices of each of the variables, whether newly created or already existing. For instance, in the example above, let us assume that gnp is the second variable in the dataset, r the 23rd, and that, in total, there are 54 variables. Returned by st_tsrevar() will be (2, 23, 55). Variable 55 is new, created by st_tsrevar(), and it contains the values of l.gnp.

Conformability

st_tsrevar(*s*):
        s: $1 \times c$
   result: $1 \times c$

Diagnostics

st_tsrevar() aborts with error if any variable name is not found or any *op.varname* combination is invalid.

Also See

[M-5] **st_varname( )** — Obtain variable names from variable indices

[M-4] **stata** — Stata interface functions

Title

[M-5] **st_updata( )** — Determine or set data-have-changed flag

Syntax

>*real scalar* st_updata()
>
>*void* st_updata(*real scalar value*)

Description

st_updata() returns 0 if the data in memory have not changed since they were last saved and returns 1 otherwise.

st_updata(*value*) sets the data-have-changed flag to 0 if *value* = 0 and 1 otherwise.

Remarks

Stata's describe command reports whether the data have changed since they were last saved. Stata's use command refuses to load a new dataset if the data currently in memory have not been saved since they were last changed. Other components of Stata also react to the data-have-changed flag.

st_updata() allows you to respect that same flag.

Also, as a Mata programmer, you must set the flag if your function changes the data in memory. Mata attempts to set the flag for you (for instance, when you add a new variable using st_addvar() [see [M-5] **st_addvar( )**]), but there are other places where the flag ought to be set, and you must do so. For instance, Mata does not set the flag every time you change a value in the dataset. Setting the flag what may be many thousands of times would reduce performance too much.

Moreover, even when Mata does set the flag, it might do so inappropriately, because the logic of your program fooled Mata. For instance, perhaps you added a variable and later dropped it. In such cases, the appropriate code is

        priorupdatavalue = st_updata()
        ...
        st_updata(priorupdatavalue)

Conformability

st_updata():
>   *result*: 1×1

st_updata(*value*):
>   *value*: 1×1
>   *result*:   *void*

Diagnostics

None.

Also See

[M-4] **stata** — Stata interface functions

Title

[M-5] **st_varformat( )** — Obtain/set format, etc., of Stata variable

Syntax

    string scalar   st_varformat(*scalar var*)

    void   st_varformat(*scalar var*, *string scalar fmt*)

    string scalar   st_varlabel(*scalar var*)

    void   st_varlabel(*scalar var*, *string scalar label*)

    string scalar   st_varvaluelabel(*scalar var*)

    void   st_varvaluelabel(*scalar var*, *string scalar labelname*)

where *var* contains a Stata variable name or a Stata variable index.

Description

st_varformat(*var*) returns the display format associated with *var*, such as "%9.0gc". st_varformat(*var*, *fmt*) changes *var*'s display format.

st_varlabel(*var*) returns the variable label associated with *var*, such as "Sex of Patient", or it returns "" if *var* has no variable label. st_varformat(*var*, *label*) changes *var*'s variable label.

st_varvaluelabel(*var*) returns the value-label name associated with *var*, such as "origin", or it returns "" if *var* has no value label. st_varvaluelabel(*var*, *labelname*) changes the value-label name associated with *var*.

Conformability

st_varformat(*var*), st_varlabel(*var*), st_varvaluelabel(*var*):
    var: 1×1
    result: 1×1

st_varformat(*var, value*), st_varlabel(*var, value*), st_varvaluelabel(*var, value*):
    var: 1×1
    value: 1×1
    result: *void*

Diagnostics

In all functions, if *var* is specified as a name, abbreviations are not allowed.

All functions abort with error if *var* is not a valid Stata variable.

st_varformat(*var, fmt*) aborts with error if *fmt* does not contain a valid display format for *var*.

st_varlabel(*var*, *label*) will truncate *label* if it is too long.

st_varvaluelabel(*var*, *labelname*) aborts with error if *var* is a Stata string variable or if *labelname* does not contain a valid name (assuming *labelname* is not ""). It is not required, however, that the label name exist.

Also See

[M-4] **stata** — Stata interface functions

Title

[M-5] **st_varindex( )** — Obtain variable indices from variable names

Syntax

>   *real rowvector*   st_varindex(*string rowvector s*)
>
>   *real rowvector*   st_varindex(*string rowvector s*, *real scalar abbrev*)
>
>   *real rowvector*   _st_varindex(*string rowvector s*)
>
>   *real rowvector*   _st_varindex(*string rowvector s*, *real scalar abbrev*)

Description

st_varindex(*s*) returns the variable index associated with each variable name recorded in *s*. st_varindex(*s*) does not allow variable-name abbreviations such as "pr" for "price".

st_varindex(*s*, *abbrev*) does the same thing but allows you to specify whether variable-name abbreviations are to be allowed. Abbreviations are allowed if *abbrev* $\neq$ 0. st_varindex(*s*) is equivalent to st_varindex(*s*, 0).

_st_varindex() does the same thing as st_varindex(). The two functions differ in how they respond when a name is not found. st_varindex() aborts with error, and _st_varindex() places missing in the appropriate element of the returned result.

Remarks

These functions require that each element of *s* contain a variable name, such as

>   *s* = ("price", "mpg", "weight")

If you have one string containing multiple names

>   *s* = ("price mpg weight")

then use tokens() to split it into the desired form, as in

>   *k* = st_varindex(tokens(*s*))

See [M-5] **tokens( )**.

Conformability

st_varindex(*s*, *abbrev*), _st_varindex(*s*, *abbrev*):
```
        s:   1 × k
   abbrev:   1 × 1   (optional)
   result:   1 × k
```

Diagnostics

st_varindex() aborts with error if any name is not found.

_st_varindex() puts missing in the appropriate element of the returned result for any name that is not found.

Also See

[M-5] **st_varname( )** — Obtain variable names from variable indices

[M-5] **tokens( )** — Obtain tokens from string

[M-4] **stata** — Stata interface functions

Title

[M-5] **st_varname( )** — Obtain variable names from variable indices

Syntax

> *string rowvector* st_varname(*real rowvector k*)
>
> *string rowvector* st_varname(*real rowvector k*, *real scalar tsmap*)

Description

st_varname(*k*) returns the Stata variable names associated with the variable indices stored in *k*. For instance, with the automobile data in memory

> names = st_varname((1..3))

results in names being ("make", "price", "mpg").

st_varname(*k*, *tsmap*) does the same thing but allows you to specify whether you want the actual or logical variable names of any time series–operated variables created by the Mata function st_tsrevar() (see [M-5] **st_tsrevar( )**) or by the Stata command tsrevar (see [TS] **tsrevar**).

st_varname(*k*) is equivalent to st_varname(*k*, 0); actual variable names are returned.

st_varname(*k*, 1) returns logical variable names.

Remarks

To understand the actions of st_varname(*k*, 1), pretend that variable 58 was created by st_tsrevar():

> k = st_tsrevar(("gnp", "r", "l.gnp"))

Pretend that k now contains (12, 5, 58). Variable 58 is a new, temporary variable, containing l.gnp values. Were you to ask for the actual names of the variables

> actualnames = st_varname(k)

actualnames would contain ("gnp", "r", "__00004a"), although the name of the last variable will vary because it is a temporary variable. Were you to ask for the logical names,

> logicalnames = st_varname(k, 1)

you would get back ("gnp", "r", "L.gnp").

Conformability

st_varname(*k*, *tsmap*)
> k: $1 \times c$
> *tsmap*: 1×1 (optional)
> *result*: $1 \times c$

Diagnostics

st_varname(*k*) and st_varname(*k*, *tsmap*) abort with error if any element of *k* is less than 1 or greater than st_nvar(); see [M-5] **st_nvar()**.

Also See

[M-5] **st_varindex()** — Obtain variable indices from variable names

[M-5] **st_tsrevar()** — Create time-series op.varname variables

[M-4] **stata** — Stata interface functions

Title

[M-5] **st_varrename( )** — Rename Stata variable

Syntax

> *void* st_varrename(*scalar var*, *string scalar newname*)

where *var* contains a Stata variable name or a Stata variable index.

Description

st_varrename(*var*, *newname*) changes the name of *var* to *newname*.

If *var* is specified as a name, abbreviations are not allowed.

Conformability

st_varrename(*var*, *newname*):
      var: 1×1
  newname: 1×1
    result: void

Diagnostics

st_varrename(*var*, *newname*) aborts with error if *var* is not a valid Stata variable or if *newname* is not a valid name or if a variable named *newname* already exists.

Also See

[M-4] **stata** — Stata interface functions

Title

[M-5] **st_vartype( )** — Storage type of Stata variable

Syntax

>   *string scalar*   st_vartype(*scalar var*)
>
>   *real scalar*   st_isnumvar(*scalar var*)
>
>   *real scalar*   st_isstrvar(*scalar var*)

where *var* contains a Stata variable name or a Stata variable index.

Description

In all the functions, if *var* is specified as a name, abbreviations are not allowed.

st_vartype(*var*) returns the storage type of the *var*, such as float, double, or str18.

st_isnumvar(*var*) returns 1 if *var* is a numeric variable and 0 otherwise.

st_isstrvar(*var*) returns 1 if *var* is a string variable and 0 otherwise.

Remarks

st_isstrvar(*var*) and st_isnumvar(*var*) are antonyms. Both functions are provided merely for convenience; they tell you nothing that you cannot discover from st_vartype(*var*).

Conformability

st_vartype(*var*):
>   *var*: 1×1
>   *result*: 1×1

st_isnumvar(*var*), st_isstrvar(*var*):
>   *var*: 1×1
>   *result*: 1×1

Diagnostics

All functions abort with error if *var* is not a valid Stata variable.

Also See

[M-4] **stata** — Stata interface functions

Title

[M-5] **st_view( )** — Make matrix that is a view onto current Stata dataset

Syntax

        void st_view(*V*, *real matrix i*, *rowvector j*)

        void st_view(*V*, *real matrix i*, *rowvector j*, *scalar selectvar*)

        void st_sview(*V*, *real matrix i*, *rowvector j*)

        void st_sview(*V*, *real matrix i*, *rowvector j*, *scalar selectvar*)

where

1. The type of *V* does not matter; it is replaced.
2. *i* may be specified in the same way as with st_data( ).
3. *j* may be specified in the same way as with st_data( ) except that time-series operators may not be specified. If time series–operated variables are needed, see [M-5] **st_tsrevar( )**.
4. *selectvar* may be specified in the same way as with st_data( ).

See [M-5] **st_data( )**.

Description

st_view( ) and st_sview( ) create a matrix that is a view onto the current Stata dataset.

Remarks

Remarks are presented under the following headings:

        Overview
        Advantages and disadvantages of views
        When not to use views
        Cautions when using views 1: Conserving memory
        Cautions when using views 2: Assignment
        Efficiency

Overview

st_view( ) does the same thing as st_data( )—and st_sview( ) does the same thing as st_sdata( )—except that, rather than returning a copy of the underlying values, st_view( ) and st_sview( ) create a matrix that is a view onto the Stata dataset itself.

To understand the distinction, consider

        X = st_data(., ("mpg", "displ", "weight"))

and

```
        st_view(X, ., ("mpg", "displ", "weight"))
```

Both commands fill in matrix X with the same data. However, were you to code

```
        X[2,1] = 123
```

after the st_data() setup, you would change the value in the matrix X, but the Stata dataset would remain unchanged. After the st_view() setup, changing the value in the matrix would cause the value of mpg in the second observation to change to 123.

Advantages and disadvantages of views

Views make it easy to change the dataset, and that can be an advantage or a disadvantage, depending on your desires.

Putting that aside: views are, in general, better than copies because (1) they take less time to set up and (2) they consume less memory; (2) is the important reason. Consider a 100,000-observation dataset on 30 variables. Coding

```
        X = st_data(., .)
```

creates a new matrix that is 24 MB in size. Meanwhile, the total storage requirement for

```
        st_view(X, ., .)
```

is roughly 128 bytes!

There is a cost: when you use the matrix X, it takes your computer longer to access the individual elements. You have to do a lot of calculation with X, however, before that extra time equals the initial savings in setup time, and even then, the extra time is probably worth it to save the extra memory.

When not to use views

Do not use views as a substitute for scalars. If you are going to loop through the data an observation at a time, and if every usage you will make of X is in scalar calculations, use _st_data(). There is nothing faster for that problem.

Putting aside that extreme: views become more efficient relative to copies the larger they are; i.e., it is more efficient to use st_data() for small amounts of data, especially if you are going to make computationally intensive calculations with it.

Cautions when using views 1: Conserving memory

If you are using views, it is probably because you are concerned about memory, and if you are, you want to be careful to avoid making copies of views. Copies of views are not views; they are copies. For instance,

```
        st_view(V, ., .)
        Y = V
```

That innocuous looking Y = V just made a copy of the entire dataset, meaning that if the dataset had 100,000 observations on 30 variables, Y now consumes 24 MB. Coding Y = V is okay in some circumstances, but in general, it is better to set up another view.

Similarly, watch out for subscripts. Consider the following code fragment

```
st_view(V, ., .)
for (i=1; i<=cols(V); i++) {
    sum = colsum(V[,i])
    ...
}
```

The problem in the above code is the V[,i]. That creates a new column vector containing the values from the ith column of V. Given 100,000 observations, that new column vector needs 800k of memory. Better to code would be

```
for (i=1; i<=cols(V); i++) {
    st_view(v, ., i)
    sum = colsum(v)
    ...
}
```

If you also need V, that is okay. You can have many views of the data setup simultaneously.

Similarly, be careful using views with operators. X'X makes a copy of X in the process of creating the transpose. Use functions such as cross() (see [M-5] **cross()**) designed to minimize the use of memory.

However, do not be overly concerned about this issue. Making a copy of a column of a view amounts to the same thing as introducing a temporary variable in a Stata program—something that is done all the time.

Cautions when using views 2: Assignment

The ability to assign to a view and so change the underlying data can be either convenient or dangerous, depending on your goals.

When making such assignments, there are two things you need be aware of.

The first is more of a Stata issue than it is a Mata issue. Assignment does not cause promotion. Coding

```
V[1,2] = 4059.125
```

might store 4059.125 in the first observation of the second variable of the view. Or, if that second variable is an int, what will be stored is 4059, or if it is a byte, what will be stored is missing.

The second caution is a Mata issue. To reassign all the values of the view, code

```
V[.,.] = matrix_expression
```

Do not code

```
V = matrix_expression
```

The second expression does not assign to the underlying dataset, it redefines V to be a regular matrix.

Efficiency

Whenever possible, specify argument i of st_view(V, i, j) and st_sview(V, i, j) as . (missing value) or as a row vector range (e.g., (i_1, i_2)) rather than as a column vector list.

Specify argument j as a real row vector rather than as a string row vector whenever st_view() and st_sview() are used inside loops with the same variables. This prevents Mata from having to look up the names over and over again.

Conformability

st_view(V, i, j), st_sview(V, i, j):

   input:
        i: $n \times 1$ or $n_2 \times 2$
        j: $1 \times k$
   output:
        V: $n \times k$

st_view$(V, i, j, selectvar)$, st_sview$(V, i, j, selectvar)$:

   input:
        i: $n \times 1$ or $n_2 \times 2$
        j: $1 \times k$
    selectvar: 1×1
   output:
        V: $(n - e) \times k$, where e is number of observations excluded by *selectvar*

Diagnostics

st_view$(i, j[, selectvar])$ and st_sview$(i, j[, selectvar])$ abort with error if any element of i is outside the range of observations or if a variable name or index recorded in j is not found. Variable-name abbreviations are allowed. If you do not want this, use st_varindex() (see [M-5] **st_varindex( )**) to translate variable names into variable indices.

st_view() and st_sview() abort with error if any element of i is out of range as described under the heading *Details of observation subscripting using st_data( ) and st_sdata( )* in [M-5] **st_data( )**.

Some functions do not allow views as arguments. If example(X) does not allow views, you can still use it by coding

      ... example(X=V) ...

because that will make a copy of view V in X. Most functions that do not allow views mention that in their *Diagnostics* section, but some do not because it was unexpected that anyone would want to use a view in that case. If a function does not allow a view, you will see in the traceback log:

```
        : myfunction(...)
              example():  3103  view found where array required
                mysub():     -  function returned error
           myfunction():     -  function returned error
                <istmt>:     -  function returned error
        r(3103);
```

The above means that function example() does not allow views.

Reference

Gould, W. 2005. Mata Matters: Using views onto the data. *Stata Journal* 5: 567–573.

Also See

[M-5] **st_subview()** — Make view from view

[M-5] **select()** — Select rows or columns

[M-5] **st_viewvars()** — Variables and observations of view

[M-5] **st_data()** — Load copy of current Stata dataset

[M-4] **stata** — Stata interface functions

Title

[M-5] **st_viewvars( )** — Variables and observations of view

Syntax

> *real rowvector* st_viewvars(*matrix V*)
>
> *real matrix* st_viewobs(*matrix V*)

where *V* is required to be a view.

Description

st_viewvars(*V*) returns the indices of the Stata variables corresponding to the columns of *V*.

st_viewobs(*V*) returns the Stata observation numbers corresponding to the rows of *V*. Returned is either a 1×2 row vector recording the observation range or an $N \times 1$ column vector recording the individual observation numbers.

Remarks

The results returned by these two functions are suitable for inclusion as arguments in subsequent calls to st_view() and st_sview(); see [M-5] **st_view( )**.

Conformability

st_viewvars(*V*):
    V: $N \times k$
  result: $1 \times k$

st_viewobs(*V*):
    V: $N \times k$
  result: 1×2 or $N \times 1$

Diagnostics

st_viewvars(*V*) and st_viewobs(*V*) abort with error if *V* is not a view.

Also See

[M-5] **st_view( )** — Make matrix that is a view onto current Stata dataset

[M-4] **stata** — Stata interface functions

Title

[M-5] **st_vlexists( )** — Use and manipulate value labels

Syntax

real scalar	st_vlexists(*name*)
void	st_vldrop(*name*)
string matrix	st_vlmap(*name*, *real matrix values*)
real matrix	st_vlsearch(*name*, *string matrix text*)
void	st_vlload(*name*, *values*, *text*)
void	st_vlmodify(*name*, *real colvector values*, *real colvector text*)

where *name* is *string scalar* and where the types of *values* and *text* in st_vlload() are irrelevant because they are replaced.

Description

st_vlexists(*name*) returns 1 if value label *name* exists and returns 0 otherwise.

st_vldrop(*name*) drops value label *name* if it exists.

st_vlmap(*name*, *values*) maps *values* through value label *name* and returns the result.

st_vlsearch(*name*, *text*) does the reverse; it returns the value corresponding to the text.

st_vlload(*name*, *values*, *text*) places value label *name* into *values* and *text*.

st_vlmodify(*name*, *values*, *text*) creates a new value label or modifies an existing one.

Remarks

Value labels are named and record a mapping from numeric values to text. For instance, a value label named sexlbl might record that 1 corresponds to male and 2 to female. Values labels are attached to Stata numeric variables. If a Stata numeric variable had the value label sexlbl attached to it, then the 1s and 2s in the variable would display as male and female. How other values would appear—if there were other values—would not be affected.

Remarks are presented under the following headings:

    Value-label mapping
    Value-label creation and editing
    Loading value labels

Value-label mapping

Let us consider value label sexlbl mapping 1 to male and 2 to female.
st_vlmap("sexlbl", *values*) would map the $r \times c$ matrix values through sexlbl and return an $r \times c$ string matrix containing the result. Any values for which there was no mapping would result in "". Thus

```
: res = st_vlmap("sexlbl", 1)
: res
male
: res = st_vlmap("sexlbl", (2,3,1))
: res
```

	1	2	3
1	female		male

st_vlsearch(*name*, *text*) performs the reverse mapping:

```
: txt = st_vlsearch("sexlbl", ("female","","male"))
: txt
```

	1	2	3
1	2	.	1

Value-label creation and editing

st_vlmodify(*name*, *values*, *text*) creates new value labels and modifies existing ones.

If value label sexlbl did not exist, coding

```
: st_vlmodify("sexlbl", (1\2), ("male"\"female"))
```

would create it. If the value label did previously exist, the above would modify it so that 1 now corresponds to male and 2 to female, regardless of what 1 or 2 previously corresponded to, if they corresponded to anything. Other mappings that might have been included in the value label remain unchanged. Thus

```
: st_vlmodify("sexlbl", 3, "unknown")
```

would add another mapping to the label. Values are deleted by specifying the text as "", so

```
: st_vlmodify("sexlbl", 3, "")
```

would remove the mapping for 3 (if there was a mapping). If you remove all the mappings, the value label itself is automatically dropped:

```
: st_vlmodify("sexlbl", (1\2), (""\""))
```

results in value label sexlbl being dropped if 1 and 2 were the final mappings in it.

Loading value labels

st_vlload(*name*, *values*, *text*) returns the value label in *values* and *text*, where you can do with it as you please. Thus you could code

    st_vlload("sexlbl", values, text)
        ...
    st_vldrop("sexlbl")
    st_vlmodify("sexlbl", values, text)

Conformability

st_vlexists(*name*):
    name: 1×1
    result: 1×1

st_vldrop(*name*):
    name: 1×1
    result: *void*

st_vlmap(*name*, *values*):
    name: 1×1
    values: $r \times c$
    result: $r \times c$

st_vlsearch(*name*, *text*):
    name: 1×1
    text: $r \times c$
    result: $r \times c$

st_vlload(*name*, *values*, *text*):

    input:
        name: 1×1

    output:
        values: $k \times 1$
        text: $k \times 1$

st_vlmodify(*name*, *values*, *text*):
    name: 1×1
    values: $m \times 1$
    text: $m \times 1$
    result: *void*

Diagnostics

The only conditions under which the above functions abort with error is when *name* is malformed or Mata is out of memory. Functions tolerate all other problems.

st_vldrop(*name*) does nothing if value label *name* does not exist.

st_vlmap(*name*, *values*) returns J(rows(values), cols(values), "") if value label *name* does not exist. When the value label does exist, individual values for which there is no recorded mapping are returned as "".

st_vlsearch(*name*, *text*): returns J(rows(values), cols(values), .) if value label *name* does not exist. When the value label does exist, individual text values for which there is no corresponding value are returned as . (missing).

st_vlload(*name*, *values*, *text*): sets *values* and *text* to be 0×1 when value label *name* does not exist.

st_vlmodify(*name*, *values*, *text*): creates the value label if it does not already exist. Value labels may map only integers and .a, .b, ..., .z. Attempts to insert a mapping for . are ignored. Noninteger values are truncated to integer values. If an element of *text* is "", then the corresponding mapping is removed.

Also See

[M-4] **stata** — Stata interface functions

Title

[M-5] **stata( )** — Execute Stata command

Syntax

> *void*   stata(*cmd*)
>
> *void*   stata(*cmd*, *nooutput*)
>
> *void*   stata(*cmd*, *nooutput*, *nomacroexpand*)
>
> *real scalar*   _stata(*cmd*)
>
> *real scalar*   _stata(*cmd*, *nooutput*)
>
> *real scalar*   _stata(*cmd*, *nooutput*, *nomacroexpand*)

where

	cmd:	*string scalar*
	nooutput:	*real scalar*
	nomacroexpand:	*real scalar*

Description

stata(*cmd*) executes the Stata command contained in the string scalar *cmd*. Output from the command appears at the terminal, and any macros contained in *cmd* are expanded.

stata(*cmd*, *nooutput*) does the same thing, but if *nooutput* $\neq$ 0, output produced by the execution is not displayed. stata(*cmd*, 0) is equivalent to stata(*cmd*).

stata(*cmd*, *nooutput*, *nomacroexpand*) does the same thing but, before execution, suppresses expansion of any macros contained in *cmd* if *nomacroexpand* $\neq$ 0. stata(*cmd*, 0, 0) is equivalent to stata(*cmd*).

_stata() repeats the syntaxes of stata(). The difference is that, whereas stata() aborts with error if the execution results in a nonzero return code, _stata() returns the resulting return code.

Remarks

The command you execute may invoke a process that causes another instance of Mata to be invoked. For instance, Stata program *A* calls Mata function *m1*(), which executes stata() to invoke Stata program *B*, which in turn calls Mata function *m2*(), which ....

stata(*cmd*) and _stata(*cmd*) execute *cmd* at the current run level. This means that any local macros refer to local macros in the caller's space. Consider the following:

```
program example
        ...
        local x = "value from A"
        mata: myfunc()
        display "`x'"
        ...
end
mata void myfunc()
{
        stata(`"local x = "new value""')
}
```

After example executes mata: myfunc(), `x' will be "new value".

That stata() and _stata() work that way was intentional: Mata functions can modify the caller's environment so that they may create temporary variables for the caller's use, etc., and you only have to exercise a little caution. Executing stata() functions to run other ado-files and programs will cause no problems because other ado-files and programs create their own new environment in which temporary variables, local macros, etc., are private.

Also, do not use stata() or _stata() to execute a multiline command or to execute the first line of what could be considered a multiline command. Once the first line is executed, Stata will fetch the remaining lines from the caller's environment. For instance, consider

─────────────────────────────────── top of myfile.do ───────
```
mata void myfunc()
{
        stata("if (1==1) {")
}
mata: myfunc()
display "hello"
}
```
─────────────────────────────────── end of myfile.do ───────

In the example above, myfunc() will consume the display "hello" and } lines.

Conformability

stata(*cmd, nooutput, nomacroexpand*):
          cmd: 1×1
    nooutput: 1×1 (optional)
  nomacroexpand: 1×1 (optional)
       result: *void*

_stata(*cmd, nooutput, nomacroexpand*):
          cmd: 1×1
    nooutput: 1×1 (optional)
  nomacroexpand: 1×1 (optional)
       result: 1×1

Diagnostics

stata() aborts with error if *cmd* is too long (exceedingly unlikely), if macro expansion fails, or if execution results in a nonzero return code.

_stata() aborts with error if *cmd* is too long.

Also See

[M-3] **mata stata** — Execute Stata command

[M-4] **stata** — Stata interface functions

Title

[M-5] **stataversion( )** — Version of Stata being used

Syntax

      real scalar   stataversion()

      real scalar   statasetversion()

      void   statasetversion(*real scalar version*)

Note: the version number is multiplied by 100: Stata 2.0 is 200, Stata 5.1 is 510, and Stata 13.0 is 1300.

Description

stataversion() returns the version of Stata/Mata that is running, multiplied by 100. For instance, if you have Stata 13 installed on your computer, stataversion() returns 1300.

statasetversion() returns the version of Stata that has been set by the user—the version of Stata that Stata is currently emulating—multiplied by 100. Usually stataversion() == statasetversion(). If the user has set a previous version—say, version 8 by typing version 8 in Stata—statasetversion() will return a number less than stataversion().

statasetversion(*version*) allows you to reset the version being emulated. Results are the same as using Stata's version command. *version*, however, is specified as an integer equal to 100 times the version you want.

Remarks

It is usually not necessary to reset statasetversion(). If you do reset statasetversion(), good form is to set it back when you are finished:

      current_version = statasetversion()
      statasetversion(*desired_version*)
      ...
      statasetversion(current_version)

Conformability

stataversion():
    result: 1×1

statasetversion():
    result: 1×1

statasetversion(*version*):
    version: 1×1
    result: *void*

Diagnostics

statasetversion(*version*) aborts with error if *version* is less than 100 or greater than stataversion().

Also See

[M-5] **bufio( )** — Buffered (binary) I/O

[M-5] **byteorder( )** — Byte order used by computer

[M-4] **programming** — Programming functions

Title

[M-5] **strdup( )** — String duplication

Syntax

> $n * s$
>
> $s * n$
>
> $n :* s$
>
> $s :* n$

where *n* is real and *s* is string.

Description

There is no `strdup()` function. Instead, the multiplication operator is used:

```
3*"example" = "exampleexampleexample"
0*"this" = ""
```

Conformability

n*s, *s**n:
n:	1×1
s:	$r \times c$
result:	$r \times c$

n:*s, *s*:*n:
n:	$r_1 \times c_1$
s:	$r_2 \times c_2$, *n* and *s* c-conformable
result:	$\max(r_1, r_2) \times \max(c_1, c_2)$

Diagnostics

If $n < 0$, the result is as if $n = 0$: "" is returned.

If *n* is not an integer, the result is as if $\text{trunc}(n)$ were specified.

Also See

[M-4] **string** — String manipulation functions

Title

[M-5] **strlen( )** — Length of string

Syntax

> *real matrix* strlen(*string matrix s*)

Description

strlen(*s*) returns the length of—the number of characters contained in—the string *s*.

When *s* is not a scalar, strlen() returns element-by-element results.

Remarks

Stata understands strlen() as a synonym for its own length() function, so you can use the function named strlen() in both your Stata and Mata code. Do not, however, use length() in Mata when you mean strlen(). Mata's length() function returns the length (number of elements) of a vector.

Conformability

strlen(*s*):
> *s*: $r \times c$
> *result*: $r \times c$

Diagnostics

strlen(*s*), when *s* is a binary string (a string containing binary 0), returns the overall length of the string, not the location of the binary 0. Use strpos(*s*, char(0)) if you want the location of the binary 0.

Also See

[M-5] **strpos( )** — Find substring in string

[M-5] **fmtwidth( )** — Width of %fmt

[M-4] **string** — String manipulation functions

Title

[M-5] **strmatch( )** — Determine whether string matches pattern

Syntax

>   *real matrix* strmatch(*string matrix s*, *string matrix pattern*)

Description

strmatch(*s*, *pattern*) returns 1 if *s* matches *pattern* and 0 otherwise.

When arguments are not scalars, strmatch() returns element-by-element results.

Remarks

In *pattern*, * means that 0 or more characters go here and ? means that exactly one character goes here. Thus *pattern*="*" matches anything and *pattern*="?p*x" matches all strings whose second character is *p* and whose last character is *x*.

Stata understands strmatch() as a synonym for its own match() function, so you can use the strmatch() function in both your Stata and Mata code.

Conformability

strmatch(*s*, *pattern*):
> *s*: $r_1 \times c_1$
> *pattern*: $r_2 \times c_2$,   *s* and *pattern* r-conformable
> *result*: $\max(r_1, r_2) \times \max(c_1, c_2)$

Diagnostics

In strmatch(*s*, *pattern*), if *s* or *pattern* contain a binary 0 (they usually would not), the strings are considered to end at that point.

Also See

[M-4] **string** — String manipulation functions

Title

> **[M-5] strofreal( )** — Convert real to string

Syntax

>   *string matrix*   strofreal(*real matrix R*)
>
>   *string matrix*   strofreal(*real matrix R*, *string matrix format*)

Description

strofreal(*R*) returns *R* as a string using Stata's %9.0g format. strofreal(*R*) is equivalent to strofreal(*R*, "%9.0g").

strofreal(*R*, *format*) returns *R* as a string formatted using *format*.

Leading blanks are trimmed from the result.

When arguments are not scalars, strofreal() returns element-by-element results.

Conformability

strofreal(*R*, *format*):
>   *R*: $r_1 \times c_1$
>   *format*: $r_2 \times c_2$,   *R* and *format* r-conformable   (optional)
>   *result*: $\max(r_1,r_2) \times \max(c_1,c_2)$

Diagnostics

strofreal(*R*, *format*) returns "." if *format* is invalid.

Also See

[M-5] **strtoreal( )** — Convert string to real

[M-4] **string** — String manipulation functions

Title

[M-5] **strpos( )** — Find substring in string

Syntax

> *real matrix* strpos(*string matrix haystack*, *string matrix needle*)

Description

strpos(*haystack*, *needle*) returns the location of the first occurrence of *needle* in *haystack* or 0 if *needle* does not occur.

When arguments are not scalars, strpos() returns element-by-element results.

Remarks

When working with binary strings, one can find the location of the binary 0 using strpos(*s*, char(0)).

Conformability

strpos(*haystack*, *needle*):
    haystack: $r_1 \times c_1$
    needle: $r_2 \times c_2$,   *haystack* and *needle* r-conformable
    result: $\max(r_1, r_2) \times \max(c_1, c_2)$

Diagnostics

strpos(*haystack*, *needle*) returns 0 if *needle* is not found in *haystack*.

Also See

[M-4] **string** — String manipulation functions

Title

[M-5] **strreverse( )** — Reverse string

Syntax

> *string matrix*   strreverse(*string matrix s*)

Description

strreverse(*s*) returns *s* with its characters in reverse order.

When *s* is not a scalar, strreverse() returns element-by-element results.

Remarks

Stata understands strreverse() as a synonym for its own reverse() function, so you can use the strreverse() name in both your Stata and Mata code.

Conformability

strreverse(*s*)
        s: $r \times c$
  result: $r \times c$

Diagnostics

None.

Also See

[M-4] **string** — String manipulation functions

Title

[M-5] **strtoreal( )** — Convert string to real

Syntax

>   *real matrix*   strtoreal(*string matrix S*)
>
>   *real scalar*   _strtoreal(*string matrix S, R*)

Description

strtoreal(*S*) returns *S* converted to real. Elements of *S* that cannot be converted are returned as . (missing value).

_strtoreal(*S, R*) does the same as above—it returns the converted values in *R*—and it returns the number of elements that could not be converted. In such cases, the corresponding value of *R* contains . (missing).

Remarks

strtoreal("1.5") returns (numeric) 1.5.

strtoreal("-2.5e+1") returns (numeric) -25.

strtoreal("not a number") returns (numeric) . (missing).

Typically, strtoreal(*S*) and _strtoreal(*S, R*) are used with scalars, but if applied to a vector or matrix *S*, element-by-element results are returned.

In performing the conversion, leading and trailing blanks are ignored: "1.5" and " 1.5 " both convert to (numeric) 1.5, but "1.5 kilometers" converts to . (missing). Use strtoreal(tokens(*S*)[1]) to convert just the first space-delimited part.

All Stata numeric formats are understood, such as 0, 1, -2, 1.5, 1.5e+2, and $-1.0\text{x}+8$, as well as the missing-value codes ., .a, .b, ..., .z.

Thus using strtoreal(*S*), if an element of *S* converts to . (missing), you cannot tell whether the element was valid and equal to "." or the element was invalid and so defaulted to . (missing), such as if *S* contained "cat" or "dog" or "1.5 kilometers".

When it is important to distinguish between these cases, use _strtoreal(*S, R*). The conversion is returned in *R* and the function returns the number of elements that were invalid. If _strtoreal() returns 0, then all values were valid.

Conformability

strtoreal(*S*):
   input:
            S: $r \times c$
   output:
           result: $r \times c$

_strtoreal(*S*, *R*):
   input:
           S: $r \times c$
   output:
           R: $r \times c$
           result: 1×1

Diagnostics

strtoreal(*S*) returns a missing value wherever an element of *S* cannot be converted to a number.

_strtoreal(*S*, *R*) does the same, but the result is returned in *R*.

Also See

[M-5] **strofreal( )** — Convert real to string

[M-4] **string** — String manipulation functions

Title

> [M-5] **strtrim()** — Remove blanks

Syntax

>> *string matrix* stritrim(*string matrix s*)
>>
>> *string matrix* strltrim(*string matrix s*)
>>
>> *string matrix* strrtrim(*string matrix s*)
>>
>> *string matrix* strtrim(*string matrix s*)

Description

stritrim(*s*) returns *s* with all consecutive, internal blanks collapsed to one blank.

strltrim(*s*) returns *s* with leading blanks removed.

strrtrim(*s*) returns *s* with trailing blanks removed.

strtrim(*s*) returns *s* with leading and trailing blanks removed.

When *s* is not a scalar, these functions return element-by-element results.

Remarks

Stata understands stritrim(), strltrim(), strrtrim(), and strtrim(), as synonyms for its own itrim(), ltrim(), rtrim(), and trim() functions, so you can use the str*() names in both your Stata and Mata code.

Conformability

> stritrim(*s*), strltrim(*s*), strrtrim(*s*), strtrim(*s*):
>> *s*: $r \times c$
>> *result*: $r \times c$

Diagnostics

None.

Also See

> [M-4] **string** — String manipulation functions

Title

[M-5] **strupper( )** — Convert string to uppercase (lowercase)

Syntax

>   *string matrix*   strupper(*string matrix s*)
>
>   *string matrix*   strlower(*string matrix s*)
>
>   *string matrix*   strproper(*string matrix s*)

Description

strupper(*s*) returns *s*, converted to uppercase.

strlower(*s*) returns *s*, converted to lowercase.

strproper(*s*) returns a string with the first letter capitalized and any other letters capitalized that immediately follow characters that are not letters; all other letters are converted to lowercase.

When *s* is not a scalar, these functions return element-by-element results.

Remarks

strproper("mR. joHn a. sMitH") returns Mr. John A. Smith.

strproper("jack o'reilly") returns Jack O'Reilly.

strproper("2-cent's worth") returns 2-Cent'S Worth.

Also, Stata understands strupper(), strlower(), and strproper() as synonyms for its own upper(), lower(), and proper() functions, so you can use the str*() names in both your Stata and Mata code.

Conformability

strupper(*s*), strlower(*s*), strproper(*s*):
>   *s*: $r \times c$
>   *result*: $r \times c$

Diagnostics

None.

Also See

[M-4] **string** — String manipulation functions

Title

[M-5] **subinstr( )** — Substitute text

Syntax

>   *string matrix*   subinstr(*string matrix s, string matrix old, string matrix new*)
>
>   *string matrix*   subinstr(*string matrix s, string matrix old, string matrix new,*
>   *real matrix cnt*)
>
>   *string matrix*   subinword(*string matrix s, string matrix old, string matrix new*)
>
>   *string matrix*   subinword(*string matrix s, string matrix old, string matrix new,*
>   *real matrix cnt*)

Description

subinstr(*s, old, new*) returns *s* with all occurrences of *old* changed to *new*.

subinstr(*s, old, new, cnt*) returns *s* with the first *cnt* occurrences of *old* changed to *new*. All occurrences are changed if *cnt* contains missing.

subinword(*s, old, new*) returns *s* with all occurrences of *old* on word boundaries changed to *new*.

subinword(*s, old, new, cnt*) returns *s* with the first *cnt* occurrences of *old* on word boundaries changed to *new*. All occurrences are changed if *cnt* contains missing.

When arguments are not scalars, these functions return element-by-element results.

Remarks

subinstr("th thin man", "th", "the") returns "the thein man".

subinword("th thin man", "th", "the") returns "the thin man".

Conformability

subinstr(*s, old, new, cnt*), subinword(*s, old, new, cnt*):

>   *s*: $r_1 \times c_1$
>   *old*: $r_2 \times c_2$
>   *new*: $r_3 \times c_3$
>   *cnt*: $r_4 \times c_4$ (optional);   *s, old, new, cnt* r-conformable
>   *result*: $\max(r_1,r_2,r_3,r_4) \times \max(c_1,c_2,c_3,c_4)$

Diagnostics

subinstr(*s, old, new, cnt*) and subinword(*s, old, new, cnt*) treat *cnt* < 0 as if *cnt* = 0 was specified; the original string *s* is returned.

Also See

[M-4] **string** — String manipulation functions

Title

[M-5] **_substr( )** — Substitute into string

Syntax

> *void* _substr(*string scalar s*, *string scalar tosub*, *real scalar pos*)

Description

_substr(*s*, *tosub*, *pos*) substitutes *tosub* into *s* at position *pos*. The first position of *s* is *pos* = 1. _substr() may be used with text or binary strings.

Do not confuse _substr() with substr(), which extracts substrings; see [M-5] **substr( )**.

Remarks

If *s* contains "abcdef", then _substr(*s*, "XY", 2) changes *s* to contain "aXYdef".

Conformability

_substr(*s*, *tosub*, *pos*):
   input:
            s: 1 × 1
        tosub: 1 × 1
          pos: 1 × 1
   output:
            s: 1 × 1

Diagnostics

_substr(*s*, *tosub*, *pos*) does nothing if *tosub*=="".

_substr(*s*, *tosub*, *pos*) may not be used to extend *s*: _substr() aborts with error if substituting *tosub* into *s* would result in a string longer than the original *s*. _substr() also aborts with error if *pos* ≤ 0 or *pos* ≥ . unless *tosub* is "".

_substr(*s*, *tosub*, *pos*) aborts with error if *s* or *tosub* are views.

Also See

[M-4] **string** — String manipulation functions

Title

[M-5] **substr( )** — Extract substring

Syntax

>   *string matrix*   substr(*string matrix si, real matrix b, real matrix l*)
>
>   *string matrix*   substr(*string matrix s, real matrix b*)

Description

substr(s, b, l) returns the substring of s starting at position b and continuing for a length of l, where

1. b specifies the starting position; the first character of the string is $b = 1$.
2. $b > 0$ is interpreted as distance from the start of the string; $b = 2$ means starting at the second character.
3. $b < 0$ is interpreted as distance from the end of string; $b = -1$ means starting at the last character; $b = -2$ means starting at the second from the last character.
4. l specifies the length; $l = 2$ means for two characters.
5. $l < 0$ is treated the same as $l = 0$: no characters are copied.
6. $l \geq .$ is interpreted to mean to the end of the string.

substr(s, b) is equivalent to substr(s, b, .) for strings that do not contain binary 0. If there is a binary 0 to the right of b, the substring from b up to but not including the binary 0 is returned.

When arguments are not scalars, strmatch() returns element-by-element results.

Conformability

substr(s, b, l):
> s: $r_1 \times c_1$
> b: $r_2 \times c_2$
> l: $r_3 \times c_3$; s, b, and l r-conformable
>   *result*: $\max(r_1, r_2, r_3) \times \max(c_1, c_2, c_3)$

substr(s, b):
> s: $r_1 \times c_1$
> b: $r_2 \times c_2$; s and b r-conformable
>   *result*: $\max(r_1, r_2) \times \max(c_1, c_2)$

Diagnostics

In $\text{substr}(s, b, l)$ and $\text{substr}(s, b)$, if b describes a position before the beginning of the string or after the end, "" is returned. If $b + l$ describes a position to the right of the end of the string, results are as if a smaller value for l were specified.

Also See

[M-4] **string** — String manipulation functions

Title

[M-5] **sum( )** — Sums

Syntax

numeric colvector	rowsum(*numeric matrix Z* [*, missing*])
numeric rowvector	colsum(*numeric matrix Z* [*, missing*])
numeric scalar	sum(*numeric matrix Z* [*, missing*])
numeric colvector	quadrowsum(*numeric matrix Z* [*, missing*])
numeric rowvector	quadcolsum(*numeric matrix Z* [*, missing*])
numeric scalar	quadsum(*numeric matrix Z* [*, missing*])

where optional argument *missing* is a real scalar that determines how missing values in Z are treated:

1. Specifying *missing* as 0 is equivalent to not specifying the argument; missing values in Z are treated as contributing 0 to the sum.

2. Specifying *missing* as 1 (or nonzero) specifies that missing values in Z are to be treated as missing values and to turn the sum to missing.

Description

rowsum(Z) and rowsum(Z, *missing*) return a column vector containing the sum over the rows of Z.

colsum(Z) and colsum(Z, *missing*) return a row vector containing the sum over the columns of Z.

sum(Z) and sum(Z, *missing*) return a scalar containing the sum over the rows and columns of Z.

quadrowsum(), quadcolsum(), and quadsum() are quad-precision variants of the above functions. The sum is accumulated in quad precision and then rounded to double precision and returned.

Argument *missing* determines how missing values are treated. If *missing* is not specified, results are the same as if *missing* = 0 were specified: missing values are treated as zero. If *missing* = 1 is specified, missing values are treated as missing values.

These functions may be used with real or complex matrix Z.

Remarks

All functions return the same type as the argument, real if argument is real, complex if complex.

Conformability

rowsum(*Z*, *missing*), quadrowsum(*Z*, *missing*):

        Z: $r \times c$
  missing: 1×1 (optional)
    result: $1 \times c$

colsum(*Z*, *missing*), quadcolsum(*Z*, *missing*):

        Z: $r \times c$
  missing: 1×1 (optional)
    result: $r \times 1$

sum(*Z*, *missing*), quadsum(*Z*, *missing*):

        Z: $r \times c$
  missing: 1×1 (optional)
    result: 1×1

Diagnostics

If *missing* = 0, missing values are treated as contributing zero to the sum; they do not turn the sum to missing. Otherwise, missing values turn the sum to missing.

Also See

[M-5] **mean()** — Means, variances, and correlations

[M-5] **runningsum()** — Running sum of vector

[M-5] **cross()** — Cross products

[M-4] **mathematical** — Important mathematical functions

[M-4] **utility** — Matrix utility functions

Title

[M-5] **svd()** — Singular value decomposition

Syntax

void	svd(*numeric matrix A, U, s, Vt*)
real colvector	svdsv(*numeric matrix A*)
void	_svd(*numeric matrix A, s, Vt*)
real colvector	_svdsv(*numeric matrix A*)
real scalar	_svd_la(*numeric matrix A, s, Vt*)

Description

svd(*A, U, s, Vt*) calculates the singular value decomposition of A: $m \times n$, $m \geq n$, returning the result in U, s, and Vt. Singular values returned in s are sorted from largest to smallest.

svdsv(*A*) returns the singular values of A: $m \times n$, $m \geq n$ or $m < n$ (i.e., no restriction), in a column vector of length $\min(m,n)$. U and Vt are not calculated.

_svd(*A, s, Vt*) does the same as svd(), except that it returns U in A. Use of _svd() conserves memory.

_svdsv(*A*) does the same as svdsv(), except that, in the process, it destroys A. Use of _svdsv() conserves memory.

_svd_la() is the interface into the [M-1] **LAPACK** SVD routines and is used in the implementation of the previous functions. There is no reason you should want to use it.

Remarks

Remarks are presented under the following headings:

> *Introduction*
> *Possibility of convergence problems*

Documented here is the thin SVD, appropriate for use with A: $m \times n$, $m \geq n$. See [M-5] **fullsvd()** for the full SVD, appropriate for use in all cases. The relationship between the two is discussed in *Relationship between the full and thin SVDs* in [M-5] **fullsvd()**.

Use of the thin SVD—the functions documented here—is preferred when $m \geq n$.

Introduction

The SVD is used to compute accurate solutions to linear systems and least-squares problems, to compute the 2-norm, and to determine the numerical rank of a matrix.

The singular value decomposition (SVD) of A: $m \times n$, $m \geq n$, is given by

$$A = U * \text{diag}(s) * V'$$

where
- U: $m \times n$ and $U'U = I(n)$
- s: $n \times 1$
- V: $n \times n$ and orthogonal (unitary)

When A is complex, the transpose operator ' is understood to mean the conjugate transpose operator.

Vector s contains the singular values, and those values are real even when A is complex. s is ordered so that the largest singular value appears first, then the next largest, and so on.

Function svd(A, U, s, Vt) returns U, s, and $Vt = V'$.

Function svdsv(A) returns s, omitting the calculation of U and Vt. Also, whereas svd() is suitable for use only in the case $m \geq n$, svdsv() may be used in all cases.

Possibility of convergence problems

It is possible, although exceedingly unlikely, that the SVD routines could fail to converge. svd(), svdsv(), _svd(), and _svdsv() then return singular values in s equal to missing.

In coding, it is perfectly acceptable to ignore this possibility because (1) it is so unlikely and (2) even if the unlikely event occurs, the missing values will properly reflect the situation. If you do wish to check, in addition to checking missing(s)>0, you must also check missing(A)==0 because that is the other reason s could contain missing values. Convergence was not achieved if missing(s) > 0 & missing(A)==0. If you are calling one of the destructive-of-A versions of SVD, remember to check missing(A)==0 before extracting singular values.

Conformability

svd(A, U, s, Vt):

  input:
  - A: $m \times n$, $m \geq n$

  output:
  - U: $m \times n$
  - s: $n \times 1$
  - Vt: $n \times n$

(Continued on next page)

svdsv(*A*):

	A:	$m \times n$, $m \geq n$ or $m < n$
	result:	$\min(m,n) \times 1$

_svd(*A*, *s*, *Vt*):

input:

	A:	$m \times n$, $m \geq n$

output:

	A:	$m \times n$, contains *U*
	s:	$n \times 1$
	Vt:	$n \times n$

_svdsv(*A*):

input:

	A:	$m \times n$, $m \geq n$ or $m < n$

output:

	A:	0×0
	result:	$\min(m,n) \times 1$

_svd_la(*A*, *s*, *Vt*):

input:

	A:	$m \times n$, $m \geq n$

output:

	A:	$m \times n$, contains *U*
	s:	$n \times 1$
	Vt:	$n \times n$
	result:	1×1

Diagnostics

svd(*A*, *U*, *s*, *Vt*) and _svd(*A*, *s*, *Vt*) return missing results if *A* contains missing. In all other cases, the routines should work, but there is the unlikely possibility of convergence problems, in which case missing results will also be returned; see *Possibility of convergence problems* above.

svdsv(*A*) and _svdsv(*A*) return missing results if *A* contains missing values or if there are convergence problems.

_svd() and _svdsv() abort with error if *A* is a view.

Direct use of _svd_la() is not recommended.

Also See

[M-5] **fullsvd( )** — Full singular value decomposition

[M-5] **svsolve( )** — Solve AX=B for X using singular value decomposition

[M-5] **pinv( )** — Moore–Penrose pseudoinverse

[M-5] **norm( )** — Matrix and vector norms

[M-5] **rank( )** — Rank of matrix

[M-4] **matrix** — Matrix functions

Title

[M-5] svsolve( ) — Solve AX=B for X using singular value decomposition

Syntax

    numeric matrix   svsolve(*A*, *B*)

    numeric matrix   svsolve(*A*, *B*, *rank*)

    numeric matrix   svsolve(*A*, *B*, *rank*, *tol*)

    real scalar   _svsolve(*A*, *B*)

    real scalar   _svsolve(*A*, *B*, *tol*)

where

    A: *numeric matrix*
    B: *numeric matrix*
    rank: irrelevant; *real scalar* returned
    tol: *real scalar*

Description

svsolve(*A*, *B*, ...), uses singular value decomposition to solve $AX = B$ and return X. When A is singular, svsolve() computes the minimum-norm least-squares generalized solution. When *rank* is specified, in it is placed the rank of A.

_svsolve(*A*, *B*, ...) does the same thing, except that it destroys the contents of A and it overwrites B with the solution. Returned is the rank of A.

In both cases, *tol* specifies the tolerance for determining whether A is of full rank. *tol* is interpreted in the standard way—as a multiplier for the default if *tol* > 0 is specified and as an absolute quantity to use in place of the default if *tol* ≤ 0 is specified.

Remarks

svsolve(*A*, *B*, ...) is suitable for use with square or nonsquare, full-rank or rank-deficient matrix A. When A is of full rank, qrsolve() returns the same solution as lusolve() (see [M-5] **lusolve( )**), ignoring roundoff error. When A is singular, svsolve() returns the minimum-norm least-squares generalized solution. qrsolve() (see [M-5] **qrsolve( )**), an alternative, returns a generalized least-squares solution that amounts to dropping rows of A.

Remarks are presented under the following headings:

    Derivation
    Relationship to inversion
    Tolerance

Derivation

We wish to solve for X

$$AX = B \tag{1}$$

Perform singular value decomposition on A so that we have $A = USV'$. Then (1) can be rewritten

$$USV'X = B$$

Premultiplying by U' and remembering that $U'U = I$, we have

$$SV'X = U'B$$

Matrix S is diagonal and thus its inverse is easily calculated, and we have

$$V'X = S^{-1}U'B$$

When we premultiply by V, remembering that $VV' = I$, the solution is

$$X = VS^{-1}U'B \tag{2}$$

See [M-5] **svd( )** for more information on the SVD.

Relationship to inversion

For a general discussion, see *Relationship to inversion* in [M-5] **lusolve( )**.

For an inverse based on the SVD, see [M-5] **pinv( )**. `pinv(A)` amounts to `svsolve(A, I(rows(A)))`, although `pinv()` has separate code that uses less memory.

Tolerance

In (2) above, we are required to calculate the inverse of diagonal matrix S. The generalized solution is obtained by substituting zero for the ith diagonal element of S^{-1}, where the ith diagonal element of S is less than or equal to *eta* in absolute value. The default value of *eta* is

$$eta = \text{epsilon}(1) * \text{rows}(A) * \max(S)$$

If you specify *tol* > 0, the value you specify is used to multiply *eta*. You may instead specify *tol* ≤ 0 and then the negative of the value you specify is used in place of *eta*; see [M-1] **tolerance**.

Conformability

svsolve(*A*, *B*, *rank*, *tol*):

>input:
>>A: $m \times n$
>>B: $m \times k$
>>*tol*: 1×1 (optional)
>
>output:
>>*rank*: 1×1 (optional)
>>*result*: $n \times k$

_svsolve(*A*, *B*, *tol*):

>input:
>>A: $m \times n$
>>B: $m \times k$
>>*tol*: 1×1 (optional)
>
>output:
>>A: 0×0
>>B: $m \times k$
>>*result*: 1×1

Diagnostics

svsolve(*A*, *B*, ...) and _svsolve(*A*, *B*, ...) return missing results if A or B contain missing.

_svsolve(*A*, *B*, ...) aborts with error if A (but not B) is a view.

Also See

[M-5] **solvelower( )** — Solve AX=B for X, A triangular

[M-5] **cholsolve( )** — Solve AX=B for X using Cholesky decomposition

[M-5] **lusolve( )** — Solve AX=B for X using LU decomposition

[M-5] **qrsolve( )** — Solve AX=B for X using QR decomposition

[M-4] **matrix** — Matrix functions

[M-4] **solvers** — Functions to solve AX=B and to obtain A inverse

Title

[M-5] **swap( )** — Interchange contents of variables

Syntax

> *void* swap(*transmorphic matrix A, transmorphic matrix B*)

Description

swap(*A*, *B*) interchanges the contents of *A* and *B*. *A* and *B* are not required to be of the same type or dimension.

Remarks

There is no faster way than swap(*A*, *B*) to assign *A*=*B* when you do not care about the contents of *B* after the assignment. For instance, you have the code

> A = B
> B = ...(*matrix expression*)...

Faster is

> swap(A, B)
> B = ...(*matrix expression*)...

The execution time of swap() is independent of the size of *A* and *B*, and swap() conserves memory to boot. Pretend that *B* is a 900 × 900 matrix. After *A*=*B* is executed, but before *B* is reassigned, two copies of the 900 × 900 matrix exist. That does not happen with swap().

Conformability

swap(*A*, *B*):
> *input*:
>> *A*: $r_1 \times c_1$
>> *B*: $r_2 \times c_2$
>
> *output*:
>> *A*: $r_2 \times c_2$
>> *B*: $r_1 \times c_1$

Diagnostics

swap(*A*, *B*) works only with variables. Do not code, for instance, swap(*A*[*i*,*j*], *A*[*j*,*i*]). It is not an error, but it will have no effect.

Also See

[M-4] **programming** — Programming functions

Title

[M-5] **Toeplitz( )** — Toeplitz matrices

Syntax

> *numeric matrix* `Toeplitz(`*numeric colvector c1, numeric rowvector r1*`)`

Description

`Toeplitz(`*c1, r1*`)` returns the Toeplitz matrix defined by *c1* being its first column and *r1* being its first row. A Toeplitz matrix T is characterized by $T[i,j] = T[i-1, j-1]$, $i,j > 1$. In a Toeplitz matrix, each diagonal is constant.

Vectors *c1* and *r1* specify the first column and first row of T.

Remarks

c1[1] is used to fill $T[1,1]$ and *r1*[1] is not used.

To obtain the symmetric (Hermitian) Toeplitz matrix, code `Toeplitz(`v, v'`)` (if v is a column vector), or `Toeplitz(`v', v`)` if v is a row vector.

Conformability

`Toeplitz(`*c1, r1*`)`:
        c1: $r \times 1$
        r1: $1 \times c$
   result: $r \times c$

Diagnostics

None. The top left element is defined by *c1*[1], and *r1*[1] is not used.

Otto Toeplitz (1881–1940) was born in Breslau, Germany (now Wrocław, Poland), and educated there in mathematics. He researched and taught at universities in Göttingen, Kiel, and Bonn, making many contributions to algebra and analysis, but he was dismissed in 1935 for being a Jew. Toeplitz emigrated to Palestine in 1939 but died a few months later in Jerusalem. He was fascinated by the history of mathematics and wrote a popular work with Hans Rademacher, *The Enjoyment of Mathematics*.

Reference

Robinson, A. 1976. Toeplitz, Otto. In *Dictionary of Scientific Biography*, vol. 13, ed. C. C. Gillispie, 428. New York: Scribner's.

Also See

[M-4] **standard** — Functions to create standard matrices

Title

[M-5] **tokenget( )** — Advanced parsing

Syntax

$t =$ tokeninit([*wchars* [, *pchars* [, *qchars* [, *allownum* [, *allowhex*]]]]])

$t =$ tokeninitstata( )

void	tokenset(*t*, *real scalar s*)
string rowvector	tokengetall(*t*)
string scalar	tokenget(*t*)
string scalar	tokenpeek(*t*)
string scalar	tokenrest(*t*)
real scalar	tokenoffset(*t*)
void	tokenoffset(*t*, *real scalar offset*)
string scalar	tokenwchars(*t*)
void	tokenwchars(*t*, *string scalar wchars*)
string rowvector	tokenpchars(*t*)
void	tokenpchars(*t*, *string rowvector pchars*)
string rowvector	tokenqchars(*t*)
void	tokenqchars(*t*, *string rowvector qchars*)
real scalar	tokenallownum(*t*)
void	tokenallownum(*t*, *real scalar allownum*)
real scalar	tokenallowhex(*t*)
void	tokenallowhex(*t*, *real scalar allowhex*)

where

t is *transmorphic* and contains the parsing environment information. You obtain a *t* from tokeninit() or tokeninitstata() and then pass *t* to the other functions.

wchars is a *string scalar* containing the characters to be treated as white space, such as " ", (" "+char(9)), or "".

pchars is a *string rowvector* containing the strings to be treated as parsing characters, such as "" and (">", "<", ">=", "<="). "" and J(1,0,"") are given the same interpretation: there are no parsing characters.

qchars is a *string rowvector* containing the character pairs to be treated as quote characters. "" (i.e., empty string) is given the same interpretation as J(1,0,""); there are no quote characters. *qchars* = (`""""'`) (i.e., the two-character string quote indicates that " is to be treated as open quote and " is to be treated as close quote. *qchars* = (`""""'`, `"«""»"`) indicates that, in addition, `"` is to be treated as open quote and `"'` as close quote. In a syntax that did not use < and > as parsing characters, *qchars* = ("<>") would indicate that < is to be treated as open quote and > as close quote.

allownum is a *string scalar* containing 0 or 1. *allownum* = 1 indicates that numbers such as 12.23 and 1.52e+02 are to be returned as single tokens even in violation of other parsing rules.

allowhex is a *string scalar* containing 0 or 1. *allowhex* = 1 indicates that numbers such as 1.921fb54442d18X+001 and 1.0x+a are to be returned as single tokens even in violation of other parsing rules.

Description

These functions provide advanced parsing. If you simply wish to convert strings into row vectors by separating on blanks, converting "mpg weight displ" into ("mpg", "weight", "displ"), see [M-5] **tokens( )**.

Remarks

Remarks are presented under the following headings:

> Concepts
> > White-space characters
> > Parsing characters
> > Quote characters
> > Overrides
> > Setting the environment to parse on blanks with quote binding
> > Setting the environment to parse full Stata syntax
> > Setting the environment to parse tab-delimited files
>
> Function overview
> > tokeninit() and tokeninitstata()
> > tokenset()
> > tokengetall()
> > tokenget(), tokenpeek(), and tokenrest()
> > tokenwchars(), tokenpchars(), and tokenqchars()
> > tokenallownum and tokenallowhex()

Concepts

Parsing refers to splitting a string into pieces, which we will call tokens. Parsing as implemented by the token*() functions is defined by (1) the white-space characters *wchars*, (2) the parsing characters *pchars*, and (3) the quote characters *qchars*.

White-space characters

Consider the string "this that what". If there are no white-space characters, no parsing characters, and no quote characters, i.e., if *wchars* = *pchars* = *qchars* = "", then the result of parsing "this that what" would be one token that would be the string just as it is: "this that what".

If *wchars* were instead " ", then parsing "this that what" results in ("this", "that", "what"). Parsing "this that   what" (note the multiple blanks) would result in the same thing. White-space characters separate one token from the next but are not otherwise significant.

Parsing characters

If we instead left *wchars* = "" and set *pchars* = " ", "this that what" parses into ("this", " ", "that", " ", "what") and parsing "this that   what" results in ("this", " ", "that", " ", " ", " ", "what").

pchars are like *wchars* except that they are themselves significant.

pchars do not usually contain space. A more reasonable definition of *pchars* is ("+", "-"). Then parsing "x+y" results in ("x", "+", "y"). Also, the parsing characters can be character combinations. If *pchars* = ("+", "-", "++", "--"), then parsing "x+y++" results in ("x", "+", "y", "++") and parsing "x+++y" results in ("x", "++", "+", "y"). Longer *pchars* are matched before shorter ones regardless of the order in which they appear in the *pchars* vector.

Quote characters

qchars specifies the quote characters. Pieces of the string being parsed that are surrounded by quotes are returned as one token, ignoring the separation that would usually occur because of the *wchars* and *pchars* definitions. Consider the string

        mystr= "x = y"

Let *wchars* = " " and *pchars* include "=". That by itself would result in the above string parsing into the five tokens

| mystr | = | "x | = | y" |

Now let *qchars* = ('""""'); i.e., *qchars* is the two-character string "". Parsing then results in the three tokens

| mystr | = | "x = y" |

Each element of *qchars* contains a character pair: the open character followed by the close character. We defined those two characters as " and " above, that is, as being the same. The two characters can differ. We might define the first as ' and the second as '. When the characters are different, quotations can nest. The quotation "he said "hello"" makes no sense because that parses into ("he said ", hello, ""). The quotation 'he said 'hello'', however, makes perfect sense and results in the single token 'he said 'hello''.

The quote characters can themselves be multiple characters. You can define open quote as '" and close as "': *qchars* = ('"""",""'). Or you can define multiple sets of quotation characters, such as *qchars* = ('""""', '"""",').

The quote characters do not even have to be quotes at all. In some context you might find it convenient to specify them as ("()"). With that definition, "(2 × (3 + 2))" would parse into one token. Specifying them like this can be useful, but in general we recommend against it. It is usually better to write your code so that quote characters really are quote characters and to push the work of handling other kinds of nested expressions back onto the caller.

Overrides

The `token*()` functions provide two overrides: *allownum* and *allowhex*. These have to do with parsing numbers. First, consider life without overrides. You have set *wchars* = " " and *pchars* = ("=", "+", "-", "*", "/"). You attempt to parse

   y = x + 1e+13

The result is

| y | = | x | + | 1e | + | 13 |

when what you wanted was

| y | = | x | + | 1e+13 |

Setting *allownum* = 1 will achieve the desired result. *allownum* specifies that, when a token could be interpreted as a number, the number interpretation is to be taken even in violation of the other parsing rules.

Setting *allownum* = 1 will not find numbers buried in the middle of strings, such as the 1e+3 in "xis1e+3", but if the number occurs at the beginning of the token according to the parsing rules set by *wchars* and *pchars*, *allownum* = 1 will continue the token in violation of those rules if that results in a valid number.

The override *allowhex* is similar and Stata specific. Stata (and Mata) provide a unique and useful way of writing hexadecimal floating-point numbers in a printable, short, and precise way: π can be written 1.921fb54442d18X+001. Setting *allowhex* = 1 allows such numbers.

Setting the environment to parse on blanks with quote binding

Stata's default rule for parsing do-file arguments is "parse on blanks and bind on quotes". The settings for duplicating that behavior are

   wchars = " "

   pchars = ("")

   qchars = (`""""' , `"`""'"')

   allownum = 0

   allowhex = 0

This behavior can be obtained by coding

   t = tokeninit(" ", "", (`""""' , `"`""'"'), 0, 0)

or by coding

   t = tokeninit()

because in `tokeninit()` the arguments are optional and "parse on blank with quote binding" is the default.

With those settings, parsing '"first second "third fourth" fifth"' results in
("first", "second", '""third fourth""', "fifth").

This result is a little different from that of Stata because the third token includes the quote binding characters. Assume that the parsed string was obtained by coding

>   res = tokengetall(t)

The following code will remove the open and close quotes, should that be desirable.

```
for (i=1; i<=cols(res); i++) {
        if (res[i]==`""""') {
                res[i] = substr(res[i], 2, strlen(res[i])-2)
        }
        else if (substr(res[i], 1, 2)=="`" + `""""') {
                res[i] = substr(res[i], 3, strlen(res[i])-4)
        }
}
```

Setting the environment to parse full Stata syntax

To parse full Stata syntax, the settings are

>   wchars = " "
>   pchars = ("\",   "~",   "!",   "=",   ":",   ";",   ",",
>   "?",   "!",   "@",   "#",   "==", "!=", ">=",
>   "<=", "<",   ">",   "&",   "|",   "&&", "||",
>   "+",   "-",   "++", "--", "*",   "/",   "^",
>   "(",   ")",   "[",   "]",   "{",   "}")
>   qchars = (`""""', `"`""'"', char(96)+char(39))
>   allownum = 1
>   allowhex = 1

The above is a slight oversimplification. Stata is an interpretive language and Stata does not require users to type filenames in quotes, although Stata does allow it. Thus "\" is sometimes a parsing character and sometimes not, and the same is true of "/". As Stata parses a line from left to right, it will change *pchars* between two tokenget() calls when the next token could be or is known to be a filename. Sometimes Stata peeks ahead to decide which way to parse. You can do the same by using the tokenpchars() and tokenpeek() functions.

To obtain the above environment, code

>   t = tokeninitstata()

Setting the environment to parse tab-delimited files

The token*() functions can be used to parse lines from tab-delimited files. A tab-delimited file contains lines of the form

>   ⟨field1⟩⟨tab⟩⟨field2⟩⟨tab⟩⟨field3⟩

The parsing environment variables are

$$wchars = \texttt{""}$$
$$pchars = (\ \texttt{char(9)}\ )\quad (\text{i.e., } tab)$$
$$qchars = (\ \texttt{""}\ )$$
$$allownum = 0$$
$$allowhex = 0$$

To set this environment, code

> t = tokeninit("", char(9), "", 0, 0)

Say that you then parse the line

> Farber, William⟨*tab*⟩   2201.00⟨*tab*⟩12

The results will be

> ("Farber, William", char(9), " 2201.00", char(9), "12")

If the line were

> Farber, William⟨*tab*⟩⟨*tab*⟩12

the result would be

> ("Farber, William", char(9), char(9), "12")

The tab-delimited format is not well defined when the missing fields occur at the end of the line. A line with the last field missing might be recorded

> Farber, William⟨*tab*⟩   2201.00⟨*tab*⟩

or

> Farber, William⟨*tab*⟩   2201.00

A line with the last two fields missing might be recorded

> Farber, William⟨*tab*⟩⟨*tab*⟩

or

> Farber, William⟨*tab*⟩

or

> Farber, William

(*Continued on next page*)

The following program would correctly parse lines with missing fields regardless of how they are recorded:

```
real rowvector readtabbed(transmorphic t, real scalar n)
{
        real scalar      i
        string rowvector res
        string scalar    token
        res = J(1, n, "")
        i = 1
        while ((token = tokenget(t))!="") {
                if (token==char(9)) i++
                else res[i] = token
        }
        return(res)
}
```

Function overview

The basic way to proceed is to initialize the parsing environment and store it in a variable,

```
t = tokeninit(...)
```

and then set the string s to be parsed,

```
tokenset(t, s)
```

and finally use tokenget() to obtain the tokens one at a time (tokenget() returns "" when the end of the line is reached), or obtain all the tokens at once using tokengetall(t). That is, either

```
while((token = tokenget(t)) != "") {
        ... process token ...
}
```

or

```
tokens = tokengetall(t)
for (i=1; i<=cols(tokens); i++) {
        ... process tokens[i] ...
}
```

After that, set the next string to be parsed,

```
tokenset(t, nextstring)
```

and repeat.

tokeninit() and tokeninitstata()

tokeninit() and tokeninitstata() are alternatives. tokeninitstata() is generally unnecessary unless you are writing a fairly complicated function.

Whichever function you use, code

> t = tokeninit(...)

or

> t = tokeninitstata()

If you declare t, declare it transmorphic. t is in fact a structure containing all the details of your parsing environment, but that is purposely hidden from you so that you cannot accidentally modify the environment.

tokeninit() allows up to five arguments:

> t = tokeninit(*wchars*, *pchars*, *qchars*, *allownum*, *allowhex*)

You may omit arguments from the end. If omitted, the default values of the arguments are

> *allowhex* = 0
>
> *allownum* = 0
>
> *qchars* = (`""""` , `"'""'"`)
>
> *pchars* = (`""`)
>
> *wchars* = " "

Notes

1. Concerning *wchars*:

   a. *wchars* is a *string scalar*. The white-space characters appear one after the other in the string. The order in which the characters appear is irrelevant.

   b. Specify *wchars* as " " to treat blank as white space.

   c. Specify *wchars* as " "+char(9) to treat blank and *tab* as white space. Including *tab* is necessary only when strings to be parsed are obtained from a file; strings obtained from Stata already have the *tab* characters removed.

   d. Any character can be treated as a white-space character, including letters.

   e. Specify *wchars* as "" to specify that there are no white-space characters.

2. Concerning *pchars*:

   a. *pchars* is a *string rowvector*. Each element of the vector is a separate parse character. The order in which the parse characters are specified is irrelevant.

   b. Specify *pchars* as ("+", "-") to make + and - parse characters.

   c. Parse characters may be character combinations such as ++ or >=. Character combinations may be up to four characters long.

   d. Specify *pchars* as "" or J(1,0,"") to specify that there are no parse characters. It makes no difference which you specify, but you will realize that J(1,0,"") is more logically consistent if you think about it.

3. Concerning *qchars*:

   a. *qchars* is a `string rowvector`. Each element of the vector contains the open followed by the close characters. The order in which sets of quote characters are specified is irrelevant.

   b. Specify *qchars* as (`'""""'`) to make " an open and close character.

   c. Specify *qchars* as (`'""""'`, `'"<"">"'`) to make "" and `'""'` quote characters.

   d. Individual quote characters can be up to two characters long.

   e. Specify *qchars* as `""` or `J(1,0,"")` to specify that there are no quote characters.

tokenset()

After `tokeninit()` or `tokeninitstata()`, you are not yet through with initialization. You must `tokenset(s)` to specify the string scalar you wish to parse. You `tokenset()` one line, parse it, and if you have more lines, you `tokenset()` again and repeat the process. Often you will need to parse only one line. Perhaps you wish to write a program to parse the argument of a complicated option in a Stata ado-file. The structure is

```
program ...
    ...
    syntax ... [, ... MYoption(string) ...]
    mata: parseoption("`myoption'")
    ...
end

mata:
void parseoption(string scalar option)
{
    transmorphic    t
    t = tokeninit(...)
    tokenset(t, option)
    ...
}
end
```

Notes

1. When you `tokenset(s)`, the contents of *s* are not stored. Instead, a pointer to *s* is stored. This approach saves memory and time, but it means that if you change *s* after setting it, you will change the subsequent behavior of the `token*()` functions.

2. Simply changing *s* is not sufficient to restart parsing. If you change *s*, you must `tokenset(s)` again.

tokengetall()

You have two alternatives in how to process the tokens. You can parse the entire line into a row vector containing all the individual tokens by using `tokengetall()`,

```
tokens = tokengetall(t)
```

or you can use `tokenget()` to process the tokens one at a time, which is discussed in the next section.

Using `tokengetall()`, `tokens[1]` will be the first token, `tokens[2]` the second, and so on. There are, in total, `cols(tokens)` tokens. If the line was empty or contained only white-space characters, `cols(tokens)` will be 0.

tokenget(), tokenpeek(), and tokenrest()

`tokenget()` returns the tokens one at a time and returns `""` when the end of the line is reached. The basic loop for processing all the tokens in a line is

```
while ( (token = tokenget( t )) != "") {
        . . .
}
```

`tokenpeek()` allows you to peek ahead at the next token without actually getting it, so whatever is returned will be returned again by the next call to `tokenget()`. `tokenpeek()` is suitable only for obtaining the next token after `tokenget()`. Calling `tokenpeek()` twice in a row will not return the next two tokens; it will return the next token twice. To obtain the next two tokens, code

```
        . . .
        current = tokenget(t)         // get the current token
        . . .
        t2 = t                        // copy parse environment
        next_1 = tokenget(t2)         // peek at next token
        next_1 = tokenget(t2)         // peek at token after that
        . . .
        current = tokenget(t)         // get next token
```

If you declare *t2*, declare it `transmorphic`.

`tokenrest()` returns the unparsed portion of the `tokenset()` string. Assume that you have just gotten the first token by using `tokenget()`. `tokenrest()` would return the rest of the original string, following the first token, unparsed. `tokenrest(t)` returns `substr`(*original_string*, `tokenoffset(t)`, .).

tokenoffset()

`tokenoffset()` is useful only when you are using the `tokenget()` rather than `tokengetall()` style of programming. Let the original string you `tokenset()` be "this is an example". Right after you have `tokenset()` this string, `tokenoffset()` is 1:

```
        this is an example
        |
  tokenoffset() = 1
```

After getting the first token (say it is `"this"`), `tokenoffset()` is 5:

```
        this is an example
            |
  tokenoffset() = 5
```

tokenoffset() is always located on the first character following the last character parsed.

The syntax of tokenoffset() is

    tokenoffset(*t*)

and

    tokenoffset(*t, newoffset*)

The first returns the current offset value. The second resets the parser's location within the string.

tokenwchars(), tokenpchars(), and tokenqchars()

tokenwchars(), tokenpchars(), and tokenqchars() allow resetting the current *wchars*, *pchars*, and *qchars*. As with tokenoffset(), they come in two syntaxes.

With one argument, *t*, they return the current value of the setting. With two arguments, *t* and *newvalue*, they reset the value.

Resetting in the midst of parsing is an advanced issue. The most useful of these functions is tokenpchars(), since for interactive grammars, it is sometimes necessary to switch on and off a certain parsing character such as /, which in one context means division and in another is a file separator.

tokenallownum and tokenallowhex()

These two functions allow obtaining the current values of *allownum* and *allowhex* and resetting them.

Conformability

tokeninit(*wchars, pchars, qchars, allownum, allowhex*):

wchars:	1×1	(optional)
pchars:	$1 \times c_p$	(optional)
qchars:	$1 \times c_q$	(optional)
allownum:	1×1	(optional)
allowhex:	1×1	(optional)
result:	*transmorphic*	

tokeninitstata():

    result: *transmorphic*

tokenset(*t, s*):

t:	*transmorphic*
s:	1×1
result:	*void*

tokengetall(*t*):

t:	*transmorphic*
result:	$1 \times k$

tokenget(*t*), tokenpeek(*t*), tokenrest(*t*):

t:	*transmorphic*
result:	1×1

tokenoffset(*t*), tokenwchars(*t*), tokenallownum(*t*), tokenallowhex(*t*):
  t: *transmorphic*
  result: 1 × 1

tokenoffset(*t*, *newvalue*), tokenwchars(*t*, *newvalue*),
  tokenallownum(*t*, *newvalue*), tokenallowhex(*t*, *newvalue*):
  t: *transmorphic*
  newvalue: 1 × 1
  result: *void*

tokenpchars(*t*), tokenqchars(*t*):
  t: *transmorphic*
  result: 1 × *c*

tokenpchars(*t*, *newvalue*), tokenqchars(*t*, *newvalue*):
  t: *transmorphic*
  newvalue: 1 × *c*
  result: *void*

Diagnostics

None.

Also See

[M-5] **tokens( )** — Obtain tokens from string

[M-4] **programming** — Programming functions

[M-4] **string** — String manipulation functions

Title

[M-5] **tokens( )** — Obtain tokens from string

Syntax

>   *string rowvector*   `tokens(`*string scalar s*`)`

>   *string rowvector*   `tokens(`*string scalar s*, *string scalar parsechars*`)`

Description

`tokens(`*s*`)` returns the contents of *s*, split into words.

`tokens(`*s*, *parsechars*`)` returns the contents of *s* split into tokens based on *parsechars*.

`tokens(`*s*`)` is equivalent to `tokens(`*s*, " "`)`.

If you need more advanced parsing, see [M-5] **tokenget( )**.

Remarks

`tokens()` is commonly used to split a string containing a sequence of variable names into a row vector, each element of which contains one variable name:

        `tokens("mpg weight displacement") = ("mpg", "weight", "displacement")`

Many of the Stata interface functions require that variable names be specified in this form. This is required, for instance, by `st_data()` and `st_view()`; see [M-5] **st_data( )** and [M-5] **st_view( )**.

Say you have an ado-file that calls private Mata function `mysub()` to perform a task. The varlist that is to be used is passed from the ado-file to the Mata function and was originally obtained from the user:

```
─────────────────────────────────────────────────── top of example.ado ──────
program example
        syntax varlist [if] [in]
        marksample touse
        ...
        mata: mysub("`varlist'", "`touse'")
        ...
end

mata void mysub(string scalar varlist, string scalar touse)
{
        ...
        st_view(V, ., tokens(varlist), touse)
        ...
}
─────────────────────────────────────────────────── end of example.ado ──────
```

st_view() requires that its third argument be a row vector of variable names. Our ado-file passed in one string that had all the variable names, listed one after the other, so we could not simply code

    st_view(V, ., varlist, touse)

Perhaps `varlist` contains "mpg weight displacement". `tokens()` converts `varlist` to the form `st_view()` requires, namely, the row vector ("mpg", "weight", "displacement").

Conformability

tokens(*s*, *parsechars*)
         s: 1 × 1
   parsechars: 1 × 1   (optional)
     result: 1 × w, w = number of words (tokens) in *s*

Diagnostics

If *s* contains "", tokens() returns J(1,0,"").

If *s* contains double-quoted or compound-double-quoted material, the quotes are stripped and that material is returned as one token. For example,

    tokens(`"this "is an" example"') = ("this", "is an", "example")

If *s* contains quoted material and the quotes do not match, results are as if the appropriate number of close quotes were added to the end of *s*. For example,

    tokens(`"this "is an example"') = ("this", "is an example")

Also See

[M-5] **invtokens( )** — Concatenate string rowvector into string scalar

[M-5] **tokenget( )** — Advanced parsing

[M-4] **string** — String manipulation functions

Title

[M-5] trace( ) — Trace of square matrix

Syntax

> *numeric scalar*   `trace(`*numeric matrix A*`)`

> *numeric scalar*   `trace(`*numeric matrix A, numeric matrix B*`)`

> *numeric scalar*   `trace(`*numeric matrix A, numeric matrix B, real scalar t*`)`

Description

trace(*A*) returns the sum of the diagonal elements of *A*. Returned result is real if *A* is real, complex if *A* is complex.

trace(*A*, *B*) returns trace(*AB*), the calculation being made without calculating or storing the off-diagonal elements of *AB*. Returned result is real if *A* and *B* are real and is complex otherwise.

trace(*A*, *B*, *t*) returns trace(*AB*) if $t = 0$ and returns trace(*A'B*) otherwise, where, if either *A* or *B* is complex, transpose is understood to mean conjugate transpose. Returned result is real if *A* and *B* are real and is complex otherwise.

Remarks

trace(*A*, *B*) returns the same result as trace(*A*∗*B*) but is more efficient if you do not otherwise need to calculate *A*∗*B*.

trace(*A*, *B*, 1) returns the same result as trace(*A'B*) but is more efficient.

For real matrices *A* and *B*,

$$\text{trace}(A') = \text{trace}(A)$$
$$\text{trace}(AB) = \text{trace}(BA)$$

and for complex matrices,

$$\text{trace}(A') = \text{conj}(\text{trace}(A))$$
$$\text{trace}(AB) = \text{trace}(BA)$$

where, for complex matrices, transpose is understood to mean conjugate transpose.

Thus for real matrices,

to calculate	code
trace(AB)	trace(A, B)
trace($A'B$)	trace(A, B, 1)
trace(AB')	trace(A, B, 1)
trace($A'B'$)	trace(A, B)

and for complex matrices,

to calculate	code
trace(AB)	trace(A, B)
trace($A'B$)	trace(A, B, 1)
trace(AB')	conj(trace(A, B, 1))
trace($A'B'$)	conj(trace(A, B))

Transpose in the first column means conjugate transpose.

Conformability

trace(A):
    A: $n \times n$
    result: 1×1

trace(A, B):
    A: $n \times m$
    B: $m \times n$
    result: 1×1

trace(A, B, t)
    A: $n \times m$ if $t = 0$, $m \times n$ otherwise
    B: $m \times n$
    t: 1×1
    result: 1×1

Diagnostics

trace(A) aborts with error if A is not square.

trace(A, B) and trace(A, B, t) abort with error if the matrices are not conformable or their product is not square.

The trace of a 0×0 matrix is 0.

Also See

[M-4] **matrix** — Matrix functions

Title

[M-5] **_transpose( )** — Transposition in place

Syntax

> *void* _transpose(*numeric matrix A*)
>
> *void* _conj(*numeric matrix A*)

Description

_transpose(*A*) replaces *A* with *A*′. Coding _transpose(*A*) is equivalent to coding *A* = *A*′, except that execution can take a little longer and less memory is used. When *A* is complex, *A* is replaced with its conjugate transpose; see [M-5] **transposeonly( )** if transposition without conjugation is desired.

_conj(*A*) replaces *A* with conj(*A*). Coding _conj(*A*) is equivalent to coding *A* = conj(*A*), except that less memory is used.

Remarks

In some calculation, you need *A*′

> *X* = ... *calculation using A*′ ...

If *A* is large, you can save considerable memory by coding

> _transpose(*A*)
> *X* = ... *calculation using A* ...
> _transpose(*A*)

Conformability

_transpose(*A*):
  input:
        A: $r \times c$
  output:
        A: $c \times r$

_conj(*A*):
  input:
        A: $r \times c$
  output:
        A: $r \times c$

Diagnostics

_transpose(A) aborts with error if A is a view.

_conj(A) does nothing if A is real (and hence, does not abort if A is a view).

Also See

[M-2] **op_transpose** — Conjugate transpose operator

[M-5] **transposeonly()** — Transposition without conjugation

[M-5] **conj()** — Complex conjugate

[M-4] **manipulation** — Matrix manipulation

Title

[M-5] **transposeonly( )** — Transposition without conjugation

Syntax

> *numeric matrix* transposeonly(*numeric matrix A*)
> *void* _transposeonly(*numeric matrix A*)

Description

transposeonly(*A*) returns *A* with its rows and columns interchanged. When *A* is real, the actions of transposeonly(*A*) are indistinguishable from coding A'; see [M-2] **op_transpose**. The returned result is the same, and the execution time is the same, too. When *A* is complex, however, transposeonly(*A*) is equivalent to coding conj(A'), but transposeonly() obtains the result more quickly.

_transposeonly(*A*) interchanges the rows and columns of *A* in place—without use of additional memory—and returns the transposed (but not conjugated) result in *A*.

Remarks

transposeonly() is useful when you are coding in the programming, rather than the mathematical, sense. Say that you have two row vectors, a and b, and you want to place the two vectors together in a matrix R, and you want to turn them into column vectors. If a and b were certain to be real, you could just code

    R = (a', b')

The above line, however, would result in not just the organization but also the values recorded in R changing if a or b were complex. The solution is to code

    R = (transposeonly(a), transposeonly(b))

The above line will work for real or complex a and b. If you were concerned about memory consumption, you could instead code

    R = (a \ b)
    _transposeonly(R)

Conformability

transposeonly(*A*):
>   *A*: $r \times c$
>   *result*: $c \times r$

_transposeonly(*A*):

input:
          A: $r \times c$

output:
          A: $c \times r$

Diagnostics

_transposeonly(*A*) aborts with error if *A* is a view.

Also See

[M-2] **op_transpose** — Conjugate transpose operator

[M-5] **_transpose( )** — Transposition in place

[M-4] **manipulation** — Matrix manipulation

Title

[M-5] **trunc( )** — Round to integer

Syntax

    real matrix   trunc(*real matrix R*)

    real matrix   floor(*real matrix R*)

    real matrix   ceil(*real matrix R*)

    real matrix   round(*real matrix R*)

    real matrix   round(*real matrix R*, *real matrix U*)

Description

These functions convert noninteger values to integers by moving toward 0, moving down, moving up, or rounding. These functions are typically used with scalar arguments, and they return a scalar in that case. When used with vectors or matrices, the operation is performed element by element.

trunc(*R*) returns the integer part of *R*.

floor(*R*) returns the largest integer *i* such that $i \leq R$.

ceil(*R*) returns the smallest integer *i* such that $i \geq R$.

round(*R*) returns the integer closest to *R*.

round(*R*, *U*) returns the values of *R* rounded in units of *U* and is equivalent to round((R:/U)):*U. For instance, round(*R*, 2) returns *R* rounded to the closest even number. round(*R*, .5) returns *R* rounded to the closest multiple of one half. round(*R*, 1) returns *R* rounded to the closest integer and so is equivalent to round(*R*).

Remarks

Remarks are presented under the following headings:

    Relationship to Stata's functions
    Examples of rounding

Relationship to Stata's functions

trunc() is equivalent to Stata's int() function.

floor(), ceil(), and round() are equivalent to Stata's functions of the same name.

Examples of rounding

x	trunc(x)	floor(x)	ceil(x)	round(x)
1	1	1	1	1
1.3	1	1	2	1
1.6	1	1	2	2
-1	-1	-1	-1	-1
-1.3	-1	-2	-1	-1
-1.6	-1	-2	-1	-2

Conformability

trunc(R), floor(R), ceil(R):
        R: $r \times c$
   result: $r \times c$

round(R):
        R: $r \times c$
   result: $r \times c$

round(R, U):
        R: $r_1 \times c_1$
        U: $r_2 \times c_2$, R and U r-conformable
   result: $\max(r_1, r_2) \times \max(c_1, c_2)$

Diagnostics

Most Stata and Mata functions return missing when arguments contain missing, and in particular, return . whether the argument is ., .a, .b, ..., .z. The logic is that performing the operation on a missing value always results in the same missing-value result. For example, sqrt(.a)==.

These functions, however, when passed a missing value, return the particular missing value. Thus trunc(.a)==.a, floor(.b)==.b, ceil(.c)==.c, and round(.d)==.d.

For round() with two arguments, this applies to the first argument and only when the second argument is not missing. If the second argument is missing (whether ., .a, ..., or .z), then . is returned.

Also See

[M-4] **scalar** — Scalar mathematical functions

Title

[M-5] **uniform( )** — Uniformly distributed random numbers

Syntax

    real matrix   uniform(*real scalar r, real scalar c*)

    string scalar   uniformseed()

    void   uniformseed(*string scalar newseed*)

    void   uniformseed(*real scalar newseed*)

Description

uniform(*r, c*) returns an $r \times c$ real matrix containing uniformly distributed random numbers $0 <$ element < 1. uniform() is the same function as Stata's uniform() function.

uniformseed() returns uniform()'s current seed in encrypted, string form. uniformseed() returns the same thing as Stata's c(seed); see [D] **generate** and [P] **creturn**.

uniformseed(*newseed*) sets the seed: a string previously obtained from uniformseed() may be specified for the argument, or an integer number may be specified. uniformseed() has the same effect as Stata's set seed command; see [D] **generate**.

Conformability

uniform(*r, c*):
        r: 1×1
        c: 1×1
   result: $r \times c$

uniformseed():
   result: 1×1

uniformseed(*newseed*):
  newseed: 1×1
   result: void

Diagnostics

uniform(*r, c*) aborts with error if $r < 0$ or $c < 0$.

uniformseed(*seed*) aborts with error if a string seed is specified and it is malformed (was not obtained from uniformseed()).

Also See

[M-4] **standard** — Functions to create standard matrices

[M-4] **statistical** — Statistical functions

Title

[M-5] **uniqrows( )** — Obtain sorted, unique values

Syntax

> *transmorphic matrix* uniqrows(*transmorphic matrix P*)

Description

uniqrows(*P*) returns a sorted matrix containing the unique rows of *P*.

Remarks

```
: x
            1   2   3
        1   4   5   7
        2   4   5   6
        3   1   2   3
        4   4   5   6

: uniqrows(x)
            1   2   3
        1   1   2   3
        2   4   5   6
        3   4   5   7
```

Conformability

uniqrows(*P*)
        P: $r_1 \times c_1$
  result: $r_2 \times c_1$, $r_2 \leq r_1$

Diagnostics

In uniqrows(*P*), if rows(*P*)==0, J(0, cols(*P*), missingof(*P*)) is returned.

If rows(*P*)>0 and cols(*P*)==0, J(1, 0, missingof(*P*)) is returned.

Also See

[M-5] **sort( )** — Reorder rows of matrix

[M-4] **manipulation** — Matrix manipulation

Title

[M-5] **unitcircle( )** — Complex vector containing unit circle

Syntax

> *complex colvector* unitcircle(*real scalar n*)

Description

unitcircle(*n*) returns a column vector containing C(cos(θ), sin(θ)) for $0 \leq \theta \leq 2\pi$ in *n* points.

Conformability

unitcircle(*n*):
        n: 1×1
   result: $n \times 1$

Diagnostics

None.

Also See

[M-4] **standard** — Functions to create standard matrices

Title

[M-5] **unlink( )** — Erase file

Syntax

> *void*   unlink(*real scalar filename*)
>
> *real scalar*   _unlink(*real scalar filename*)

Description

unlink(*filename*) erases *filename* if it exists, does nothing if *filename* does not exist, and aborts with error if *filename* exists but cannot be erased.

_unlink(*filename*) does the same, except that, if *filename* cannot be erased, rather than aborting with error, _unlink() returns a negative error code. _unlink() returns 0 if *filename* was erased or *filename* did not exist.

Remarks

To remove directories, see rmdir() in [M-5] **chdir( )**.

Conformability

> unlink(*filename*)
>   *filename*: 1×1
>   *result*:   *void*
>
> _unlink(*filename*)
>   *filename*: 1×1
>   *result*: 1×1

Diagnostics

unlink(*filename*) aborts with error when _unlink() would give a negative result.

_unlink(*filename*) returns a negative result if the file cannot be erased and returns 0 otherwise. If the file did not exist, 0 is returned. When there is an error, most commonly returned are −3602 (filename invalid) or −3621 (file is read-only).

Also See

[M-4] **io** — I/O functions

Title

> **[M-5] valofexternal( )** — Obtain value of external global

Syntax

> *transmorphic matrix*   valofexternal(*string scalar name*)

Description

valofexternal(*name*) returns the contents of the external global matrix, vector, or scalar whose name is specified by *name*; it returns J(0,0,.) if the external global is not found.

Also see *Linking to external globals* in [M-2] **declarations**.

Remarks

Also see [M-5] **findexternal( )**. Rather than returning a pointer to the external global, as does findexternal(), valofexternal() returns the contents of the external global. This is useful when the external global contains a scalar:

```
tol = valofexternal("tolerance")
if (tol==J(0,0,.)) tol = 1e-6
```

Using findexternal(), one alternative would be

```
if ((p = findexternal("tolerance"))==NULL) tol = 1e-6
else tol = *p
```

For efficiency reasons, use of valofexternal() should be avoided with nonscalar objects; see [M-5] **findexternal( )**.

Conformability

valofexternal(*name*):
> *name*: 1×1
> *result*: $r \times c$ or 0×0 if not found

Diagnostics

valofexternal() aborts with error if *name* contains an invalid name.

valofexternal(*name*) returns J(0,0,.) if *name* does not exist.

Also See

[M-5] **findexternal( )** — Find, create, and remove external globals

[M-4] **programming** — Programming functions

Title

[M-5] **Vandermonde( )** — Vandermonde matrices

Syntax

>   *numeric matrix* Vandermonde(*numeric colvector x*)

Description

Vandermonde(*x*) returns the Vandermonde matrix containing the geometric progression of *x* in each row

$$\begin{bmatrix} 1 & x_1 & x_1^2 & x_1^3 & \cdots & x_1^{n-1} \\ 1 & x_2 & x_2^2 & x_2^3 & \cdots & x_2^{n-1} \\ \vdots & \vdots & \vdots & \vdots & \ddots & \vdots \\ 1 & x_n & x_n^2 & x_n^3 & \cdots & x_n^{n-1} \end{bmatrix}$$

where $n = \text{rows}(x)$. Some authors use the transpose of the above matrix.

Remarks

Vandermonde matrices are useful in polynomial interpolation.

Conformability

Vandermonde(*x*):
>   *x*: $n \times 1$
>   *result*: $n \times n$

Diagnostics

None.

Alexandre-Théophile Vandermonde (1735–1796) was born in Paris. His first passion was music (particularly the violin) and he turned to mathematics only at the age of 35. Four papers dated 1771 and 1772 are his entire mathematical output, although all contain good work. He also worked in experimental science and the theory of music, arguing that musicians should ignore all theory and trust their trained ears, and was busy with various committees and other administration. Vandermonde was a strong supporter of the French Revolution. He is now best known for the Vandermonde determinant, even though it does not appear in any of his papers, and for the associated matrix. Lebesgue later conjectured that the attribution arises from a misreading of Vandermonde's notation.

Reference

Jones, P. S. 1976. Vandermonde, Alexandre-Théophile. In *Dictionary of Scientific Biography*, vol. 13, ed. C. C. Gillispie, 571–572. New York: Scribner's.

Also See

[M-4] **standard** — Functions to create standard matrices

Title

[M-5] **vec( )** — Stack matrix columns

Syntax

> *transmorphic colvector* vec(*transmorphic matrix T*)
>
> *transmorphic colvector* vech(*transmorphic matrix T*)
>
> *transmorphic matrix* invvech(*transmorphic colvector v*)

Description

vec(*T*) returns *T* transformed into a column vector with one column stacked onto the next.

vech(*T*) returns square and typically symmetric matrix *T* transformed into a column vector; only the lower half of the matrix is recorded.

invvech(*v*) returns vech()-style column vector *v* transformed into a symmetric (Hermitian) matrix.

Remarks

Remarks are presented under the following headings:

> Example of vec( )
> Example of vech( ) and invvech( )

Example of vec( )

```
: x
         1   2   3
    1    1   2   3
    2    4   5   6

: vec(x)
         1
    1    1
    2    4
    3    2
    4    5
    5    3
    6    6
```

Example of vech( ) and invvech( )

```
: x
[symmetric]
       1   2   3
    ┌─────────────┐
  1 │ 1           │
  2 │ 2   4       │
  3 │ 3   6   9   │
    └─────────────┘

: v = vech(x)
: v
       1
    ┌─────┐
  1 │ 1   │
  2 │ 2   │
  3 │ 3   │
  4 │ 4   │
  5 │ 6   │
  6 │ 9   │
    └─────┘

: invvech(v)
[symmetric]
       1   2   3
    ┌─────────────┐
  1 │ 1           │
  2 │ 2   4       │
  3 │ 3   6   9   │
    └─────────────┘
```

Conformability

vec(T):

$\qquad T: \quad r \times c$
$\qquad result: \quad r*c \times 1$

vech(T):

$\qquad T: \quad n \times n$
$\qquad result: \quad (n(n+1)/2 \times 1)$

invvech(v):

$\qquad v: \quad (n(n+1)/2 \times 1)$
$\qquad result: \quad n \times n$

Diagnostics

vec(T) cannot fail.

vech(T) aborts with error if T is not square. vech( ) records only the lower triangle of T; it does not require T be symmetric.

invvech(v) aborts with error if v does not have 0, 1, 3, 6, 10, ... rows.

Also See

[M-4] **manipulation** — Matrix manipulation

[M-6] Mata glossary of common terms

Title

[M-6] **glossary** — Mata glossary of common terms

Description

Commonly used terms are defined here.

Mata glossary

arguments

The values a function receives are called the function's arguments. For instance, in lud(A, L, U), A, L, and U are the arguments.

broad type

Two matrices are said to be of the same broad type if the elements in each are numeric, are string, or are pointers. Mata provides two numeric types, real and complex. The term *broad type* is used to mask the distinction within numeric and is often used when discussing operators or functions. One might say, "The comma operator can be used to join the rows of two matrices of the same broad type," and the implication of that is that one could join a real to a complex. The result would be complex. Also see *type*.

c-conformability

Matrix, vector, or scalar A is said to be c-conformable with matrix, vector, or scalar B if they have the same number of rows and columns (they are *p-conformable*), or if they have the same number of rows and one is a vector, or if they have the same number of columns and one is a vector, or if one or the other is a scalar. c stands for colon; c-conformable matrices are suitable for being used with Mata's :*op* operators. A and B are c-conformable if and only if

A	B
$r \times c$	$r \times c$
$r \times 1$	$r \times c$
$1 \times c$	$r \times c$
1×1	$r \times c$
$r \times c$	$r \times 1$
$r \times c$	$1 \times c$
$r \times c$	1×1

The idea behind c-conformability is generalized elementwise operation. Consider $C=A:*B$. If A and B have the same number of rows and have the same number of columns, then $||C_{ij}|| = ||A_{ij}*B_{ij}||$. Now say that A is a column vector and B is a matrix. Then $||C_{ij}|| = ||A_i*B_{ij}||$: each element of A is applied to the entire row of B. If A is a row vector, each column of A is applied to the entire column of B. If A is a scalar, A is applied to every element of B. And then all the rules repeat, with the roles of A and B interchanged. See [M-2] **op_colon** for a complete definition.

colon operators

Colon operators are operators preceded by a colon, and the colon indicates that the operator is to be performed elementwise. $A:*B$ indicates element-by-element multiplication, whereas $A*B$ indicates matrix multiplication. Colons may be placed in front of any operator. Usually one thinks of elementwise as meaning $c_{ij} = a_{ij} <op> b_{ij}$, but in Mata, elementwise is also generalized to include c-conformability. See [M-2] **op_colon**.

colvector

See *vector*.

complex

A matrix is said to be complex if its elements are complex numbers. Complex is one of two numeric types in Stata, the other being real. Complex is generally used to describe how a matrix is stored and not the kind of numbers that happen to be in it: complex matrix Z might happen to contain real numbers. Also see *type*.

condition number

The condition number associated with a numerical problem is a measure of that quantity's amenability to digital computation. A problem with a low condition number is said to be well conditioned, whereas a problem with a high condition number is said to be ill conditioned.

Sometimes reciprocals of condition numbers are reported and yet authors will still refer to them sloppily as condition numbers. Reciprocal condition numbers are often scaled between 0 and 1, with values near epsilon(1) indicating problems.

conformability

Conformability refers to row-and-column matching between two or more matrices. For instance, to multiply $A*B$, A must have the same number of columns as B has rows. If that is not true, then the matrices are said to be nonconformable (for multiplication).

Three kinds of conformability are often mentioned in the Mata documentation: *p-conformability*, *c-conformability*, and *r-conformability*.

conjugate

If $z = a + bi$, the conjugate of z is $\text{conj}(z) = a - bi$. The conjugate is obtained by reversing the sign of the imaginary part. The conjugate of a real number is the number itself.

conjugate transpose

See *transpose*.

data matrix

A dataset containing n observations on k variables in often stored in an $n \times k$ matrix. An observation refers to a row of that matrix; a variable refers to a column. When the rows are observations and the columns are variables, the matrix is called a data matrix.

declarations

Declarations state the *eltype* and *orgtype* of functions, arguments, and variables. In

```
real matrix myfunc(real vector A, complex scalar B)
{
        real scalar i
        ...
}
```

the `real matrix` is a function declaration, the `real vector` and `complex scalar` are argument declarations, and `real scalar i` is a variable declaration. The `real matrix` states the function returns a real matrix. The `real vector` and `complex scalar` state the kind of arguments `myfunc()` expects and requires. The `real scalar i` helps Mata to produce more efficient compiled code.

Declarations are optional, so the above could just as well have read

```
function myfunc(A, B)
{
        ...
}
```

When you omit the function declaration, you must substitute the word `function`.

When you omit the other declarations, `transmorphic matrix` is assumed, which is fancy jargon for a matrix that can hold anything. The advantages of explicit declarations are that they reduce the chances you make a mistake either in coding or in using the function, and they assist Mata in producing more efficient code. Working interactively, most people omit the declarations.

See [M-2] **declarations** for more information.

defective matrix

An $n \times n$ matrix is defective if it does not have n linearly independent eigenvectors.

diagonal matrix

A matrix is diagonal if its off-diagonal elements are zero; A is diagonal if $A[i,j]==0$ for $i!=j$. Usually, diagonal matrices are also *square*. Some definitions require that a diagonal matrix also be a square matrix.

diagonal of a matrix

The diagonal of a matrix is the set of elements $A[i,j]$.

eltype

See *type*.

epsilon(1), etc.

epsilon(1) refers to the unit roundoff error associated with a computer, also informally called machine precision. It is the smallest amount by which a number may differ from 1. For IEEE double-precision variables, epsilon(1) is approximately $2.22045e-16$.

$\texttt{epsilon}(x)$ is the smallest amount by which a real number can differ from x, or an approximation thereof; see [M-5] **epsilon( )**.

exp

exp is used in syntax diagrams to mean "any valid expression may appear here"; see [M-2] **exp**.

external variable

See *global variable*.

function

The words *program* and *function* are used interchangeably. The programs that you write in Mata are in fact functions. Functions receive arguments and optionally return results.

Examples of functions that are included with Mata are $\texttt{sqrt()}$, $\texttt{ttail()}$, and $\texttt{substr()}$. Such functions are often referred to as the built-in functions or the library functions. Built-in functions refer to functions implemented in the C code that implements Mata, and library functions refer to functions written in the Mata programming language, but many users use the words interchangeably because how functions are implemented is of little importance. If you have a choice between using a built-in function and a library function, however, the built-in function will usually execute more quickly and the library function will be easier to use. Mostly, however, features are implemented one way or the other and you have no choice.

Also see *underscore functions*.

For a list of the functions that Mata provides, see [M-4] **intro**.

global variable

Global variables, also known as external variables and as global external variables, refer to variables that are common across programs and which programs may access without the variable being passed as an argument.

The variables you create interactively are global variables. Even so, programs cannot access those variables without engaging in another step, and global variables can be created without your creating them interactively.

To access (and create if necessary) global external variables, you declare the variable in the body of your program:

```
function myfunction(...)
{
    external real scalar globalvar
    ...
}
```

See *Linking to external globals* in [M-2] **declarations**.

There are other ways of creating and accessing global variables, but the declaration method is recommended. The alternatives are crexternal(), findexternal(), and rmexternal() documented in [M-5] **findexternal( )** and valofexternal() documented in [M-5] **valofexternal( )**.

Hermitian matrix

Matrix A is Hermitian if it is equal to its conjugate transpose; $A = A'$; see *transpose*. This means that each off-diagonal element a_{ij} must equal the conjugate of a_{ji}, and that the diagonal elements must be real. The following matrix is Hermitian:

$$\begin{bmatrix} 2 & 4+5i \\ 4-5i & 6 \end{bmatrix}$$

The definition $A = A'$ is the same as the definition for a symmetric matrix, although usually the word *symmetric* is reserved for real matrices and Hermitian, for complex matrices. In this manual, we use the word *symmetric* for both; see *symmetric matrix*.

istmt

An *istmt* is an interactive statement, a statement typed at Mata's colon prompt.

J(r, c, value)

J() is the function that returns an $r \times c$ matrix with all elements set to *value*; see [M-5] **J( )**. Also, J() is often used in the documentation to describe the various types of *void* matrices; see *void*. Thus the documentation might say that such-and-such returns J(0, 0, .) under certain conditions. That is another way of saying that such-and-such returns a 0×0 real matrix.

When r or c is 0, there are no elements to be filled in with *value*, but even so, *value* is used to determine the type of the matrix. Thus J(0, 0, 1i) refers to a 0×0 complex matrix, J(0, 0, "") refers to a 0×0 string matrix, and J(0, 0, NULL) refers to a 0×0 *pointer* matrix.

In the documentation, J() is used for more than describing 0×0 matrices. Sometimes, the matrices being described are $r \times 0$ or are $0 \times c$. Say that a function example(X) is supposed to return a column vector; perhaps it returns the last column of X. Now say that X is 0×0. Function example() still should return a column vector, and so it returns a 0×1 matrix. This would be documented by noting that example() returns J(0, 1, .) when X is 0×0.

LAPACK

LAPACK stands for Linear Algebra PACKage and forms the basis for many of Mata's linear algebra capabilities; see [M-1] **LAPACK**.

lval

lval stands for left-hand-side value and is defined as the property of being able to appear on the left-hand side of an equal-assignment operator. Matrices are *lvals* in Mata, and thus

    X = ...

is valid. Functions are not *lvals*; thus, you cannot code

    substr(mystr,1,3) = "abc"

lvals would be easy to describe except that *pointers* can also be lvals. Few people ever use pointers. See [M-2] **op_assignment** for a complete definition.

.mata file

By convention, we store the Mata source code for function *function()* in file *function*.mata; see [M-1] **source**.

.mlib library

The object code of functions can be collected and stored in a library. Most Mata functions, in fact, are located in the official libraries provided with Stata. You can create your own libraries. See [M-3] **mata mlib**.

.mo file

The object code of a function can be stored in a .mo file, where it can be later reused. See [M-1] **how** and [M-3] **mata mosave**.

machine precision

See *epsilon(1)*.

matrix

The most general organization of data, containing r rows and c columns. Vectors, column vectors, row vectors, and scalars are special cases of matrices.

norm

A norm is a real-valued function $f(x)$ satisfying

$$\begin{aligned} f(0) &= 0 \\ f(x) &> 0 \quad \text{for all } x \neq 0 \\ f(cx) &= |c| f(x) \\ f(x+y) &\leq f(x) + f(y) \end{aligned}$$

The word *norm* applied to a vector x usually refers to its Euclidean norm, $p = 2$ norm, or length: the square root of the sum of its squared elements. The are other norms, the popular ones being $p = 1$ (the sum of the absolute values of its elements) and $p = $ infinity (the maximum element). Norms can also be generalized to deal with matrices. See [M-5] **norm( )**.

NULL

A special value for a *pointer* that means "points to nothing". If you list the contents of a pointer variable that contains NULL, the address will show as 0x0. See *pointer*.

numeric

A matrix is said to be numeric if its elements are real or complex; see *type*.

object code

Object code refers to the binary code that Mata produces from the source code you type as input. See [M-1] **how**.

observations and variables

A dataset containing n observations on k variables in often stored in an $n \times k$ matrix. An observation refers to a row of that matrix; a variable refers to a column.

operator

An operator is +, -, and the like. Most operators are binary (or dyadic), such as + in $A+B$ and * in $C*D$. Binary operators also include logicial operators such as & and | ("and" and "or") in $E\&F$ and $G|H$. Other operators are unary (or monadic), such as ! (not) in $!J$, or both unary and binary, such as - in $-K$ and in $L-M$. When we say "operator" without specifying which, we mean binary operator. Thus colon operators are in fact colon binary operators. See [M-2] **exp**.

optimization

Mata compiles the code that you write. After compilation, Mata performs an *optimization* step, the purpose of which is to make the compiled code execute more quickly. You can turn off the optimization step—see [M-3] **mata set**—but doing so is not recommended.

orgtype

See *type*.

orthogonal matrix and unitary matrix

A is orthogonal if *A* is *square* and $A'A==I$. The word orthogonal is usually reserved for real matrices; if the matrix is complex, it is said to be unitary (and then transpose means conjugate-transpose). We use the word orthogonal for both real and complex matrices.

If *A* is orthogonal, then $\det(A) = \pm 1$.

p-conformability

Matrix, vector, or scalar *A* is said to be p-conformable with matrix, vector, or scalar *B* if rows(A)==rows(B) and cols(A)==cols(B). p stands for plus; p-conformability is one of the properties necessary to be able to add matrices together. p-conformability, however, does not imply that the matrices are of the same type. Thus (1,2,3) is p-conformable with (4,5,6) and with ("this","that","what") but not with (4\5\6).

permutation matrix and permutation vector

A *permutation matrix* is an $n \times n$ matrix that is a row (or column) permutation of the identity matrix. If *P* is a permutation matrix, then *P*∗*A* permutes the rows of *A* and *A*∗*P* permutes the columns of *A*. Permutation matrices also have the property that $P^{-1} = P'$.

A *permutation vector* is a $1 \times n$ or $n \times 1$ vector that contains a permutation of the integers 1, 2, ..., *n*. Permutation vectors can be used with subscripting to reorder the rows or columns of a matrix. Permutation vectors are a memory-conserving way of recording permutation matrices; see [M-1] **permutation**.

pointer

A matrix is said to be a pointer matrix if its elements are pointers.

A pointer is the address of a *variable*. Say that variable *X* contains a matrix. Another variable *p* might contain 137,799,016 and, if 137,799,016 were the address at which *X* were stored, then *p* would be said to point to *X*. Addresses are seldom written in base 10, and so rather than saying *p* contains 137,799,016, we would be more likely to say that *p* contains 0x836a568, which is the way we write numbers in base 16. Regardless of how we write addresses, however, *p* contains a number and that number corresponds to the address of another variable.

In our program, if we refer to *p*, we are referring to *p*'s contents, the number 0x836a568. The monadic operator * is defined as "refer to the address" or "dereference": *∗p* means *X*. We could code Y = *p or Y = X, and either way, we would obtain the same result. In our program, we could refer to X[i,j] or (*p)[i,j], and either way, we would obtain the *i*, *j* element of *X*.

The monadic operator & is how we put addresses into *p*. To load *p* with the address of *X*, we code p = &X.

The special address 0 (zero, written in hexadecimal as 0x0), also known as NULL, is how we record that a pointer variable points to nothing. A pointer variable contains NULL or it contains a valid address of another variable.

See [M-2] **pointers** for a complete description of pointers and their use.

pragma

"(Pragmatic information) A standardised form of comment which has meaning to a compiler. It may use a special syntax or a specific form within the normal comment syntax. A pragma usually conveys non-essential information, often intended to help the compiler to optimise the program." See *The Free On-line Dictionary of Computing*, http://www.foldoc.org/, Editor Denis Howe. For Mata, see [M-2] **pragma**.

r-conformability

A set of two or more matrices, vectors, or scalars $A, B, \ldots$, are said to be r-conformable if each is *c-conformable* with a matrix of max(rows(A), rows(B), ...) rows and max(cols(A), cols(B), ...) columns.

r-conformability is a more relaxed form of *c-conformability* in that, if two matrices are c-conformable, they are r-conformable, but not vice versa. For instance, $A: 1 \times 3$ and $B: 3 \times 1$ are r-conformable but not c-conformable. Also, c-conformability is defined with respect to a pair of matrices only; r-conformability can be applied to a set of matrices.

r-conformability is often required of the arguments for functions that would otherwise naturally be expected to require scalars. See *r-conformability* in [M-5] **normal( )** for an example.

rank

Terms in common use are rank, row rank, and column rank. The row rank of a matrix $A: m \times n$ is the number of rows of A that are linearly independent. The column rank is defined similarly, as the number of columns that are linearly independent. The terms *row rank* and *column rank*, however, are used merely for emphasis; the ranks are equal and the result is simply called the rank of A.

For a square matrix A (where $m==n$), the matrix is invertible if and only if rank(A)==n. One often hears that A is of full rank in this case and rank deficient in the other. See [M-5] **rank( )**.

real

A matrix is said to be a real matrix if its elements are all reals and it is stored in a real matrix. Real is one of the two numeric types in Mata, the other being complex. Also see *type*.

rowvector

See *vector*.

scalar

A special case of a *matrix* with one row and one column. A scalar may be substituted anywhere a matrix, vector, column vector, or row vector is required, but not vice versa.

source code

Source code refers to the human-readable code that you type into Mata to define a function. Source code is compiled into object code, which is binary. See [M-1] **how**.

square matrix

A matrix is square if it has the same number of rows and columns. A 3×3 matrix is square; a 3×4 matrix is not.

string

A matrix is said to be a string matrix if its elements are strings (text); see *type*. In Mata, a string may be text or binary and may be up to 2,147,483,647 characters (bytes) long.

structure

A structure is an *eltype*, indicating a set of variables tied together under one name. struct mystruct might be

```
struct mystruct {
        real scalar     n1, n2
        real matrix     X
}
```

If variable a was declared a struct mystruct scalar, then the scalar a would contain three pieces: two real scalars and one real matrix. The pieces would be referred to as a.n1, a.n2, and a.X. If variable b were also declared a struct mystruct scalar, it too would contain three pieces, b.n1, b.n2, and b.X. The advantage of structures is that they can be referred to as a whole. You can code a.n1=b.n1 to copy one piece, or you can code a=b if you wanted to copy all three pieces. In all ways, a and b are variables. You may pass a to a subroutine, for instance, which amounts to passing all three values.

Structures variables are usually scalar, but they are not limited to being so. If A were a struct mystruct matrix, then each element of A would contain three pieces, and one could refer, for instance, to A[2,3].n1, A[2,3].n2, and A[2,3].X, and even to A[2,3].X[3,2].

See [M-2] **struct**.

subscripts

Subscripts are how you refer to an element or even a submatrix of a matrix.

Mata provides two kinds of subscripts, known as list subscripts and range subscripts.

In list subscripts, A[2,3] refers to the (2,3) element of A. A[(2\3), (4,6)] refers to the submatrix made up of the second and third rows, fourth and sixth columns, of A.

In range subscripts, A[|2,3|] also refers to the (2,3) element of A. A[|2,3\4,6|] refers to the submatrix beginning at the (2,3) element and ending at the (4,6) element.

See [M-2] **subscripts** for more information.

symmetric matrix

Matrix A is symmetric if $A = A'$. The word *symmetric* is usually reserved for real matrices, and in that case, a symmetric matrix is a square matrix with a_{ij}==a_{ji}.

Matrix A is said to be Hermitian if $A = A'$, where the transpose operator is understood to mean the conjugate-transpose operator; see *Hermitian matrix*. In Mata, the ' operator is the conjugate-transpose operator, and thus, in this manual, we will use the word *symmetric* both to refer to real, symmetric matrices and to refer to complex, Hermitian matrices.

Sometimes, you will see us follow the word *symmetric* with a parenthesized Hermitian, as in, "the resulting matrix is symmetric (Hermitian)". That is done only for emphasis.

The inverse of a symmetric (Hermitian) matrix is symmetric (Hermitian).

symmetriconly

Symmetriconly is a word we have coined to refer to a square matrix whose corresponding off-diagonal elements are equal to each other, whether the matrix is real or complex. Symmetriconly matrices have no mathematical significance, but sometimes, in data-processing and memory-management routines, it is useful to be able to distinguish such matrices.

time series–operated variable

Time-series operated variables are a Stata concept. The term refers to *op.varname* combinations such as L.gnp to mean the lagged value of variable gnp. Mata's [M-5] **st_data( )** function works with time series–operated variables just as it works with other variables, but many other Stata-interface functions do not allow *op.varname* combinations. In those cases, you must use [M-5] **st_tsrevar( )**.

traceback log

When a function fails—either because of a programming error or because it was used incorrectly—it produces a traceback log:

```
       : myfunction(2,3)
                  solve():    3200  conformability error
                  mysub():       -  function returned error
             myfunction():       -  function returned error
                  <istmt>:       -  function returned error
       r(3200);
```

The log says that solve() detected the problem—arguments are not conformable—and that solve() was called by mysub() was called by myfunction() was called by what you typed at the keyboard. See [M-2] **errors** for more information.

transmorphic

Transmorphic is an *eltype*. A scalar, vector, or matrix can be transmorphic, which indicates that its elements may be real, complex, string, pointer, or even a structure. The elements are all of the same type; you are just not saying which they are. Variables that are not declared are assumed to be transmorphic, or a variable can be explicitly declared to be transmorphic. Transmorphic is just fancy jargon for saying that the elements of the scalar, vector, or matrix can be anything and that, from one instant to the next, the scalar, vector, or matrix might change from holding elements of one type to elements of another.

See [M-2] **declarations**.

transpose

The transpose operator is written different ways in different books, including ′, superscript *, superscript *T*, and superscript *H*. Here we use the ′ notation: A' means the transpose of A, A with its rows and columns interchanged.

In complex analysis, the transpose operator, however it is written, is usually defined to mean the conjugate transpose; that is, one interchanges the rows and columns of the matrix and then one takes the conjugate of each element, or one does it in the opposite order—it makes no difference. Conjugation simply means reversing the sign of the imaginary part of a complex number: the conjugate of 1+2i is 1-2i. The conjugate of a real is the number itself; the conjugate of 2 is 2.

In Mata, ′ is defined to mean conjugate transpose. Since the conjugate of a real is the number itself, A' is regular transposition when A is real. Similarly, we have defined ′ so that it performs regular transposition for string and pointer matrices. For complex matrices, however, ′ also performs conjugation.

If you have a complex matrix and simply want to transpose it without taking the conjugate of its elements, see [M-5] **transposeonly( )**. Or code conj(A'). The extra conj() will undo the undesired conjugation performed by the transpose operator.

Usually, however, you want transposition and conjugation to go hand in hand. Most mathematical formulas, generalized to complex values, work that way.

triangular matrix

A triangular matrix is a matrix with all elements equal to zero above the diagonal or all elements equal to zero below the diagonal.

A matrix A is *lower triangular* if all elements are zero above the diagonal, i.e., if $A[i,j]==0, j>i$.

A matrix A is *upper triangular* if all elements are zero below the diagonal, i.e., if $A[i,j]==0, j<i$.

A *diagonal matrix* is both lower and upper triangular. That is worth mentioning because any function suitable for use with triangular matrices is suitable for use with diagonal matrices.

A triangular matrix is usually *square*.

The inverse of a triangular matrix is a triangular matrix. The determinant of a triangular matrix is the product of the diagonal elements. The eigenvalues of a triangular matrix are the diagonal elements.

type, eltype, and orgtype

The *type* of a matrix (or vector or scalar) is formally defined as the matrix's *eltype* and *orgtype*, listed one after the other—such as real vector—but it can also mean just one or the other—such as the *eltype* real or the *orgtype* vector.

eltype refers to the type of the elements. The *eltypes* are

real	numbers such as 1, 2, 3.4
complex	numbers such as 1+2i, 3+0i
string	strings such as "bill"
pointer	pointers such as &*varname*
struct	structures
numeric	meaning real or complex
transmorphic	meaning any of the above

orgtype refers to the organizational type. *orgtype* specifies how the elements are organized. The *orgtypes* are

matrix	two-dimensional arrays
vector	one-dimensional arrays
colvector	one-dimensional column arrays
rowvector	one-dimensional row arrays
scalar	single items

The fully specified type is the element and organization types combined, as in real vector.

underscore functions

Functions whose names start with an underscore are called underscore functions, and when an underscore function exists, usually a function without the underscore prefix also exists. In those cases, the function is usually implemented in terms of the underscore function, and the underscore function is harder to use but is faster or provides greater control. Usually, the difference is in the handling of errors.

For instance, function fopen() opens a file. If the file does not exist, execution of your program is aborted. Function _fopen() does the same thing, but if the file cannot be opened, it returns a special value indicating failure, and it is the responsibility of your program to check the indicator and to take the appropriate action. This can be useful when the file might not exist, and if it does not, you wish to take a different action. Usually, however, if the file does not exist, you will wish to abort, and use of fopen() will allow you to write less code.

unitary matrix

See *orthogonal matrix*.

variable

In a program, the entities that store values (*a*, *b*, *c*, ..., *x*, *y*, *z*) are called variables. Variables are given names of 1 to 32 characters long. To be terribly formal about it: a variable is a container; it contains a matrix, vector, or scalar and is referred to by its variable name or by another variable containing a *pointer* to it.

Also, *variable* is sometimes used to refer to columns of data matrices; see *data matrix*.

vector, colvector, and rowvector

A special case of a matrix with either one row or one column. A vector may be substituted anywhere a matrix is required. A matrix, however, may not be substituted for a vector.

A colvector is a vector with one column.

A rowvector is a vector with one row.

A vector is either a rowvector or colvector, without saying which.

view

A view is a special type of matrix that appears to be an ordinary matrix, but in fact the values in the matrix are the values of certain or all variables and observations in the Stata dataset that is currently in memory. Its values are not just equal to the dataset's values; they are the dataset's values: if an element of the matrix is changed, the corresponding variable and observation in the Stata dataset also changes. Views are obtained by st_view() and are efficient; see [M-5] **st_view()**.

void function

A function is said to be void if it returns nothing. For instance, the function [M-5] **printf()** is a void function; it prints results, but it does not return anything in the sense that, say, [M-5] **sqrt()** does. It would not make any sense to code x = printf("hi there"), but coding x = sqrt(2) is perfectly logical.

void matrix

A matrix is said to be void if it is 0×0, $r \times 0$, or $0 \times c$; see [M-2] **void**.

Also See

[M-0] **intro** — Introduction to the Mata manual

[M-1] **intro** — Introduction and advice

Subject and author index

This is the subject and author index for the *Mata Reference Manual*. Readers interested in topics other than Mata should see the combined subject index (and the combined author index) in the *Stata Quick Reference and Index*. The combined index indexes the *Getting Started* manuals, the *User's Guide*, and all the Reference manuals except this one.

-> operator, [M-2] **struct**
/* */ comment delimiter, [M-2] **comments**
// comment indicator, [M-2] **comments**

A

abs() function, [M-5] **abs()**
acos() function, [M-5] **sin()**
acosh() function, [M-5] **sin()**
acosr() function, [M-5] **sin()**
addition, [M-2] **op_arith**, [M-2] **op_colon**
adjoint matrix, [M-2] **op_transpose**, [M-5] **conj()**
adjugate matrix, [M-2] **op_transpose**, [M-5] **conj()**
ado-files, [M-1] **ado**
ado-path, [M-5] **adosubdir()**
adosubdir() function, [M-5] **adosubdir()**
all() function, [M-5] **all()**
allof() function, [M-5] **all()**
Anderson, E., [M-1] **LAPACK**
any() function, [M-5] **all()**
anyof() function, [M-5] **all()**
arg() function, [M-5] **sin()**
args() function, [M-5] **args()**
arguments,
    program, [M-2] **declarations**, [M-6] **glossary**
    values returned in, [M-1] **returnedargs**
    varying number, [M-2] **optargs**, [M-5] **args()**
arithmetic operators, [M-2] **op_arith**
ASCII codes, [M-5] **ascii()**
ascii() function, [M-5] **ascii()**
asin() function, [M-5] **sin()**
asinh() function, [M-5] **sin()**
asinr() function, [M-5] **sin()**
assert() function, [M-5] **assert()**
asserteq() function, [M-5] **assert()**
assignment operator, [M-2] **op_assignment**
atan() function, [M-5] **sin()**
atan2() function, [M-5] **sin()**
atanh() function, [M-5] **sin()**
atanr() function, [M-5] **sin()**

B

Bai, Z., [M-1] **LAPACK**
base conversion, [M-5] **inbase()**
Berndt, E. K., [M-5] **optimize()**
Berndt–Hall–Hall–Hausman method, [M-5] **optimize()**
beta function, [M-5] **normal()**
betaden() function, [M-5] **normal()**
binary I/O, [M-5] **bufio()**
binomial() function, [M-5] **normal()**
binomialtail() function, [M-5] **normal()**
binormal() function, [M-5] **normal()**
Bischof, C., [M-1] **LAPACK**
Blackford, S., [M-1] **LAPACK**
block diagonal matrix, [M-5] **blockdiag()**
blockdiag() function, [M-5] **blockdiag()**
break, [M-2] **break**
break key processing, [M-5] **setbreakintr()**
breakkey() function, [M-5] **setbreakintr()**
breakkeyreset() function, [M-5] **setbreakintr()**
broad type, [M-6] **glossary**
Broyden–Fletcher–Goldfarb–Shanno, [M-5] **optimize()**
bufbfmtisnum() function, [M-5] **bufio()**
bufbfmtlen() function, [M-5] **bufio()**
bufbyteorder() function, [M-5] **bufio()**
buffered I/O, [M-5] **bufio()**
bufget() function, [M-5] **bufio()**
bufio() function, [M-5] **bufio()**
bufmissingvalue() function, [M-5] **bufio()**
bufput() function, [M-5] **bufio()**
byteorder() function, [M-5] **byteorder()**

C

c-conformability, [M-2] **op_colon**, [M-6] **glossary**
C() function, [M-5] **C()**
c() function, [M-5] **c()**
callersversion() function, [M-5] **callersversion()**
cat() function, [M-5] **cat()**
Cdhms() function, [M-5] **date()**
ceil() function, [M-5] **trunc()**
Chabert, J.-L., [M-5] **cholesky()**
char() function, [M-5] **ascii()**
characteristic roots, [M-5] **eigensystem()**
_chdir() function, [M-5] **chdir()**
chdir() function, [M-5] **chdir()**
chi2() function, [M-5] **normal()**
chi2tail() function, [M-5] **normal()**
Chms() function, [M-5] **date()**
Choi, M.-D., [M-5] **Hilbert()**
Cholesky, A.-L., [M-5] **cholesky()**
Cholesky decomposition, [M-5] **cholesky()**
_cholesky() function, [M-5] **cholesky()**
cholesky() function, [M-5] **cholesky()**
_cholinv() function, [M-5] **cholinv()**
cholinv() function, [M-5] **cholinv()**
_cholsolve() function, [M-5] **cholsolve()**
cholsolve() function, [M-5] **cholsolve()**
clear, [M-3] **mata clear**

Clock() function, [M-5] **date( )**
clock() function, [M-5] **date( )**
cloglog() function, [M-5] **logit( )**
Cmdyhms() function, [M-5] **date( )**
Cofc() function, [M-5] **date( )**
cofC() function, [M-5] **date( )**
Cofd() function, [M-5] **date( )**
cofd() function, [M-5] **date( )**
_collate() function, [M-5] **sort( )**
colmax() function, [M-5] **minmax( )**
colmaxabs() function, [M-5] **minmax( )**
colmin() function, [M-5] **minmax( )**
colminmax() function, [M-5] **minmax( )**
colmissing() function, [M-5] **missing( )**
colnonmissing() function, [M-5] **missing( )**
colon operators, [M-2] **op_colon**, [M-6] **glossary**
cols() function, [M-5] **rows( )**
colscalefactors() function, [M-5] **_equilrc( )**
colshape() function, [M-5] **rowshape( )**
colsum() function, [M-5] **sum( )**
column-join operator, [M-2] **op_join**
column of matrix, selecting, [M-5] **select( )**
colvector, [M-2] **declarations**, [M-6] **glossary**
comb() function, [M-5] **comb( )**
combinatorial function, [M-5] **comb( )**
comments, [M-2] **comments**
complex, [M-2] **declarations**, [M-6] **glossary**
cond() function, [M-5] **cond( )**
condition number, [M-5] **cond( )**, [M-6] **glossary**
conditional operator, [M-2] **op_conditional**
conformability, [M-2] **void**, [M-6] **glossary**, *also see* c-conformability, r-conformability, and p-conformability
_conj() function, [M-5] **_transpose( )**
conj() function, [M-5] **conj( )**
conjugate, [M-5] **conj( )**, [M-5] **_transpose( )**, [M-6] **glossary**
conjugate transpose, [M-2] **op_transpose**, [M-5] **conj( )**, [M-5] **_transpose( )**, [M-6] **glossary**
convolve() function, [M-5] **fft( )**
copysource, [M-1] **source**
Corr() function, [M-5] **fft( )**
_corr() function, [M-5] **corr( )**
corr() function, [M-5] **corr( )**
correlation, [M-5] **corr( )**, [M-5] **mean( )**, [M-5] **fft( )**
correlation() function, [M-5] **mean( )**
cos() function, [M-5] **sin( )**
cosh() function, [M-5] **sin( )**
crexternal() function, [M-5] **findexternal( )**
cross() function, [M-5] **cross( )**
cross product, [M-5] **cross( )**, [M-5] **crossdev( )**, [M-5] **quadcross( )**
crossdev() function, [M-5] **crossdev( )**
cubic natural splines, [M-5] **spline3( )**
cvpermute() function, [M-5] **cvpermute( )**
cvpermutesetup() function, [M-5] **cvpermute( )**

D

data matrix, [M-5] **st_data( )**, [M-5] **st_view( )**, [M-6] **glossary**
date and time, [M-5] **c( )**
date() function, [M-5] **date( )**
date functions, [M-5] **date( )**
Davidon, W. C., [M-5] **optimize( )**
Davidon–Fletcher–Powell method, [M-5] **optimize( )**
day() function, [M-5] **date( )**
declarations, [M-2] **declarations**, [M-6] **glossary**
decomposition, [M-5] **cholesky( )**, [M-5] **lud( )**, [M-5] **qrd( )**, [M-5] **svd( )**, [M-5] **fullsvd( )**
deconvolve() function, [M-5] **fft( )**
decrement operator, [M-2] **op_increment**
defective matrix, [M-6] **glossary**
delete, [M-5] **unlink( )**
#delimit, [M-2] **semicolons**
Demmel, J., [M-1] **LAPACK**
density functions, [M-5] **normal( )**
dereferencing, [M-2] **pointers**, [M-2] **ftof**
describe, [M-3] **mata describe**
design matrix, [M-5] **designmatrix( )**, [M-5] **I( )**
designmatrix() function, [M-5] **designmatrix( )**
det() function, [M-5] **det( )**
determinant of matrix, [M-5] **det( )**
dettriangular() function, [M-5] **det( )**
deviation cross product, [M-5] **crossdev( )**, [M-5] **quadcross( )**
dgammapda() function, [M-5] **normal( )**
dgammapdada() function, [M-5] **normal( )**
dgammapdadx() function, [M-5] **normal( )**
dgammapdx() function, [M-5] **normal( )**
dgammapdxdx() function, [M-5] **normal( )**
dhms() function, [M-5] **date( )**
_diag() function, [M-5] **_diag( )**
diag() function, [M-5] **diag( )**
diag0cnt() function, [M-5] **diag0cnt( )**
diagonal, [M-5] **diagonal( )**, [M-6] **glossary**
diagonal() function, [M-5] **diagonal( )**
diagonal matrix, [M-5] **_diag( )**, [M-5] **diag( )**, [M-5] **diagonal( )**, [M-5] **isdiagonal( )**, [M-6] **glossary**
digamma() function, [M-5] **factorial( )**
dir() function, [M-5] **dir( )**
directories, [M-5] **chdir( )**, [M-5] **dir( )**, [M-5] **direxists( )**
direxists() function, [M-5] **direxists( )**
direxternal() function, [M-5] **direxternal( )**
display
  as error, [M-5] **displayas( )**, [M-5] **errprintf( )**
  as text, as result, etc., [M-5] **displayas( )**
display() function, [M-5] **display( )**
displayas() function, [M-5] **displayas( )**
displayflush() function, [M-5] **displayflush( )**
distribution functions, [M-5] **normal( )**
division, [M-2] **op_arith**, [M-2] **op_colon**
do...while, [M-2] **do**, [M-2] **continue**, [M-2] **break**

dofC() function, [M-5] **date( )**
dofc() function, [M-5] **date( )**
dofh() function, [M-5] **date( )**
dofm() function, [M-5] **date( )**
dofq() function, [M-5] **date( )**
dofw() function, [M-5] **date( )**
dofy() function, [M-5] **date( )**
Dongarra, J., [M-1] **LAPACK**
dow() function, [M-5] **date( )**
doy() function, [M-5] **date( )**
drop, [M-3] **mata drop**
dsign() function, [M-5] **dsign( )**, [M-5] **sign( )**
Du Croz, J., [M-1] **LAPACK**

E

e() function, [M-5] **e( )**
_editmissing() function, [M-5] **editmissing( )**
editmissing() function, [M-5] **editmissing( )**
_edittoint() function, [M-5] **edittoint( )**
edittoint() function, [M-5] **edittoint( )**
_edittointtol() function, [M-5] **edittoint( )**
edittointtol() function, [M-5] **edittoint( )**
_edittozero() function, [M-5] **edittozero( )**
edittozero() function, [M-5] **edittozero( )**
_edittozerotol() function, [M-5] **edittozero( )**
edittozerotol() function, [M-5] **edittozero( )**
_editvalue() function, [M-5] **editvalue( )**
editvalue() function, [M-5] **editvalue( )**
_eigen_la() function, [M-5] **eigensystem( )**
_eigensystem() function, [M-5] **eigensystem( )**
eigensystem() function, [M-5] **eigensystem( )**
eigenvalues, [M-5] **eigensystem( )**
_eigenvalues() function, [M-5] **eigensystem( )**
eigenvalues() function, [M-5] **eigensystem( )**
eigenvectors, [M-5] **eigensystem( )**
eltype, [M-2] **declarations**, [M-6] **glossary**
eltype() function, [M-5] **eltype( )**
end, [M-3] **end**
epsilon, [M-6] **glossary**
epsilon() function, [M-5] **epsilon( )**
_equilc() function, [M-5] **_equilrc( )**
equilibration, [M-5] **_equilrc( )**
_equilr() function, [M-5] **_equilrc( )**
_equilrc() function, [M-5] **_equilrc( )**
erase, [M-5] **unlink( )**
error codes, [M-2] **errors**
_error() function, [M-5] **error( )**
error() function, [M-5] **error( )**
errprintf() function, [M-5] **errprintf( )**
exit() function, [M-5] **exit( )**
exit mata, [M-3] **end**
exp, [M-2] **exp**, [M-6] **glossary**
exp() function, [M-5] **exp( )**
exponentiation, [M-5] **exp( )**, [M-5] **matexpsym( )**
external, [M-2] **declarations**

externals, [M-2] **declarations**, [M-5] **findexternal( )**, [M-5] **valofexternal( )**, [M-5] **direxternal( )**, [M-6] **glossary**
extract diagonal, [M-5] **diagonal( )**, [M-5] **diag( )**

F

F() function, [M-5] **normal( )**
factorial() function, [M-5] **factorial( )**
favorspeed() function, [M-5] **favorspeed( )**
fbufget() function, [M-5] **bufio( )**
fbufput() function, [M-5] **bufio( )**
_fclose() function, [M-5] **fopen( )**
fclose() function, [M-5] **fopen( )**
Fden() function, [M-5] **normal( )**
ferrortext() function, [M-5] **ferrortext( )**
_fft() function, [M-5] **fft( )**
fft() function, [M-5] **fft( )**
_fget() function, [M-5] **fopen( )**
fget() function, [M-5] **fopen( )**
_fgetmatrix() function, [M-5] **fopen( )**
fgetmatrix() function, [M-5] **fopen( )**
_fgetnl() function, [M-5] **fopen( )**
fgetnl() function, [M-5] **fopen( )**
file processing, [M-5] **bufio( )**, [M-4] **io**, [M-5] **findfile( )**, [M-5] **fileexists( )**, [M-5] **ferrortext( )**, [M-5] **fopen( )**, [M-5] **cat( )**, [M-5] **unlink( )**
fileexists() function, [M-5] **fileexists( )**
filename manipulation, [M-5] **adosubdir( )**, [M-5] **pathjoin( )**
_fillmissing() function, [M-5] **_fillmissing( )**
findexternal() function, [M-5] **findexternal( )**
findfile() function, [M-5] **findfile( )**
Fletcher, R., [M-5] **optimize( )**
float, [M-5] **floatround( )**
floatround() function, [M-5] **floatround( )**
floor() function, [M-5] **trunc( )**
fmtwidth() function, [M-5] **fmtwidth( )**
folders, *see* directories
_fopen() function, [M-5] **fopen( )**
fopen() function, [M-5] **fopen( )**
for, [M-2] **for**, [M-2] **continue**, [M-2] **break**, [M-2] **semicolons**
format width, [M-5] **fmtwidth( )**
FORTRAN, [M-2] **goto**, [M-5] **dsign( )**
Fourier transform, [M-5] **fft( )**
_fput() function, [M-5] **fopen( )**
fput() function, [M-5] **fopen( )**
_fputmatrix() function, [M-5] **fopen( )**
fputmatrix() function, [M-5] **fopen( )**
_fread() function, [M-5] **fopen( )**
fread() function, [M-5] **fopen( )**
freturncode() function, [M-5] **ferrortext( )**
frombase() function, [M-5] **inbase( )**
_fseek() function, [M-5] **fopen( )**
fseek() function, [M-5] **fopen( )**

fstatus() function, [M-5] **fopen( )**
Ftail() function, [M-5] **normal( )**
_ftell() function, [M-5] **fopen( )**
ftell() function, [M-5] **fopen( )**
ftfreqs() function, [M-5] **fft( )**
ftpad() function, [M-5] **fft( )**
ftperiodogram() function, [M-5] **fft( )**
ftretime() function, [M-5] **fft( )**
_ftruncate() function, [M-5] **fopen( )**
ftruncate() function, [M-5] **fopen( )**
ftunwrap() function, [M-5] **fft( )**
ftwrap() function, [M-5] **fft( )**
fullsdiag() function, [M-5] **fullsvd( )**
_fullsvd() function, [M-5] **fullsvd( )**
fullsvd() function, [M-5] **fullsvd( )**
function, [M-2] **declarations**, [M-6] **glossary**
function arguments, [M-1] **returnedargs**, *also see* arguments
function naming convention, [M-1] **naming**
functions,
    passing to functions, [M-2] **ftof**
    underscore, [M-6] **glossary**
_fwrite() function, [M-5] **fopen( )**
fwrite() function, [M-5] **fopen( )**

G

gamma() function, [M-5] **factorial( )**
gammaden() function, [M-5] **normal( )**
gammap() function, [M-5] **normal( )**
gammaptail() function, [M-5] **normal( )**
generalized inverse, [M-5] **invsym( )**, [M-5] **pinv( )**, [M-5] **qrinv( )**
Geweke–Hajivassiliou–Keane multivariate normal simulator, [M-5] **ghk( )**, [M-5] **ghkfast( )**
ghalton() function, [M-5] **halton( )**
ghk() function, [M-5] **ghk( )**
ghkfast() function, [M-5] **ghkfast( )**
Gleick, J., [M-5] **optimize( )**
global variable, [M-2] **declarations**, [M-5] **direxternal( )**, [M-5] **findexternal( )**, [M-5] **valofexternal( )**, [M-6] **glossary**
Goldfarb, D., [M-5] **optimize( )**
goto, [M-2] **goto**
Gould, W., [M-1] **how**, [M-1] **interactive**, [M-2] **exp**, [M-2] **goto**, [M-2] **subscripts**, [M-2] **syntax**, [M-5] **inbase( )**, [M-5] **st_addvar( )**, [M-5] **st_view( )**
grammar, [M-2] **syntax**
Greenbaum, A., [M-1] **LAPACK**

H

halfyear() function, [M-5] **date( )**
halfyearly() function, [M-5] **date( )**
Hall, B. H., [M-5] **optimize( )**
Hall, R. E., [M-5] **optimize( )**
_halton() function, [M-5] **halton( )**

halton() function, [M-5] **halton( )**
Halton set, [M-5] **halton( )**
Hammarling, S., [M-1] **LAPACK**
Hammersley set, [M-5] **halton( )**
Hausman, J. A., [M-5] **optimize( )**
help, [M-1] **help**, [M-3] **mata help**
Hermite, C., [M-5] **issymmetric( )**
Hermitian
    adjoin, [M-2] **op_transpose**, [M-5] **conj( )**
    matrices, [M-5] **issymmetric( )**, [M-5] **makesymmetric( )**, [M-6] **glossary**
    transpose, [M-2] **op_transpose**, [M-5] **conj( )**
Herriot, J. G., [M-5] **spline3( )**
hh() function, [M-5] **date( )**
hhC() function, [M-5] **date( )**
Hilbert, D., [M-5] **Hilbert( )**
Hilbert() function, [M-5] **Hilbert( )**
HILO, [M-5] **byteorder( )**
hms() function, [M-5] **date( )**
hofd() function, [M-5] **date( )**
hours() function, [M-5] **date( )**
Householder, A. S., [M-5] **qrd( )**
_hqrd() function, [M-5] **qrd( )**
hqrd() function, [M-5] **qrd( )**
hqrdmultq() function, [M-5] **qrd( )**
hqrdmultq1t() function, [M-5] **qrd( )**
_hqrdp() function, [M-5] **qrd( )**
hqrdp() function, [M-5] **qrd( )**
_hqrdp_la() function, [M-5] **qrd( )**
hqrdq() function, [M-5] **qrd( )**
hqrdq1() function, [M-5] **qrd( )**
hqrdr() function, [M-5] **qrd( )**
hqrdr1() function, [M-5] **qrd( )**
hyperbolic functions, [M-5] **sin( )**

I

I() function, [M-5] **I( )**
ibeta() function, [M-5] **normal( )**
ibetatail() function, [M-5] **normal( )**
identity matrix, [M-5] **I( )**
if, [M-2] **if**
Im() function, [M-5] **Re( )**
imaginary part, [M-5] **Re( )**
inbase() function, [M-5] **inbase( )**
incomplete
    beta function, [M-5] **normal( )**
    gamma function, [M-5] **normal( )**
increment operator, [M-2] **op_increment**
index,
    mathematical functions, [M-4] **statistical**
    matrix functions, [M-4] **utility**
    statistical functions, [M-4] **statistical**
    utility functions, [M-4] **utility**
indexnot() function, [M-5] **indexnot( )**
input/output functions, [M-4] **io**
integers, [M-5] **trunc( )**

invbinomial() function, [M-5] **normal( )**
invbinomialtail() function, [M-5] **normal( )**
invchi2() function, [M-5] **normal( )**
invchi2tail() function, [M-5] **normal( )**
invcloglog() function, [M-5] **logit( )**
inverse matrix, [M-4] **solvers**, [M-5] **invsym( )**,
    [M-5] **cholinv( )**, [M-5] **luinv( )**, [M-5] **qrinv( )**,
    [M-5] **pinv( )**, [M-5] **solve_tol( )**
invF() function, [M-5] **normal( )**
_invfft() function, [M-5] **fft( )**
invfft() function, [M-5] **fft( )**
invFtail() function, [M-5] **normal( )**
invgammap() function, [M-5] **normal( )**
invgammaptail() function, [M-5] **normal( )**
invHilbert() function, [M-5] **Hilbert( )**
invibeta() function, [M-5] **normal( )**
invibetatail() function, [M-5] **normal( )**
invlogit() function, [M-5] **logit( )**
invnchi2() function, [M-5] **normal( )**
invnFtail() function, [M-5] **normal( )**
invnibeta() function, [M-5] **normal( )**
invnormal() function, [M-5] **normal( )**
invorder() function, [M-5] **invorder( )**
_invsym() function, [M-5] **invsym( )**
invsym() function, [M-5] **invsym( )**
invtokens() function, [M-5] **invtokens( )**
invttail() function, [M-5] **normal( )**
invvech() function, [M-5] **vec( )**
I/O functions, [M-4] **io**
iscomplex() function, [M-5] **isreal( )**
isdiagonal() function, [M-5] **isdiagonal( )**
isfleeting() function, [M-5] **isfleeting( )**
ispointer() function, [M-5] **isreal( )**
isreal() function, [M-5] **isreal( )**
isrealvalues() function, [M-5] **isrealvalues( )**
isstring() function, [M-5] **isreal( )**
issymmetric() function, [M-5] **issymmetric( )**
issymmetriconly() function, [M-5] **issymmetric( )**
istmt, [M-1] **how**, [M-6] **glossary**
isview() function, [M-5] **isview( )**

J

J() function, [M-5] **J( )**, [M-2] **void**, [M-6] **glossary**
James, I., [M-2] **op_kronecker**, [M-5] **issymmetric( )**,
    [M-5] **pinv( )**
join operator, [M-2] **op_join**
Jones, P. S., [M-5] **Vandermonde( )**
_jumble() function, [M-5] **sort( )**
jumble() function, [M-5] **sort( )**

K

Kronecker direct product, [M-2] **op_kronecker**
Kronecker, L., [M-2] **op_kronecker**

L

LAPACK, [M-1] **LAPACK**, [M-5] **cholesky( )**,
    [M-5] **cholinv( )**, [M-5] **cholsolve( )**,
    [M-5] **eigensystem( )**, [M-5] **fullsvd( )**,
    [M-5] **lud( )**, [M-5] **luinv( )**, [M-5] **lusolve( )**,
    [M-5] **qrd( )**, [M-5] **qrinv( )**, [M-5] **qrsolve( )**,
    [M-5] **svd( )**, [M-5] **svsolve( )**, [M-6] **glossary**
latent roots, [M-5] **eigensystem( )**
left eigenvectors, [M-5] **eigensystem( )**
_lefteigensystem() function, [M-5] **eigensystem( )**
lefteigensystem() function, [M-5] **eigensystem( )**
length, [M-5] **abs( )**, [M-5] **rows( )**, [M-5] **strlen( )**
length() function, [M-5] **rows( )**
libraries, [M-1] **how**, [M-3] **mata mlib**,
    [M-3] **mata which**
limits, [M-1] **limits**
list subscripts, *see* subscripts
liststruct() function, [M-5] **liststruct( )**
ln() function, [M-5] **exp( )**
lnfactorial() function, [M-5] **factorial( )**
lngamma() function, [M-5] **factorial( )**
lnnormal() function, [M-5] **normal( )**
lnnormalden() function, [M-5] **normal( )**
log() function, [M-5] **exp( )**
log10() function, [M-5] **exp( )**
logarithms, [M-5] **exp( )**, [M-5] **matexpsym( )**
logical operators, [M-2] **op_logical**
logit() function, [M-5] **logit( )**
LOHI, [M-5] **byteorder( )**
lower-triangular matrix, *see* triangular matrix
lowercase, [M-5] **strupper( )**
_lowertriangle() function, [M-5] **lowertriangle( )**
lowertriangle() function, [M-5] **lowertriangle( )**
LU decomposition, [M-5] **lud( )**
_lud() function, [M-5] **lud( )**
lud() function, [M-5] **lud( )**
_lud_la() function, [M-5] **lud( )**
_luinv() function, [M-5] **luinv( )**
luinv() function, [M-5] **luinv( )**
_luinv_la() function, [M-5] **luinv( )**
_lusolve() function, [M-5] **lusolve( )**
lusolve() function, [M-5] **lusolve( )**
_lusolve_la() function, [M-5] **lusolve( )**
lval, [M-2] **op_assignment**, [M-6] **glossary**

M

machine precision, [M-5] **epsilon( )**, [M-6] **glossary**
_makesymmetric() function, [M-5] **makesymmetric( )**
makesymmetric() function, [M-5] **makesymmetric( )**
Marquardt algorithm, [M-5] **optimize( )**
Marquardt, D. W., [M-5] **optimize( )**
Mata
    commands, [M-3] **intro**
    error messages, [M-5] **error( )**, *also see* traceback
        log

mata, [M-3] **mata clear**, [M-3] **mata describe**,
    [M-3] **mata drop**, [M-3] **mata help**,
    [M-3] **mata matsave**, [M-3] **mata memory**,
    [M-3] **mata mlib**, [M-3] **mata mosave**,
    [M-3] **mata rename**, [M-3] **mata set**,
    [M-3] **mata stata**, [M-3] **mata which**,
    [M-3] **namelists**
.mata file, [M-1] **source**, [M-3] **mata mlib**,
    [M-6] **glossary**
mata invocation command, [M-3] **mata**
matacache, [M-3] **mata set**
matafavor, [M-3] **mata set**, [M-5] **favorspeed( )**
matalibs, [M-3] **mata set**
matalnum, [M-3] **mata set**
matamofirst, [M-3] **mata set**
mataoptimize, [M-3] **mata set**
matastrict, [M-3] **mata set**, [M-2] **declarations**,
    [M-1] **ado**
matdescribe, [M-3] **mata matsave**
_matexpsym( ) function, [M-5] **matexpsym( )**
matexpsym( ) function, [M-5] **matexpsym( )**
mathematical functions, [M-4] **mathematical**,
    [M-4] **matrix**, [M-4] **scalar**, [M-4] **solvers**,
    [M-4] **standard**
_matlogsym( ) function, [M-5] **matexpsym( )**
matlogsym( ) function, [M-5] **matexpsym( )**
_matpowersym( ) function, [M-5] **matpowersym( )**
matpowersym( ) function, [M-5] **matpowersym( )**
matrix, [M-4] **intro**, [M-6] **glossary**
    functions, [M-4] **manipulation**, [M-4] **matrix**,
        [M-4] **solvers**, [M-4] **standard**
    norm, [M-5] **norm( )**
matrix, [M-2] **declarations**
matsave, [M-3] **mata matsave**
matsize, [M-1] **limits**
matuse, [M-3] **mata matsave**
max( ) function, [M-5] **minmax( )**
maxdouble( ) function, [M-5] **mindouble( )**
maximization, [M-5] **optimize( )**
maximum
    length of string, [M-1] **limits**
    size of matrix, [M-1] **limits**
maximums, [M-5] **minindex( )**
maxindex( ) function, [M-5] **minindex( )**
McKenney, A., [M-1] **LAPACK**
mdy( ) function, [M-5] **date( )**
mdyhms( ) function, [M-5] **date( )**
Mead, R., [M-5] **optimize( )**
mean( ) function, [M-5] **mean( )**
meanvariance( ) function, [M-5] **mean( )**
memory, [M-3] **mata memory**
memory utilization, [M-1] **limits**, [M-3] **mata memory**
min( ) function, [M-5] **minmax( )**
mindouble( ) function, [M-5] **mindouble( )**
minimization, [M-5] **optimize( )**
minimums, [M-5] **minindex( )**
minindex( ) function, [M-5] **minindex( )**
minmax( ) function, [M-5] **minmax( )**

minutes( ) function, [M-5] **date( )**
missing( ) function, [M-5] **missing( )**
missing values, [M-5] **missing( )**, [M-5] **missingof( )**,
    [M-5] **editmissing( )**, [M-5] **_fillmissing( )**
missingof( ) function, [M-5] **missingof( )**
_mkdir( ) function, [M-5] **chdir( )**
mkdir( ) function, [M-5] **chdir( )**
mlib, [M-3] **mata mlib**
.mlib library files, [M-1] **how**, [M-3] **mata describe**,
    [M-3] **mata mlib**, [M-3] **mata set**,
    [M-3] **mata which**, [M-6] **glossary**
mm( ) function, [M-5] **date( )**
.mmat files, [M-3] **mata matsave**
mmC( ) function, [M-5] **date( )**
.mo file, [M-1] **how**, [M-3] **mata mosave**,
    [M-3] **mata which**, [M-6] **glossary**
mod( ) function, [M-5] **mod( )**
modulus( ) function, [M-5] **mod( )**
mofd( ) function, [M-5] **date( )**
month( ) function, [M-5] **date( )**
monthly( ) function, [M-5] **date( )**
Moore, E. H., [M-5] **pinv( )**
Moore–Penrose inverse, [M-5] **pinv( )**
more( ) function, [M-5] **more( )**
mosave, [M-3] **mata mosave**
mreldif( ) function, [M-5] **reldif( )**
mreldifre( ) function, [M-5] **reldif( )**
mreldifsym( ) function, [M-5] **reldif( )**
msofhours( ) function, [M-5] **date( )**
msofminutes( ) function, [M-5] **date( )**
msofseconds( ) function, [M-5] **date( )**
multiplication, [M-2] **op_arith**, [M-2] **op_colon**
multivariate normal simulator, [M-5] **ghk( )**,
    [M-5] **ghkfast( )**

N

nameexternal( ) function, [M-5] **findexternal( )**
namelists, [M-3] **namelists**
naming convention, [M-1] **naming**
nbetaden( ) function, [M-5] **normal( )**
nchi2( ) function, [M-5] **normal( )**
negation, [M-2] **op_arith**
Nelder, J. A., [M-5] **optimize( )**
Nelder–Mead method, [M-5] **optimize( )**
Newton, I., [M-5] **optimize( )**
Newton–Raphson method, [M-5] **optimize( )**
nFden( ) function, [M-5] **normal( )**
nFtail( ) function, [M-5] **normal( )**
nibeta( ) function, [M-5] **normal( )**
nonmissing( ) function, [M-5] **missing( )**
norm, [M-6] **glossary**
norm( ) function, [M-5] **norm( )**
normal( ) function, [M-5] **normal( )**
normalden( ) function, [M-5] **normal( )**
npnchi2( ) function, [M-5] **normal( )**
NULL, [M-2] **pointers**, [M-6] **glossary**
numeric, [M-2] **declarations**, [M-6] **glossary**

O

object code, [M-1] **how**, [M-6] **glossary**
online help, [M-1] **help**, [M-3] **mata help**
operators, [M-2] **op_arith**, [M-2] **op_assignment**,
  [M-2] **op_colon**, [M-2] **op_conditional**,
  [M-2] **op_increment**, [M-2] **op_join**,
  [M-2] **op_kronecker**, [M-2] **op_logical**,
  [M-2] **op_range**, [M-2] **op_transpose**,
  [M-6] **glossary**
optimization, [M-3] **mata set**, [M-5] **optimize( )**,
  [M-6] **glossary**
_optimize( ) function, [M-5] **optimize( )**
optimize( ) function, [M-5] **optimize( )**
_optimize_evaluate( ) function, [M-5] **optimize( )**
optimize_evaluate( ) function, [M-5] **optimize( )**
optimize_init_*( ) functions, [M-5] **optimize( )**
optimize_query( ) functions, [M-5] **optimize( )**
optimize_result_*( ) functions, [M-5] **optimize( )**
order( ) function, [M-5] **sort( )**
orgtype, [M-2] **declarations**, [M-6] **glossary**
orgtype( ) function, [M-5] **eltype( )**
orthogonal matrix, [M-6] **glossary**

P

p-conformability, [M-6] **glossary**
panel data, [M-5] **panelsetup( )**
panelsetup( ) function, [M-5] **panelsetup( )**
panelstats( ) function, [M-5] **panelsetup( )**
panelsubmatrix( ) function, [M-5] **panelsetup( )**
panelsubview( ) function, [M-5] **panelsetup( )**
parsing, [M-5] **tokenget( )**, [M-5] **tokens( )**
pathasciisuffix( ) function, [M-5] **pathjoin( )**
pathbasename( ) function, [M-5] **pathjoin( )**
pathisabs( ) function, [M-5] **pathjoin( )**
pathisurl( ) function, [M-5] **pathjoin( )**
pathjoin( ) function, [M-5] **pathjoin( )**
pathlist( ) function, [M-5] **pathjoin( )**
pathrmsuffix( ) function, [M-5] **pathjoin( )**
pathsearchlist( ) function, [M-5] **pathjoin( )**
pathsplit( ) function, [M-5] **pathjoin( )**
pathstatasuffix( ) function, [M-5] **pathjoin( )**
pathsubsysdir( ) function, [M-5] **pathjoin( )**
pathsuffix( ) function, [M-5] **pathjoin( )**
pattern matching, [M-5] **strmatch( )**
Penrose, R., [M-5] **pinv( )**
_perhapsequilc( ) function, [M-5] **_equilrc( )**
_perhapsequilr( ) function, [M-5] **_equilrc( )**
_perhapsequilrc( ) function, [M-5] **_equilrc( )**
permutation matrix and vector, [M-1] **permutation**,
  [M-5] **invorder( )**, [M-6] **glossary**
permutations, [M-5] **cvpermute( )**
pi( ) function, [M-5] **sin( )**
_pinv( ) function, [M-5] **pinv( )**
pinv( ) function, [M-5] **pinv( )**
pointers, [M-2] **pointers**, [M-2] **ftof**,
  [M-5] **findexternal( )**, [M-6] **glossary**

polyadd( ) function, [M-5] **polyeval( )**
polyderiv( ) function, [M-5] **polyeval( )**
polydiv( ) function, [M-5] **polyeval( )**
polyeval( ) function, [M-5] **polyeval( )**
polyinteg( ) function, [M-5] **polyeval( )**
polymult( ) function, [M-5] **polyeval( )**
polynomials, [M-5] **polyeval( )**
polyroots( ) function, [M-5] **polyeval( )**
polysolve( ) function, [M-5] **polyeval( )**
polytrim( ) function, [M-5] **polyeval( )**
Powell, M. J. D., [M-5] **optimize( )**
power, [M-2] **op_arith**, [M-2] **op_colon**,
  [M-5] **matpowersym( )**
pragma, [M-2] **pragma**, [M-6] **glossary**
printf( ) function, [M-5] **printf( )**
product, [M-2] **op_arith**, [M-2] **op_colon**,
  [M-2] **op_kronecker**, [M-5] **cross( )**,
  [M-5] **crossdev( )**, [M-5] **quadcross( )**
programming
  functions, [M-4] **programming**
  use, [M-1] **ado**
proper values, [M-5] **eigensystem( )**
pseudoinverse, [M-5] **pinv( )**
pwd( ) function, [M-5] **chdir( )**

Q

qofd( ) function, [M-5] **date( )**
QR decomposition, [M-5] **qrd( )**
qrd( ) function, [M-5] **qrd( )**
qrdp( ) function, [M-5] **qrd( )**
_qrinv( ) function, [M-5] **qrinv( )**
qrinv( ) function, [M-5] **qrinv( )**
_qrsolve( ) function, [M-5] **qrsolve( )**
qrsolve( ) function, [M-5] **qrsolve( )**
quad precision, [M-5] **runningsum( )**, [M-5] **sum( )**,
  [M-5] **mean( )**, [M-5] **quadcross( )**
quadcolsum( ) function, [M-5] **sum( )**
quadcorrelation( ) function, [M-5] **mean( )**
quadcross( ) function, [M-5] **quadcross( )**
quadcrossdev( ) function, [M-5] **quadcross( )**
quadmeanvariance( ) function, [M-5] **mean( )**
quadrant( ) function, [M-5] **sign( )**
quadrowsum( ) function, [M-5] **sum( )**
_quadrunningsum( ) function, [M-5] **runningsum( )**
quadrunningsum( ) function, [M-5] **runningsum( )**
quadsum( ) function, [M-5] **sum( )**
quadvariance( ) function, [M-5] **mean( )**
quarter( ) function, [M-5] **date( )**
quarterly( ) function, [M-5] **date( )**
query, [M-3] **mata set**
querybreakintr( ) function, [M-5] **setbreakintr( )**

R

r-conformability, [M-5] **normal( )**, [M-6] **glossary**
random numbers, [M-5] **uniform( )**

range
  operators, [M-2] op_range
  subscripts, see subscripts
  vector, [M-5] range( )
range( ) function, [M-5] range( )
rangen( ) function, [M-5] range( )
rank, [M-5] rank( ), [M-6] glossary
rank( ) function, [M-5] rank( )
Raphson, J., [M-5] optimize( )
Re( ) function, [M-5] Re( )
real, [M-2] declarations, [M-6] glossary
real part, [M-5] Re( )
Reid, C., [M-5] Hilbert( )
Reinsch, C. H., [M-5] spline3( )
reldif( ) function, [M-5] reldif( )
rename, [M-3] mata rename
reserved words, [M-2] reswords
return, [M-2] return
revorder( ) function, [M-5] invorder( )
right eigenvectors, [M-5] eigensystem( )
_rmdir( ) function, [M-5] chdir( )
rmdir( ) function, [M-5] chdir( )
rmexternal( ) function, [M-5] findexternal( )
Robinson, A., [M-5] Toeplitz( )
roots of polynomials, [M-5] polyeval( )
round( ) function, [M-5] trunc( )
roundoff error, [M-5] epsilon( ), [M-5] edittozero( ), [M-5] edittoint( )
row-join operator, [M-2] op_join
row of matrix, selecting, [M-5] select( )
rowmax( ) function, [M-5] minmax( )
rowmaxabs( ) function, [M-5] minmax( )
rowmin( ) function, [M-5] minmax( )
rowminmax( ) function, [M-5] minmax( )
rowmissing( ) function, [M-5] missing( )
rownonmissing( ) function, [M-5] missing( )
rows( ) function, [M-5] rows( )
rowscalefactors( ) function, [M-5] _equilrc( )
rowshape( ) function, [M-5] rowshape( )
rowsum( ) function, [M-5] sum( )
rowvector, [M-2] declarations, [M-6] glossary
_runningsum( ) function, [M-5] runningsum( )
runningsum( ) function, [M-5] runningsum( )

S

scalar, [M-2] declarations, [M-6] glossary
scalar functions, [M-4] scalar
seconds( ) function, [M-5] date( )
select( ) function, [M-5] select( )
semicolons, [M-2] semicolons
set, [M-3] mata set
setbreakintr( ) function, [M-5] setbreakintr( )
setmore( ) function, [M-5] more( )
setmoreonexit( ) function, [M-5] more( )
Shanno, D. F., [M-5] optimize( )

sign( ) function, [M-5] dsign( ), [M-5] sign( )
Simpson, T., [M-5] optimize( )
sin( ) function, [M-5] sin( )
singular value decomposition, [M-5] svd( ), [M-5] fullsvd( )
sinh( ) function, [M-5] sin( )
sizeof( ) function, [M-5] sizeof( )
smallestdouble( ) function, [M-5] mindouble( )
SMCL, see Stata Markup and Control Language
solve AX=B, [M-4] solvers, [M-5] cholsolve( ), [M-5] lusolve( ), [M-5] qrsolve( ), [M-5] svsolve( ), [M-5] solvelower( ), [M-5] solve_tol( )
solve_tol( ) function, [M-5] solve_tol( )
_solvelower( ) function, [M-5] solvelower( )
solvelower( ) function, [M-5] solvelower( )
_solvetolerance, [M-5] solve_tol( )
_solveupper( ) function, [M-5] solvelower( )
solveupper( ) function, [M-5] solvelower( )
Sorensen, D., [M-1] LAPACK
_sort( ) function, [M-5] sort( )
sort( ) function, [M-5] sort( )
source code, [M-1] how, [M-1] source, [M-6] glossary
spline3( ) function, [M-5] spline3( )
spline3eval( ) function, [M-5] spline3( )
sprintf( ) function, [M-5] printf( )
sqrt( ) function, [M-5] sqrt( )
square
  matrix, [M-6] glossary
  root, [M-5] sqrt( ), [M-5] cholesky( )
ss( ) function, [M-5] date( )
ssC( ) function, [M-5] date( )
_st_addobs( ) function, [M-5] st_addobs( )
st_addobs( ) function, [M-5] st_addobs( )
_st_addvar( ) function, [M-5] st_addvar( )
st_addvar( ) function, [M-5] st_addvar( )
_st_data( ) function, [M-5] st_data( )
st_data( ) function, [M-5] st_data( )
st_dropobsif( ) function, [M-5] st_dropvar( )
st_dropobsin( ) function, [M-5] st_dropvar( )
st_dropvar( ) function, [M-5] st_dropvar( )
st_eclear( ) function, [M-5] st_rclear( )
st_global( ) function, [M-5] st_global( )
st_isfmt( ) function, [M-5] st_isfmt( )
st_islmname( ) function, [M-5] st_isname( )
st_isname( ) function, [M-5] st_isname( )
st_isnumfmt( ) function, [M-5] st_isfmt( )
st_isnumvar( ) function, [M-5] st_vartype( )
st_isstrfmt( ) function, [M-5] st_isfmt( )
st_isstrvar( ) function, [M-5] st_vartype( )
st_keepobsif( ) function, [M-5] st_dropvar( )
st_keepobsin( ) function, [M-5] st_dropvar( )
st_keepvar( ) function, [M-5] st_dropvar( )
st_local( ) function, [M-5] st_local( )
_st_macroexpand( ) function, [M-5] st_macroexpand( )

st_macroexpand() function,
 [M-5] **st_macroexpand( )**
st_matrix() function, [M-5] **st_matrix( )**
st_matrixcolstripe() function, [M-5] **st_matrix( )**
st_matrixrowstripe() function, [M-5] **st_matrix( )**
st_nobs() function, [M-5] **st_nvar( )**
st_numscalar() function, [M-5] **st_numscalar( )**
st_nvar() function, [M-5] **st_nvar( )**
st_rclear() function, [M-5] **st_rclear( )**
st_replacematrix() function, [M-5] **st_matrix( )**
st_sclear() function, [M-5] **st_rclear( )**
_st_sdata() function, [M-5] **st_data( )**
st_sdata() function, [M-5] **st_data( )**
st_select() function, [M-5] **select( )**
_st_sstore() function, [M-5] **st_store( )**
st_sstore() function, [M-5] **st_store( )**
_st_store() function, [M-5] **st_store( )**
st_store() function, [M-5] **st_store( )**
st_strscalar() function, [M-5] **st_numscalar( )**
st_subview() function, [M-5] **st_subview( )**
st_tempfilename() function, [M-5] **st_tempname( )**
st_tempname() function, [M-5] **st_tempname( )**
st_tsrevar() function, [M-5] **st_tsrevar( )**
st_updata() function, [M-5] **st_updata( )**
st_varformat() function, [M-5] **st_varformat( )**
_st_varindex() function, [M-5] **st_varindex( )**
st_varindex() function, [M-5] **st_varindex( )**
st_varlabel() function, [M-5] **st_varformat( )**
st_varname() function, [M-5] **st_varname( )**
st_varrename() function, [M-5] **st_varrename( )**
st_vartype() function, [M-5] **st_vartype( )**
st_varvaluelabel() function, [M-5] **st_varformat( )**
st_view() function, [M-5] **st_view( )**
st_viewobs() function, [M-5] **st_viewvars( )**
st_viewvars() function, [M-5] **st_viewvars( )**
st_vldrop() function, [M-5] **st_vlexists( )**
st_vlexists() function, [M-5] **st_vlexists( )**
st_vlload() function, [M-5] **st_vlexists( )**
st_vlmap() function, [M-5] **st_vlexists( )**
st_vlmodify() function, [M-5] **st_vlexists( )**
st_vlsearch() function, [M-5] **st_vlexists( )**
Stata
 c()-class results, [M-5] **st_global( )**
 characteristics, [M-5] **st_global( )**, [M-5] **st_dir( )**
 e()-class results, [M-5] **st_global( )**, [M-5] **st_dir( )**,
  [M-5] **st_rclear( )**
 error message, [M-5] **error( )**
 execute command, [M-3] **mata stata**, [M-5] **stata( )**
 macros, [M-5] **st_global( )**, [M-5] **st_local( )**,
  [M-5] **st_dir( )**
 matrices, [M-5] **st_matrix( )**, [M-5] **st_dir( )**
 op.varname, see Stata, time series–operated variable
 r()-class results, [M-5] **st_global( )**, [M-5] **st_dir( )**,
  [M-5] **st_rclear( )**
 s()-class results, [M-5] **st_global( )**, [M-5] **st_dir( )**,
  [M-5] **st_rclear( )**
 scalars, [M-5] **st_numscalar( )**, [M-5] **st_dir( )**

Stata, continued
 temporary
  filenames, [M-5] **st_tempname( )**
  names, [M-5] **st_tempname( )**
 time series–operated variable, [M-5] **st_tsrevar( )**,
  [M-6] **glossary**
 value labels, [M-5] **st_varformat( )**,
  [M-5] **st_vlexists( )**
 variable
  formats, [M-5] **st_varformat( )**
  labels, [M-5] **st_varformat( )**
stata, [M-3] **mata stata**
_stata() function, [M-5] **stata( )**
stata() function, [M-5] **stata( )**
Stata Markup and Control Language, [M-5] **display( )**,
 [M-5] **printf( )**, [M-5] **errprintf( )**
statasetversion() function, [M-5] **stataversion( )**
stataversion() function, [M-5] **stataversion( )**
statistical
 density functions, [M-5] **normal( )**
 distribution functions, [M-5] **normal( )**
st_dir() function, [M-5] **st_dir( )**
steepest descent (ascent), [M-5] **optimize( )**
string
 duplication, [M-5] **strdup( )**
 functions, [M-4] **string**
 pattern matching, [M-5] **strmatch( )**
string, [M-2] **declarations**, [M-6] **glossary**
string concatenation, [M-5] **invtokens( )**
string to real, convert, [M-5] **strtoreal( )**
stritrim() function, [M-5] **strtrim( )**
strlen() function, [M-5] **strlen( )**
strlower() function, [M-5] **strupper( )**
strltrim() function, [M-5] **strtrim( )**
strmatch() function, [M-5] **strmatch( )**
strofreal() function, [M-5] **strofreal( )**
strpos() function, [M-5] **strpos( )**
strproper() function, [M-5] **strupper( )**
strreverse() function, [M-5] **strreverse( )**
strrtrim() function, [M-5] **strtrim( )**
_strtoreal() function, [M-5] **strtoreal( )**
strtoreal() function, [M-5] **strtoreal( )**
strtrim() function, [M-5] **strtrim( )**
struct, [M-2] **struct**
structures, [M-2] **struct**, [M-5] **liststruct( )**,
 [M-6] **glossary**
strupper() function, [M-5] **strupper( )**
subinstr() function, [M-5] **subinstr( )**
subinword() function, [M-5] **subinstr( )**
subscripts, [M-2] **subscripts**, [M-6] **glossary**
_substr() function, [M-5] **_substr( )**
substr() function, [M-5] **substr( )**
subtraction, [M-2] **op_arith**, [M-2] **op_colon**
sum() function, [M-5] **sum( )**
sum of vector, [M-5] **runningsum( )**
SVD, see singular value decomposition
_svd() function, [M-5] **svd( )**

svd() function, [M-5] **svd( )**
_svd_la() function, [M-5] **svd( )**, [M-5] **fullsvd( )**
_svdsv() function, [M-5] **svd( )**
svdsv() function, [M-5] **svd( )**
_svsolve() function, [M-5] **svsolve( )**
svsolve() function, [M-5] **svsolve( )**
swap() function, [M-5] **swap( )**
_symeigen_la() function, [M-5] **eigensystem( )**
_symeigensystem() function, [M-5] **eigensystem( )**
symeigensystem() function, [M-5] **eigensystem( )**
_symeigenvalues() function, [M-5] **eigensystem( )**
symeigenvalues() function, [M-5] **eigensystem( )**
symmetric matrices, [M-5] **issymmetric( )**, [M-5] **makesymmetric( )**, [M-6] **glossary**
symmetriconly, [M-6] **glossary**
syntax, [M-2] **syntax**

T

tan() function, [M-5] **sin( )**
tanh() function, [M-5] **sin( )**
tden() function, [M-5] **normal( )**
time and date, [M-5] **c( )**
time series–operated variable, [M-5] **st_data( )**, [M-5] **st_tsrevar( )**, [M-6] **glossary**
Toeplitz() function, [M-5] **Toeplitz( )**
Toeplitz, O., [M-5] **Toeplitz( )**
tokenallowhex() function, [M-5] **tokenget( )**
tokenallownum() function, [M-5] **tokenget( )**
tokenget() function, [M-5] **tokenget( )**
tokengetall() function, [M-5] **tokenget( )**
tokeninit() function, [M-5] **tokenget( )**
tokeninitstata() function, [M-5] **tokenget( )**
tokenoffset() function, [M-5] **tokenget( )**
tokenpchars() function, [M-5] **tokenget( )**
tokenpeek() function, [M-5] **tokenget( )**
tokenqchars() function, [M-5] **tokenget( )**
tokenrest() function, [M-5] **tokenget( )**
tokens() function, [M-5] **tokens( )**
tokenset() function, [M-5] **tokenget( )**
tokenwchars() function, [M-5] **tokenget( )**
tolerances, [M-1] **tolerance**, [M-5] **solve_tol( )**
trace() function, [M-5] **trace( )**
trace of matrix, [M-5] **trace( )**
traceback log, [M-2] **errors**, [M-5] **error( )**, [M-6] **glossary**
transmorphic, [M-2] **declarations**, [M-6] **glossary**
transpose, [M-2] **op_transpose**, [M-5] **_transpose( )**, [M-5] **transposeonly( )**, [M-6] **glossary**, *also see* conjugate transpose
_transpose() function, [M-5] **_transpose( )**
_transposeonly() function, [M-5] **transposeonly( )**
transposeonly() function, [M-5] **transposeonly( )**
transposition, [M-2] **op_transpose**, [M-5] **_transpose( )**, [M-5] **transposeonly( )**
triangular matrix, [M-5] **solvelower( )**, [M-6] **glossary**
trigamma() function, [M-5] **factorial( )**
trigonometric functions, [M-5] **sin( )**
trunc() function, [M-5] **trunc( )**
ttail() function, [M-5] **normal( )**
type, [M-2] **declarations**, [M-6] **glossary**
type, broad, [M-6] **glossary**

U

underscore functions, [M-1] **naming**, [M-6] **glossary**
uniform() function, [M-5] **uniform( )**
uniformly distributed random numbers, [M-5] **uniform( )**
uniformseed() function, [M-5] **uniform( )**
uniqrows() function, [M-5] **uniqrows( )**
unit vectors, [M-5] **e( )**
unitary matrix, [M-6] **glossary**
unitcircle() function, [M-5] **unitcircle( )**
_unlink() function, [M-5] **unlink( )**
unlink() function, [M-5] **unlink( )**
unorder() function, [M-5] **sort( )**
upper-triangular matrix, *see* triangular matrix
uppercase, [M-5] **strupper( )**
_uppertriangle() function, [M-5] **lowertriangle( )**
uppertriangle() function, [M-5] **lowertriangle( )**

V

valofexternal() function, [M-5] **valofexternal( )**
Vandermonde, A.-T., [M-5] **Vandermonde( )**
Vandermonde() function, [M-5] **Vandermonde( )**
variable, [M-2] **declarations**, [M-5] **st_data( )**, [M-6] **glossary**
variable naming convention, [M-1] **naming**
variance() function, [M-5] **mean( )**
vec() function, [M-5] **vec( )**
vech() function, [M-5] **vec( )**
vector, [M-2] **declarations**, [M-6] **glossary**
vector norm, [M-5] **norm( )**
version, [M-2] **version**
version control, [M-2] **version**, [M-5] **callersversion( )**
version of Stata, [M-5] **stataversion( )**
view matrix, [M-5] **isview( )**, [M-5] **st_subview( )**, [M-5] **st_view( )**, [M-5] **st_viewvars( )**, [M-6] **glossary**
viewsource, [M-1] **source**
void
    function, [M-2] **declarations**, [M-6] **glossary**
    matrix, [M-2] **void**, [M-6] **glossary**

W

warning messages, [M-2] **pragma**
week() function, [M-5] **date( )**
weekly() function, [M-5] **date( )**
Westfall, R. S., [M-5] **optimize( )**

which, [M-3] **mata which**
while, [M-2] **while**, [M-2] **continue**, [M-2] **break**, [M-2] **semicolons**
width of %*fmt*, [M-5] **fmtwidth( )**
wofd( ) function, [M-5] **date( )**

Y

year( ) function, [M-5] **date( )**
yearly( ) function, [M-5] **date( )**
yh( ) function, [M-5] **date( )**
ym( ) function, [M-5] **date( )**
yofd( ) function, [M-5] **date( )**
Ypma, T. J., [M-5] **optimize( )**
yq( ) function, [M-5] **date( )**
yw( ) function, [M-5] **date( )**